Exploring
Anatomy & Physiology
in the
Laboratory

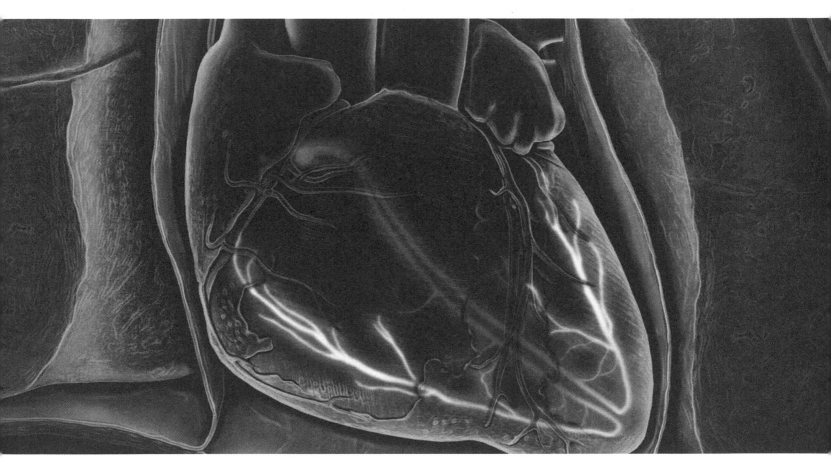

Erin C. Amerman

Morton Publishing Company
925 W. Kenyon Ave., Unit 12
Englewood, Colorado 80110
http://www.morton-pub.com

Book Team

Publisher:	Douglas N. Morton
Biology Editor:	David Ferguson
Production Manager:	Joanne Saliger
Production Assistant:	Desiree Coscia
	Patricia Billiot
Typography:	Ash Street Typecrafters, Inc.
Copyediting:	Carolyn Acheson
Illustrations:	Imagineering Media Services, Inc.
Cover Design:	Imagineering Media Services, Inc.

EXPLORING ANATOMY & PHYSIOLOGY IN THE LABORATORY

Preface

Many years ago, when I first started teaching anatomy and physiology, my biggest frustration with the course came from the laboratory. It seemed as if I were pulling teeth to get my students interested in the lab material and most students were unable to make the connections between lecture and lab. As I attended conferences and interacted with other instructors, I found that I wasn't the only one facing this challenge. Many of my colleagues were also frustrated with the lab. And they were not only frustrated with the lab, but also with the available lab manuals, which were too long, were too expensive, lacked focused activities for the students, and required equipment that most colleges simply do not have the funds to purchase in today's era of budget cuts.

In an attempt to solve these problems, I began writing lab exercises for my students. Over time, I found that my students not only learned from these exercises, but also enjoyed them because they were active for the entire lab period. Consequently, their grades on practical examinations improved dramatically. And their grades on classroom examinations also improved, as the exercises in lab helped them to better understand the lecture.

Seeing how much my own students benefitted from these exercises, I wondered if there was a way to create a full lab manual with these exercises so I could share them with my fellow instructors. In 2003 I was fortunate enough to be provided with an opportunity to do just that. The finished product was the black-and-white text *Exercises for the Anatomy and Physiology Laboratory*, which featured innovative pedagogy that proved to be popular with students and instructors alike.

When the time approached to do a second edition of the *Exercises*, my editor and I decided to expand on its innovative pedagogy, develop a full-color art program, add some more unique features, and provide just the right amount of text to walk the students through the material. The result is the book you are holding: *Exploring Anatomy & Physiology in the Laboratory*.

As you read through this text, you will find an assortment of tools to help you and your students navigate the A&P lab. Many of these tools were adapted from the original *Exercises*, and others are brand-new and found only in this text. Some of the highlights include:

Pre-Lab Exercises

Each unit opens with a list of key terms that students should define before coming to the lab. Additionally, the Pre-Lab Exercises of most units feature labeling and coloring exercises for anatomical structures, exercises that help students review material from previous units, and/or questions regarding basic physiological principles that apply to the material in the unit. The Pre-Lab Exercises in effect allow this text to function as both a laboratory manual and a study guide.

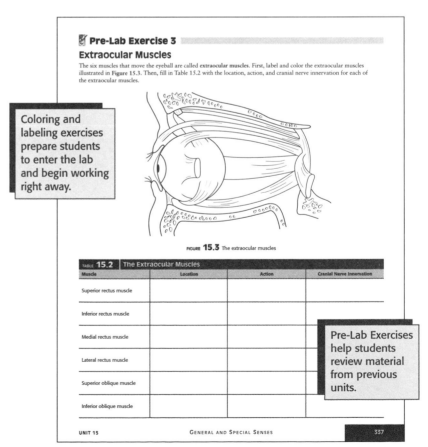

Pre-Lab Exercise 3
Extraocular Muscles

The six muscles that move the eyeball are called **extraocular muscles**. First, label and color the extraocular muscles illustrated in **Figure 15.3**. Then, fill in Table 15.2 with the location, action, and cranial nerve innervation for each of the extraocular muscles.

FIGURE **15.3** The extraocular muscles

Coloring and labeling exercises prepare students to enter the lab and begin working right away.

Pre-Lab Exercises help students review material from previous units.

TABLE 15.2	The Extraocular Muscles		
Muscle	Location	Action	Cranial Nerve Innervation
Superior rectus muscle			
Inferior rectus muscle			
Medial rectus muscle			
Lateral rectus muscle			
Superior oblique muscle			
Inferior oblique muscle			

UNIT 15   GENERAL AND SPECIAL SENSES   337

Organized Anatomy and Model Inventories

For each unit in which anatomy is a component, the anatomical structures are organized in a way that provides a centralized list for students that is easy for instructors to customize based upon preference.

   Each anatomy list is followed by a Model Inventory, in which students assign descriptive names to their anatomical models and then list the structures that they are able to locate on the model. This helps students to focus more on the anatomy and to engage more parts of their brains as they examine, pronounce, and write down the names of the anatomical structures.

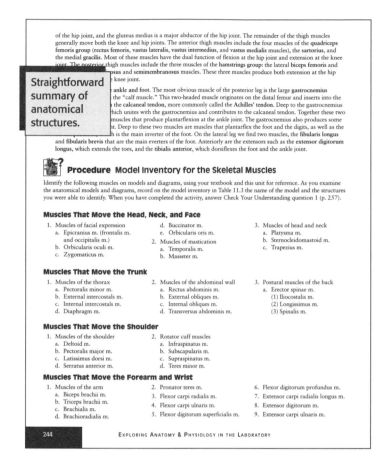

of the hip joint, and the gluteus medius is a major abductor of the hip joint. The remainder of the thigh muscles generally move both the knee and hip joints. The anterior thigh muscles include the four muscles of the **quadriceps femoris group** (**rectus femoris, vastus lateralis, vastus intermedius,** and **vastus medialis** muscles), the **sartorius,** and the medial **gracilis.** Most of these muscles have the dual function of flexion at the hip joint and extension at the knee joint. The posterior thigh muscles include the three muscles of the **hamstrings group:** the lateral **biceps femoris** and [...]osus and **semimembranosus** muscles. These three muscles produce both extension at the hip [...] knee joint.

Straightforward summary of anatomical structures.

[...] ankle and foot. The most obvious muscle of the posterior leg is the large **gastrocnemius** [...] the "calf muscle." This two-headed muscle originates on the distal femur and inserts into the [...] the **calcaneal tendon,** more commonly called the **Achilles' tendon.** Deep to the gastrocnemius [...] which unites with the gastrocnemius and contributes to the calcaneal tendon. Together these two [...] muscles that produce plantarflexion at the ankle joint. The gastrocnemius also produces some [...]. Deep to these two muscles are muscles that plantarflex the foot and the digits, as well as the [...] which is the main inverter of the foot. On the lateral leg we find two muscles, the **fibularis longus** and **fibularis brevis** that are the main everters of the foot. Anteriorly are the extensors such as the **extensor digitorum longus,** which extends the toes, and the **tibialis anterior,** which dorsiflexes the foot and the ankle joint.

Procedure Model Inventory for the Skeletal Muscles

Identify the following muscles on models and diagrams, using your textbook and this unit for reference. As you examine the anatomical models and diagrams, record on the model inventory in Table 11.3 the name of the model and the structures you were able to identify. When you have completed the activity, answer Check Your Understanding question 1 (p. 257).

Muscles That Move the Head, Neck, and Face

1. Muscles of facial expression
   a. Epicranius m. (frontalis m. and occipitalis m.)
   b. Orbicularis oculi m.
   c. Zygomaticus m.
   d. Buccinator m.
   e. Orbicularis oris m.
2. Muscles of mastication
   a. Temporalis m.
   b. Masseter m.
3. Muscles of head and neck
   a. Platysma m.
   b. Sternocleidomastoid m.
   c. Trapezius m.

Muscles That Move the Trunk

1. Muscles of the thorax
   a. Pectoralis minor m.
   b. External intercostals m.
   c. Internal intercostals m.
   d. Diaphragm m.
2. Muscles of the abdominal wall
   a. Rectus abdominis m.
   b. External obliques m.
   c. Internal obliques m.
   d. Transversus abdominis m.
3. Postural muscles of the back
   a. Erector spinae m.
      (1) Iliocostalis m.
      (2) Longissimus m.
      (3) Spinalis m.

Muscles That Move the Shoulder

1. Muscles of the shoulder
   a. Deltoid m.
   b. Pectoralis major m.
   c. Latissimus dorsi m.
   d. Serratus anterior m.
2. Rotator cuff muscles
   a. Infraspinatus m.
   b. Subscapularis m.
   c. Supraspinatus m.
   d. Teres minor m.

Muscles That Move the Forearm and Wrist

1. Muscles of the arm
   a. Biceps brachii m.
   b. Triceps brachii m.
   c. Brachialis m.
   d. Brachioradialis m.
2. Pronator teres m.
3. Flexor carpi radialis m.
4. Flexor carpi ulnaris m.
5. Flexor digitorum superficialis m.
6. Flexor digitorum profundus m.
7. Extensor carpi radialis longus m.
8. Extensor digitorum m.
9. Extensor carpi ulnaris m.

Muscles That Move the Hip and Knee

1. Muscles of the pelvic girdle
   a. Iliopsoas m. (Iliacus m. and Psoas major m.)
   b. Gluteus maximus m.
   c. Gluteus medius m.
2. Muscles of the thigh
   a. Sartorius m.
   b. Tensor fascia lata m.
   c. Pectineus m.
   d. Quadriceps femoris group
      (1) Rectus femoris m.
      (2) Vastus medialis m.
      (3) Vastus intermedius m.
      (4) Vastus lateralis m.

Model Inventories allow students to identify and describe the anatomical models they see in the lab.

Muscles That Move the Ankle and Foot

1. Gastrocnemius m.
2. Soleus m.
3. Flexor digitorum longus m.
4. Fibularis (peroneus) longus m.
5. Fibularis (peroneus) brevis m.
6. Tibialis posterior m.
7. Tibialis anterior m.
8. Extensor digitorum longus m.

Your instructor may wish to omit certain muscles included above or add muscles not included in these lists. List any additional structures below:

TABLE **11.3** Model Inventory for Skeletal Muscles	
Model/Diagram	Structures Identified

Focused Activities and Tracing Exercises

Some of the more popular features from the original *Exercises* were the focused activities and tracing exercises. We expanded upon these features extensively in this text so every unit contains activities for the students to perform. These activities were written with cost concerns in mind, and seldom require special equipment or materials. We also added more tracing exercises, in which students trace the pathway of a certain substance (e.g., a molecule of glucose or an erythrocyte) throughout the body to develop a "big picture" view of both anatomy and physiology.

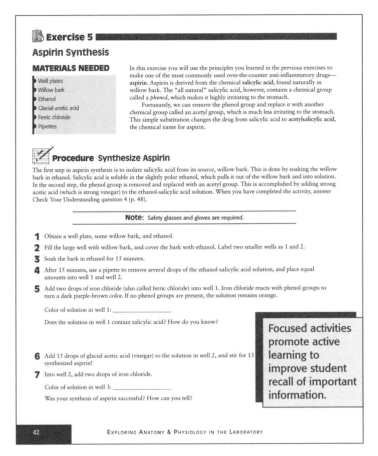

Exercise 5
Aspirin Synthesis

MATERIALS NEEDED

- Well plates
- Willow bark
- Ethanol
- Glacial acetic acid
- Ferric chloride
- Pipettes

In this exercise you will use the principles you learned in the previous exercises to make one of the most commonly used over-the-counter anti-inflammatory drugs—**aspirin**. Aspirin is derived from the chemical **salicylic acid**, found naturally in willow bark. The "all natural" salicylic acid, however, contains a chemical group called a *phenol*, which makes it highly irritating to the stomach.

Fortunately, we can remove the phenol group and replace it with another chemical group called an *acetyl* group, which is much less irritating to the stomach. This simple substitution changes the drug from salicylic acid to **acetylsalicylic acid**, the chemical name for aspirin.

Procedure Synthesize Aspirin

The first step in aspirin synthesis is to isolate salicylic acid from its source, willow bark. This is done by soaking the willow bark in ethanol. Salicylic acid is soluble in the slightly polar ethanol, which pulls it out of the willow bark and into solution. In the second step, the phenol group is removed and replaced with an acetyl group. This is accomplished by adding strong acetic acid (which is strong vinegar) to the ethanol-salicylic acid solution. When you have completed the activity, answer Check Your Understanding question 4 (p. 48).

Note: Safety glasses and gloves are required.

---

1 Obtain a well plate, some willow bark, and ethanol.

2 Fill the large well with willow bark, and cover the bark with ethanol. Label two smaller wells as 1 and 2.

3 Soak the bark in ethanol for 15 minutes.

4 After 15 minutes, use a pipette to remove several drops of the ethanol-salicylic acid solution, and place equal amounts into well 1 and well 2.

5 Add two drops of iron chloride (also called ferric chloride) into well 1. Iron chloride reacts with phenol groups to turn a dark purple-brown color. If no phenol groups are present, the solution remains orange.

Color of solution in well 1: _____

Does the solution in well 1 contain salicylic acid? How do you know?

6 Add 15 drops of glacial acetic acid (vinegar) to the solution in well 2, and stir for 15 synthesized aspirin!

7 Into well 2, add two drops of iron chloride.

Color of solution in well 3: _____

Was your synthesis of aspirin successful? How can you tell?

> Focused activities promote active learning to improve student recall of important information.

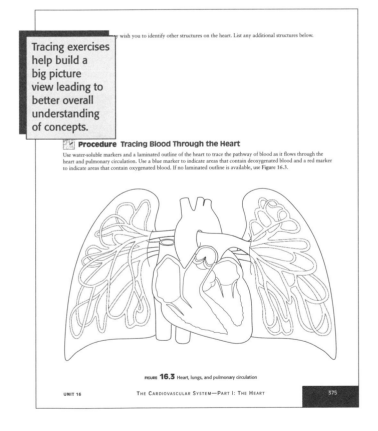

> Tracing exercises help build a big picture view leading to better overall understanding of concepts.

...y wish you to identify other structures on the heart. List any additional structures below.

Procedure Tracing Blood Through the Heart

Use water-soluble markers and a laminated outline of the heart to trace the pathway of blood as it flows through the heart and pulmonary circulation. Use a blue marker to indicate areas that contain deoxygenated blood and a red marker to indicate areas that contain oxygenated blood. If no laminated outline is available, use Figure 16.3.

FIGURE 16.3 Heart, lungs, and pulmonary circulation

Engaging Art Program

We directed efforts toward developing full-color, professionally-rendered figures for every unit that are both informative and engaging. We also included full-color photomicrographs with units that contain histology exercises. The photos were taken using the same slides that your students use and with the objective lenses that most student microscopes contain. This is intended to help students find the right tissues and structures on their microscope slides and minimize their frustration.

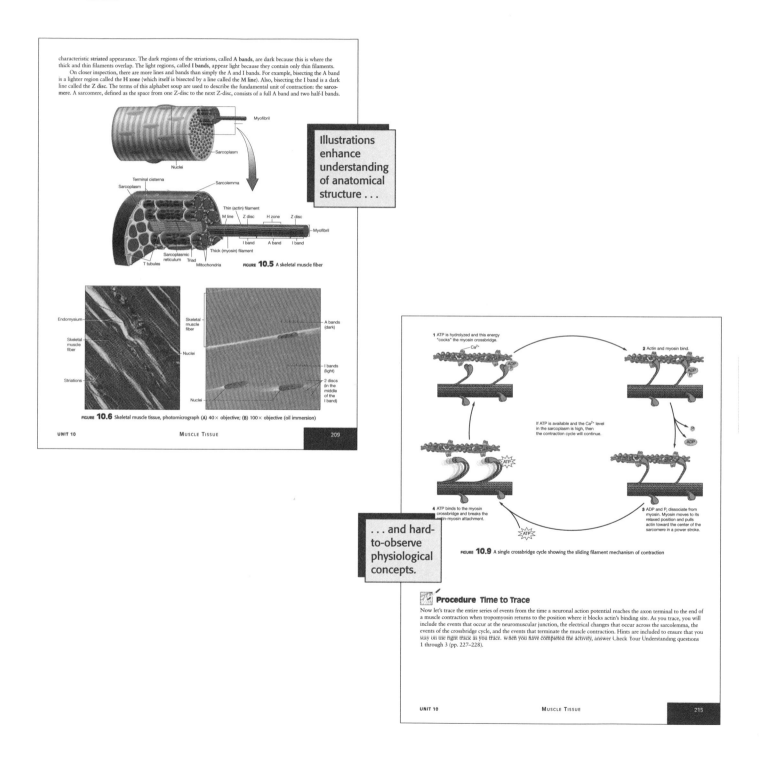

End-of-Unit Quizzes

Every unit-ending quiz contains two parts: a set of ten *Check Your Recall* questions to measure students' ability to recall the pertinent facts of the lab, and a set of three to six *Check Your Understanding* critical thinking questions that test their comprehension of the material. The quizzes are designed so that students may hand them in for grading without permanently removing pages from their lab material.

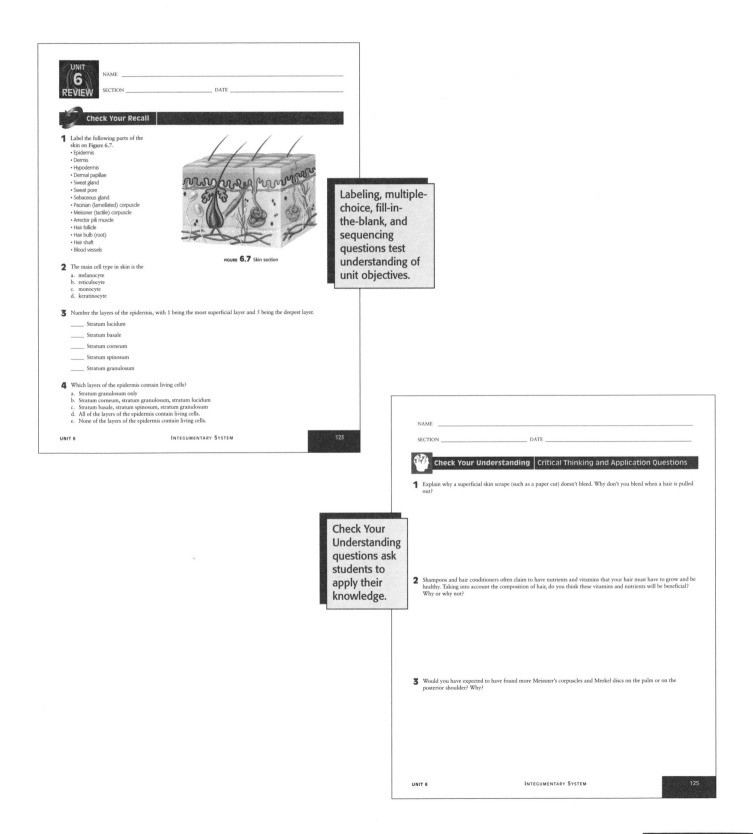

UNIT 6 REVIEW

NAME _____

SECTION _____ DATE _____

Check Your Recall

1 Label the following parts of the skin on Figure 6.7.
• Epidermis
• Dermis
• Hypodermis
• Dermal papillae
• Sweat gland
• Sweat pore
• Sebaceous gland
• Pacinian (lamellated) corpuscle
• Meissner (tactile) corpuscle
• Arrector pili muscle
• Hair follicle
• Hair bulb (root)
• Hair shaft
• Blood vessels

FIGURE **6.7** Skin section

Labeling, multiple-choice, fill-in-the-blank, and sequencing questions test understanding of unit objectives.

2 The main cell type in skin is the
a. melanocyte
b. reticulocyte
c. monocyte
d. keratinocyte

3 Number the layers of the epidermis, with 1 being the most superficial layer and 5 being the deepest layer.

_____ Stratum lucidum

_____ Stratum basale

_____ Stratum corneum

_____ Stratum spinosum

_____ Stratum granulosum

4 Which layers of the epidermis contain living cells?
a. Stratum granulosum only
b. Stratum corneum, stratum granulosum, stratum lucidum
c. Stratum basale, stratum spinosum, stratum granulosum
d. All of the layers of the epidermis contain living cells.
e. None of the layers of the epidermis contain living cells.

UNIT 6   INTEGUMENTARY SYSTEM   123

NAME _____

SECTION _____ DATE _____

Check Your Understanding | **Critical Thinking and Application Questions**

1 Explain why a superficial skin scrape (such as a paper cut) doesn't bleed. Why don't you bleed when a hair is pulled out?

Check Your Understanding questions ask students to apply their knowledge.

2 Shampoos and hair conditioners often claim to have nutrients and vitamins that your hair must have to grow and be healthy. Taking into account the composition of hair, do you think these vitamins and nutrients will be beneficial? Why or why not?

3 Would you have expected to have found more Meissner's corpuscles and Merkel discs on the palm or on the posterior shoulder? Why?

UNIT 6   INTEGUMENTARY SYSTEM   125

Hints and Tips Boxes

Students nearly always have difficulty with certain topics in A&P. For these topics we added a boxed feature called *Hints and Tips*. These boxes are scattered throughout the text to help students tackle these particularly difficult topics.

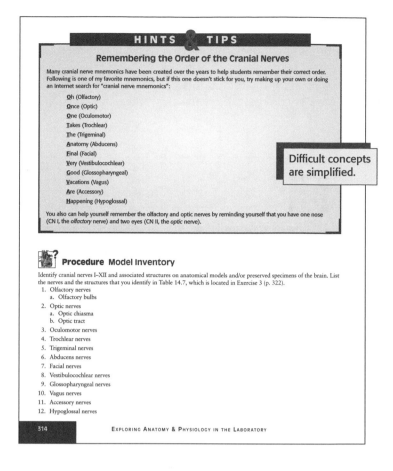

Affordability

Textbooks are expensive, and the last thing the average student needs is to purchase a lab manual that is one hundred dollars or more. This text is unique in that it provides high-quality material priced with student budgets in mind.

It is my sincere hope that *Exploring Anatomy & Physiology in the Laboratory* will provide you and your students with the tools necessary for a productive and interesting laboratory experience. I welcome all comments and suggestions for future editions of this book. Please feel free to contact me at **eapl@morton-pub.com**

Acknowledgments

Although it is my name on the cover of this text, textbooks are never a solo effort. Many people were integral to the production and development of this book, and I would like to take this brief opportunity to express my gratitude.

First and foremost I would like to thank my family, particularly Elise, my mother Cathy, and Chris. Without your unwavering support, this text would not have been possible. And to Elise, thank you especially for being patient with me being behind my computer screen so often, even if it meant I didn't get to play princess with you as often as I would have liked. Also, I can't forget my animals, particularly my cats, who unfailingly managed to be completely in the way of whatever I was doing.

Next I would like to extend my gratitude to the talented book team with whom I was fortunate enough to work: Joanne Saliger at Ash Street Typecrafters, Inc., who expertly designed and produced the book; Carolyn Acheson, who skillfully copyedited the text; the team at Imagineering Media Services, Inc., who provided the beautiful illustrations; and John Crawley and Michael Leboffe, who allowed me to use several of their excellent photos and photomicrographs. I truly appreciate all of your hard work and generosity.

I would also like to thank the following reviewers for their valuable suggestions that helped to shape the contents of this book:

Darren Mattone – Muskegon Community College
Cathy Whiting – Gainesville State College
Michele Robichaux – Nicholls State University
Angela Corbin – Nicholls State University
Justin Moore – American River College
Lori Smith– American River College
Elizabeth Hodgson – York College of Pennsylvania

The acknowledgements would be incomplete without thanking Doug Morton, who has been kind enough to provide me with another opportunity to publish with his company. And finally, I extend a special thank you to Biology Editor David Ferguson for his support, patience, friendship, and willingness to cheer for the Florida Gators.

About the Author

Erin C. Amerman has been involved in anatomy and physiology education for over 10 years as an author and professor, most recently at Santa Fe College in Gainesville, Florida. She received a B.S. in Cellular and Molecular Biology from the University of West Florida and a Doctorate in Podiatric medicine from Des Moines University. *Exploring Anatomy & Physiology in the Laboratory* is her second book with Morton Publishing.

Contents

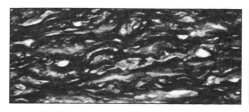  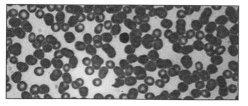

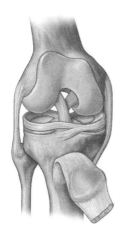

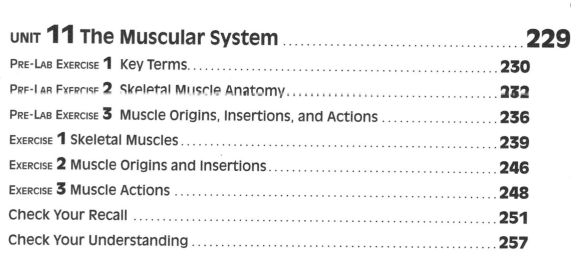

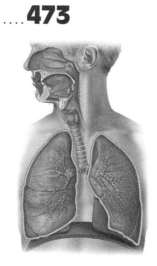

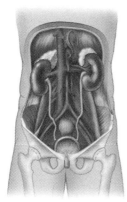

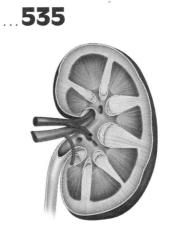
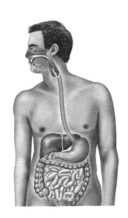
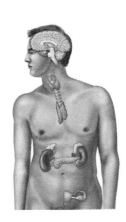

Introduction to Anatomical Terms

OBJECTIVES

Once you have completed this unit, you should be able to:

- Demonstrate and describe anatomical position.

- Apply directional terms to descriptions of anatomical parts.

- Use regional terms to describe anatomical locations.

- Locate and describe the divisions of the major body cavities and the membranes lining each cavity.

- Demonstrate and describe anatomical planes of section.

- Identify the organ systems, their functions, and the major organs in each system.

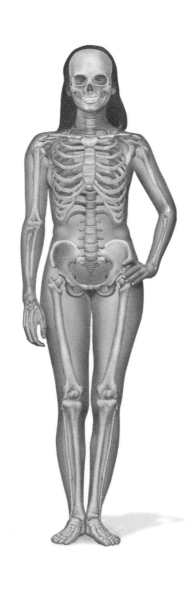

Pre-Lab Exercise 1
Key Terms

Table 1.1 lists the key terms with which you should be familiar before coming to lab.

TABLE **1.1**	Key Terms

Term | **Definition**

Directional Terms

Anterior (ventral) _____

_____

Posterior (dorsal) _____

_____

Superior (cranial) _____

_____

Inferior (caudal) _____

_____

Proximal _____

_____

Distal _____

_____

Superficial _____

_____

Deep _____

_____

Body Cavities and Membranes

Dorsal body cavity _____

_____

Ventral body cavity _____

_____

Serous membrane _____

_____

Planes of Section

Sagittal plane _____

_____

Frontal (coronal) plane _____

_____

Transverse plane _____

_____

✎ Pre-Lab Exercise 2

Organ Systems

The body has 11 organ systems, each of which contains certain organs, and each with a specific subset of functions. Note that some organs are part of more than one system.

In this exercise you will identify the 11 organ systems, their major organ(s), and the basic function(s) of each system (Table 1.2). Use your textbook and Exercise 6 from this unit for reference.

TABLE **1.2** Organ Systems		
Organ System	**Major Organs**	**Organ System Functions**
Integumentary System		
Skeletal System		
Muscular System		
Nervous System		
Cardiovascular System		
Respiratory System		
Lymphatic System		
Urinary System		
Digestive System		
Endocrine System		
Reproductive System		

Exercises

The bullet entered the right posterior scapular region, 3 centimeters lateral to the vertebral region, 4 centimeters inferior to the cervical region, and penetrated deep to the muscle and bone, but superficial to the parietal pleura . . .

Would you believe that by the end of this unit, you will be able to translate the above sentence and also locate the hypothetical wound? Unit 1 will introduce you to the world of anatomy and physiology. We will begin with an introduction to the unique language of anatomy and physiology. Like learning any new language, this may seem overwhelming at first. The key to success is repetition and application: The more you use the terms, the easier it will be for them to become part of your normal vocabulary.

From the terminology we will move on to the organization of the body into body cavities and organ systems. Once you have completed this unit, return to the above sentence and challenge yourself to locate the precise position of the bullet wound on an anatomical model.

Exercise 1

Anatomical Position

In the study of anatomy and physiology, most anatomical specimens are presented in a standard position termed **anatomical position**. In anatomical position the specimen is presented facing forward, with the toes pointing forward, the feet slightly apart, and the palms facing forward, as shown in **Figure 1.1**. This presentation of anatomical specimens creates a standard point of reference that facilitates communication among scientists and health care professionals.

 Procedure Anatomical Position

Have your lab partner stand in a normal, relaxed way, then adjust his or her position so it matches anatomical position. When you have completed this exercise, answer Check Your Understanding question 1 (p. 25).

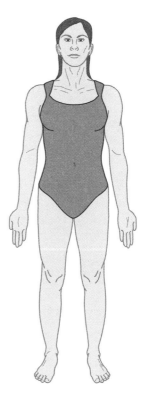

FIGURE 1.1 Anatomical position

Exercise 2

Directional Terms

Another method that makes communication easier and less prone to errors is to use certain terms to define the location of body parts and body markings. For example, when describing a wound on the chest, we could say:

▶ The wound is near the middle and top of the chest; or

▶ The wound is on the right *anterior* thoracic region, 4 centimeters *lateral* to the sternum, and 3 centimeters *inferior* to the acromial region.

The second option is precise and allows the reader to locate the wound exactly. Note that these descriptions are referring to a figure in anatomical position.

Review the definitions of the directional terms that you completed in Pre-Lab Exercise 1 and in **Figure 1.2**. Use these terms to fill in the correct directional term in the following practice procedure.

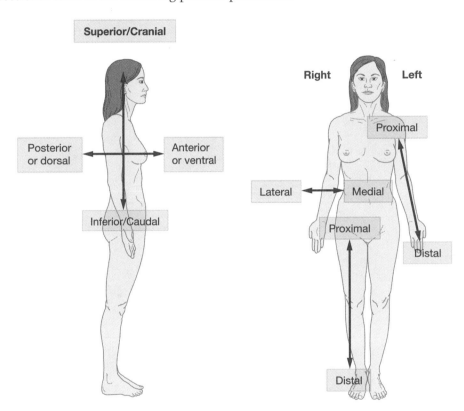

FIGURE 1.2 Directional terms

 Procedure Directional Terms

Fill in the correct directional term for the items below:

The elbow is _proximal_ to the wrist.

The chin is _inferior/caudal_ to the nose.

The shoulder is _lateral_ to the sternum (breastbone).

The forehead is _superior/cranial_ to the mouth.

The skin is _superficial_ to the muscle.

The spine is _posterior/dorsal_ to the esophagus.

The mouth is _medial_ to the ear.

The spine is on the _Dorsal_ side of the body.

The arm is _lateral_ to the torso.

The knee is _Distal_ to the hip.

Exercise 3

Regional Terms

MATERIALS NEEDED

- Laminated outline of the human body
- Water-soluble marking pens

You may have noticed in the previous exercise that we used the anatomical terms *thoracic region* and *acromial region* instead of using generic terms such as "chest" and "shoulder." This is a standard practice, again intended to make descriptions as specific as possible and to reduce the potential for errors in communication. For example, "shoulder" could consist of quite a large anatomical area, whereas the "acromial region" refers to one specific location on the shoulder.

The following regional terms, illustrated in **Figure 1.3**, are among the more common terms you will encounter in your study of anatomy and physiology. Note that most of these terms are *adjectives* rather than nouns. This means that the term is not complete unless it is paired with the term "region." For example, we cannot say, "The wound is in the antebrachial;" we instead must say, "The wound is in the antebrachial *region*."

The following list may look daunting, but you are probably familiar with several of the terms already. For example, you likely know the locations of the "oral," "nasal," and "abdominal" regions. Watch for other terms that you may know.

1. Cephalic
2. Frontal
3. Orbital
4. Nasal
5. Buccal
6. Otic
7. Oral
8. Occipital
9. Mental
10. Cervical
11. Acromial
12. Scapular
13. Sternal
14. Thoracic
15. Mammary
16. Axillary
17. Arm
18. Brachial
19. Antecubital
20. Vertebral
21. Abdominal

22. Umbilical
23. Antebrachial
24. Forearm
25. Carpal
26. Palmar
27. Digital
28. Pelvic
29. Lumbar
30. Gluteal
31. Inguinal
32. Pubic
33. Thigh
34. Femoral
35. Patellar
36. Popliteal
37. Crural
38. Sural
39. Leg
40. Tarsal
41. Calcaneal
42. Plantar

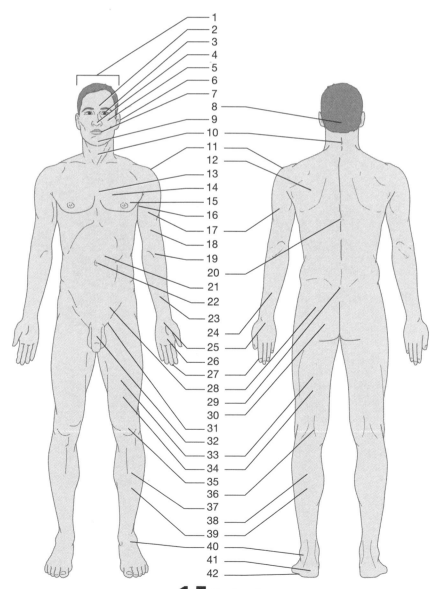

FIGURE 1.3 Regional terms

Procedure **Labeling Body Regions**

Use water-soluble markers to locate and label each of the following regions on laminated outlines of the human body. If outlines are unavailable, label the regions on **Figure 1.4**. When you have completed this exercise, answer Check Your Understanding question 2 (p. 25).

Adjectives

Abdominal	Crural	Oral	Tarsal
Acromial	Digital	Orbital	Thoracic
Antebrachial	Femoral	Otic	Umbilical
Antecubital	Frontal	Palmar	Vertebral
Axillary	Gluteal	Patellar	
Brachial	Inguinal	Pelvic	**Nouns**
Buccal	Lumbar	Plantar	Arm
Calcaneal	Mammary	Popliteal	Forearm
Carpal	Mental	Pubic	Leg
Cephalic	Nasal	Scapular	Lower limb
Cervical	Nuchal	Sternal	Thigh
Cranial	Occipital	Sural	Upper limb

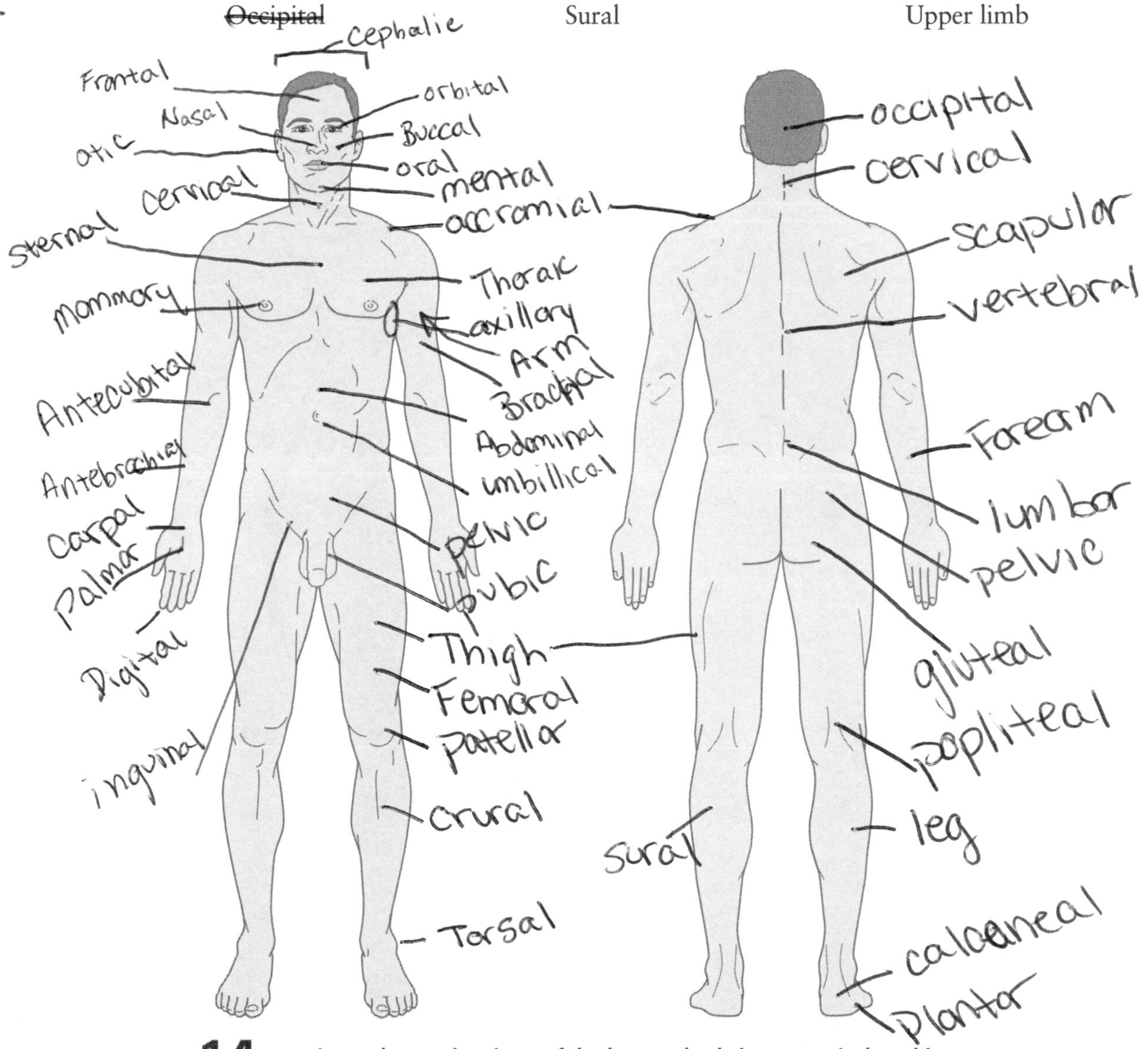

FIGURE 1.4 Anterior and posterior views of the human body in anatomical position

Exercise 4

Body Cavities and Membranes

MATERIALS NEEDED

- Human torso models
- Fetal pigs (or other preserved small mammal)
- Dissection kits / dissection trays

The body is divided into several fluid-filled cavities, each of which contains specific organs. In this exercise you will identify the body cavities and the organs contained within each cavity.

As you can see in **Figure 1.5**, there are two major body cavities, each of which is subdivided into smaller cavities, as follows:

1. **Dorsal (posterior) cavity.** As implied by its name, the dorsal body cavity is largely on the posterior, or dorsal, side of the body. It contains two smaller cavities:

a. **Cranial cavity.** The cranial cavity is the area encased by the skull. It contains the brain and the special sense organs such as the eyes and the organs for hearing.

b. **Vertebral (or spinal) cavity.** The vertebral cavity is the area encased by the vertebrae. It contains the spinal cord.

2. **Ventral (anterior) cavity.** The ventral body cavity is on the anterior, or ventral, side of the body. The divisions of the ventral cavity are surrounded by **serous membranes,** which produce a watery liquid called **serous fluid.** Serous membranes are composed of two layers—the outer **parietal layer,** which is attached to the surrounding structures, and an inner **visceral layer,** which is attached to the organ or organs. The cavities are the narrow spaces between the parietal layer and the visceral layer, where the serous fluid is located. Serous fluid lubricates the organs in the cavity and allows them to move and slide past one another without friction.

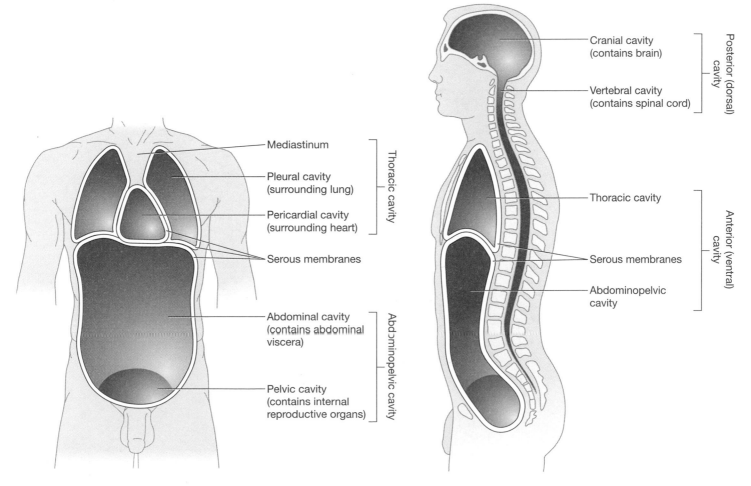

FIGURE 1.5 Anterior and lateral views of the body cavities

EXPLORING ANATOMY & PHYSIOLOGY IN THE LABORATORY

The following are the divisions of the ventral cavity:

a. **Thoracic cavity.** The thoracic cavity encompasses the area that is superior to the diaphragm and encased by the ribs. Its two subdivisions are:

   (1) **Pleural cavities.** Each pleural cavity surrounds one of the lungs. The pleural cavities are housed between the **pleural membranes.** The **parietal pleura** is attached to the body wall, and the **visceral pleura** is attached to the surface of the lung.

   (2) **Mediastinum.** The area between the pleural cavities, called the mediastinum, contains the great vessels, the esophagus, the trachea and bronchi, and other structures. It also houses another cavity called the **pericardial cavity,** which surrounds the heart. The pericardial cavity is between the **pericardial membranes;** the **parietal pericardium** is attached to surrounding structures, and the inner **visceral pericardium** is attached to the heart muscle.

b. **Abdominopelvic cavity.** The abdominopelvic cavity encompasses the area inferior to the diaphragm and extends into the bony pelvis. It is lined by the **peritoneal membranes;** the **parietal peritoneum** is attached to the body wall and surrounding structures, and the inner **visceral peritoneum** is attached to the surface of many of the organs in the cavity. Between these two layers of peritoneal membranes we find the **peritoneal cavity.** Note that some organs are posterior to the peritoneal cavity; such organs are said to be **retroperitoneal.** The abdominopelvic cavity is divided into two smaller cavities:

   (1) **Abdominal cavity.** The area superior to the bony pelvis, called the abdominal cavity, houses many of the organs of the digestive and lymphatic systems.

   (2) **Pelvic cavity.** The cavity housed within the bony pelvis, the pelvic cavity, contains certain organs of the reproductive system as well as certain organs of the digestive and urinary systems.

We often divide the abdominopelvic cavity into four quadrants or nine regions based on a series of lines drawn over the surface of the abdomen. The regions are listed and labeled in **Figure 1.6.**

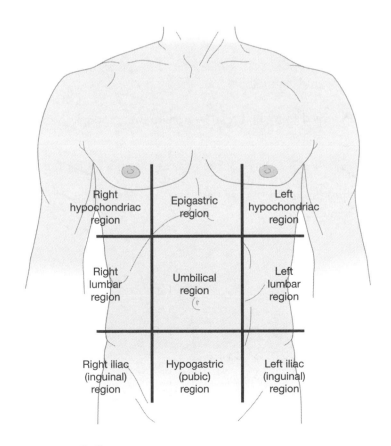

FIGURE 1.6 Regions of the abdominopelvic cavity

 **Procedure** Body Cavities

In this procedure you will use a human torso model or a preserved small mammal such as a cat, a fetal pig, or a rat to examine the body cavities.

Note: Safety glasses and gloves are required!

If you are using a preserved small mammal, you may use the following procedure to open the body cavities. Note that you will not open the animal fully in this procedure to preserve structures for future dissections.

1 Place the animal on a dissecting tray with its dorsal (back) side facing you. Use your scalpel to make a shallow, +-shaped incision. Make the first incision through the skin along the nape of the neck. Make the second incision from the anterior part of the skull down along the animal's midline to about 4 inches inferior to the first incision. Ensure that your incisions are shallow so you do not damage structures deep to the skin. List the organs you are able to see in Table 1.3.

2 Gently peel back the skin using a blunt dissection probe, and note the appearance of the skin and the exposed muscles, bones, and joints. If your instructor wants you to expose the brain and spinal cord, use either a scalpel or a saw to carefully cut through the skull and a section of the vertebral column.

3 When you have finished your examination of the animal's dorsal side, close the skin and wrap the area with wet towels soaked with a preservative solution.

4 Flip the animal over and place it in the dissecting tray with its ventral side facing you. Use a scalpel to make a shallow incision along the animal's midline from the superior neck down to its groin. Make a second incision across the animal's chest and a third incision across the animal's abdomen.

5 Using a blunt dissection probe, peel back the skin of the abdomen carefully to expose the abdominopelvic cavity and examine its contents.

6 Gently peel back the skin of the chest and neck incisions. Use scissors or a scalpel to carefully cut through the sternum and expose the thoracic cavity and examine its contents.

When you have opened your preserved small mammal or human torso model, locate and identify each cavity and do the following:

1 List the organs you are able to see in Table 1.3. See Figures 1.7 and 1.8 for reference.

2 Mark each region of the abdominopelvic cavity with a pin or marking tape (if working with a torso). Note which organs are visible in each region.

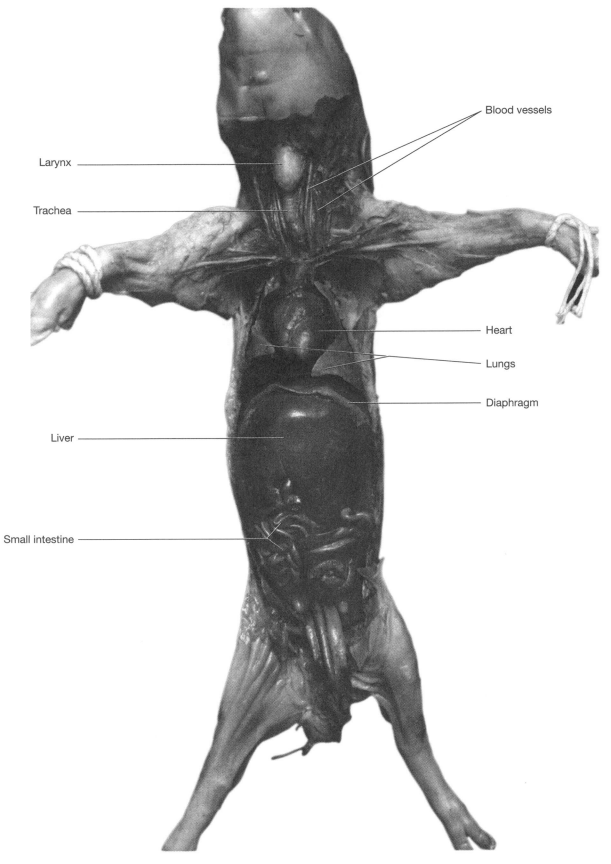

Blood vessels

Larynx

Trachea

Heart

Lungs

Diaphragm

Liver

Small intestine

FIGURE **1.7** Ventral view of the fetal pig

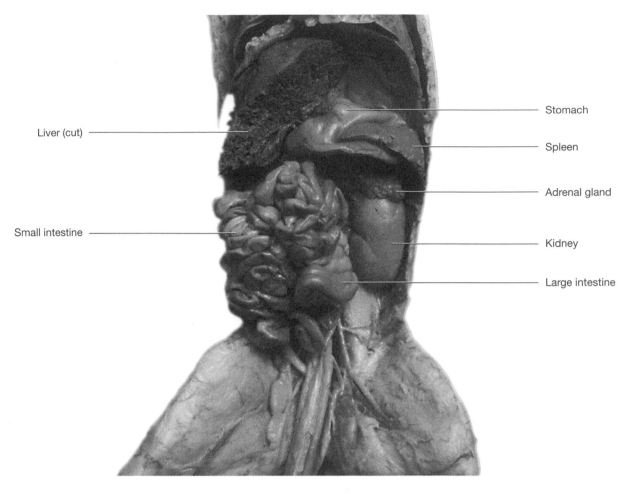

Liver (cut)

Small intestine

Stomach

Spleen

Adrenal gland

Kidney

Large intestine

FIGURE **1.8** Abdominopelvic cavity of the fetal pig

TABLE 1.3 — Body Cavities and Regions of the Abdominopelvic Cavity

Cavity	Organ(s)
Dorsal cavity:	
1. Cranial cavity	Brain, eyes, ears/hearing organs
2. Vertebral cavity	Spinal cord
Ventral cavity:	
1. Thoracic cavity	
a. Pleural cavities	lungs
b. Mediastinum	esophagus, bronchi, trachea
(1) Pericardial cavity	heart
2. Abdominopelvic cavity	
a. Subdivisions:	
(1) Abdominal cavity	Digestive / lymphatic organs
(2) Pelvic cavity	Reproductive, urinary and some digestive
b. Regions:	
(1) Right hypochondriac region	liver
(2) Epigastric region	liver, stomach
(3) Left hypochondriac region	Stomach, spleen
(4) Right lumbar region	Adrenal / Kidney, Intest.
(5) Umbilical region	Intestines
(6) Left lumbar region	Adrenal / Kidney ''
(7) Right iliac region	''
(8) Hypogastric region	
(9) Left iliac region	

Procedure Serous Membranes

Part 1: Serous membranes are best examined on a preserved specimen such as the fetal pig or a cat, as their structure is difficult to appreciate on a model. As you dissect the fetal pig or cat, look for the serous membranes listed in Table 1.4. Take care not to tear the fragile membranes, which consist of just a few layers of cells. As you identify each membrane, record in the table where you found the membrane and the structure to which the membrane is attached (the lungs, heart, abdominal wall, etc.).

TABLE **1.4**	Serous Membranes	
Membrane	**Cavity**	**Structure**
Parietal pleura		
Visceral pleura		
Parietal pericardium		
Visceral pericardium		
Parietal peritoneum		
Visceral peritoneum		

Part 2: Draw in the body cavities on your laminated outlines of the human body or Figure 1.9. Label the serous membranes surrounding the cavity where applicable.

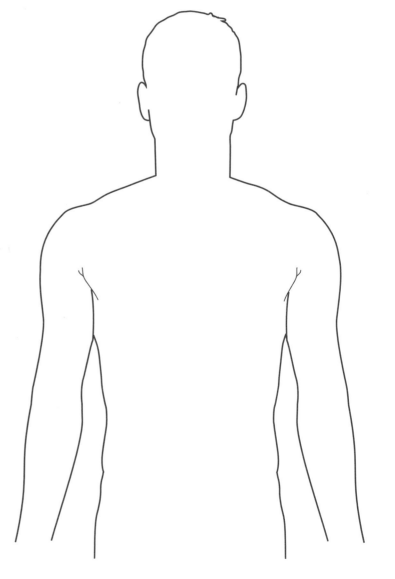

FIGURE 1.9 Anterior view of the human torso

EXPLORING ANATOMY & PHYSIOLOGY IN THE LABORATORY

Procedure Applications of Terms, Cavities, and Membranes

Assume that you are acting as coroner and you have a victim with three gunshot wounds. In this scenario your "victim" will be your fetal pig or a human torso model that your instructor has "shot." For each "bullet wound," state the anatomical region and/or body cavity in which the "bullet" was found and any serous membranes involved, and describe the location of the wound using at least three directional terms. As coroner, you have to be as specific as possible and keep your patient in anatomical position!

Shot 1	
Shot 2	
Shot 3	

Exercise 5

Planes of Section

MATERIALS NEEDED

▶ Various anatomical models demonstrating planes of section
▶ Modeling clay
▶ Knife or scalpel

Often in science and in the medical field, it is necessary to obtain different views of the internal anatomy of an organ or a body cavity. These views are obtained by making an anatomical section along a specific plane. The four commonly used planes of section, shown in Figure 1.10, are:

1. **Sagittal plane.** A section along the sagittal plane is parallel to the body's longitudinal axis and divides the body part into right and left parts. The sagittal section has two variations:

   a. **Midsagittal sections** divide the body part into equal right and left halves.

   b. **Parasagittal sections** divide the body part into unequal right and left parts.

2. **Frontal plane.** The frontal plane, also known as the **coronal plane,** is also parallel to the body's longitudinal axis. It divides the body part into an anterior (front) part and a posterior (back) part.

3. **Transverse plane.** The transverse plane, also known as a **cross-section** or the **horizontal plane,** is perpendicular to the body's longitudinal axis. It divides the body part into a superior (or proximal) part and an inferior (or distal) part.

4. **Oblique section.** An oblique section cuts at an angle and is intended to allow examination of structures that are difficult to see with standard angles. Note that an oblique section is not illustrated in Figure 1.10.

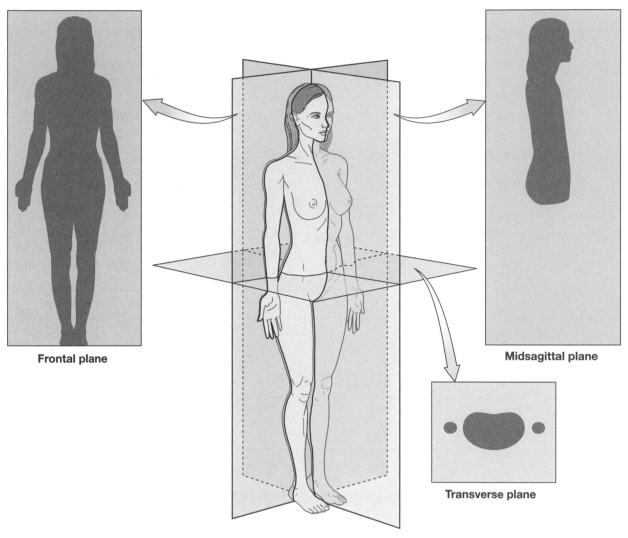

Frontal plane

Midsagittal plane

Transverse plane

FIGURE **1.10** Anatomical planes of section

Procedure Sectioning Along Anatomical Planes

Use a scalpel to cut a ball of modeling clay in each of following anatomical planes. Before you make your cuts, you may wish to mold your clay into the shape of a head and to draw eyes on the head to denote anterior and posterior sides. Use **Figure 1.10** for reference. When you have completed this exercise, answer Check Your Understanding questions 3 and 4 (p. 26).

1. Sagittal
   a. Midsagittal
   b. Parasagittal
2. Frontal
3. Transverse
4. Oblique

Procedure **Identifying Examples of Anatomical Planes of Section**

Identify at least two examples of each plane of section listed in Table 1.5 on anatomical models that your instructor provides. In addition, state which organs are visible in the section.

TABLE **1.5**	Anatomical Models and Planes of Section
Examples of midsagittal and/or parasagittal sections:	
Model	**Organ(s) Visible**
1.	
2.	
Examples of frontal sections:	
Model	**Organ(s) Visible**
1.	
2.	
Examples of transverse sections:	
Model	**Organ(s) Visible**
1.	
2.	

Exercise 6

Organs and Organ Systems

MATERIALS NEEDED

- Human torso models
- Fetal pigs (or other preserved small mammal)
- Dissection kits / dissection trays

The human body has 11 organ systems, each with specific organs and functions (**Figure 1.11**). In this exercise you will examine the organ systems and identify their major anatomical structures.

 Procedure Organs

Identify the following organs on your preserved mammal specimen or human torso models. Check off each organ as you identify it, and record the organ system to which it belongs in Table 1.6.

Adrenal glands	Intestines	Pancreas	Thymus
Blood vessels	Joints	Skin	Thyroid gland
Bones	Kidneys	Spinal cord	Trachea
Brain	Larynx	Spleen	Urinary bladder
Esophagus	Liver	Stomach	
Gallbladder	Lungs	Testes (male) or	
Heart	Muscles	ovaries (female)	

TABLE **1.6**	Organs and Organ Systems
Organ System	**Major Organ(s)**
Integumentary system	
Skeletal system	
Muscular system	
Nervous system	
Cardiovascular system	
Lymphatic system	
Respiratory system	
Digestive system	
Urinary system	
Endocrine system	
Reproductive system	

Procedure Organ Systems

The body's organ systems are illustrated in Figure 1.11. Fill in the blanks next to each organ system to identify the major organs and principal functions of each system. When you have completed this exercise, answer Check Your Understanding question 5 (p. 26).

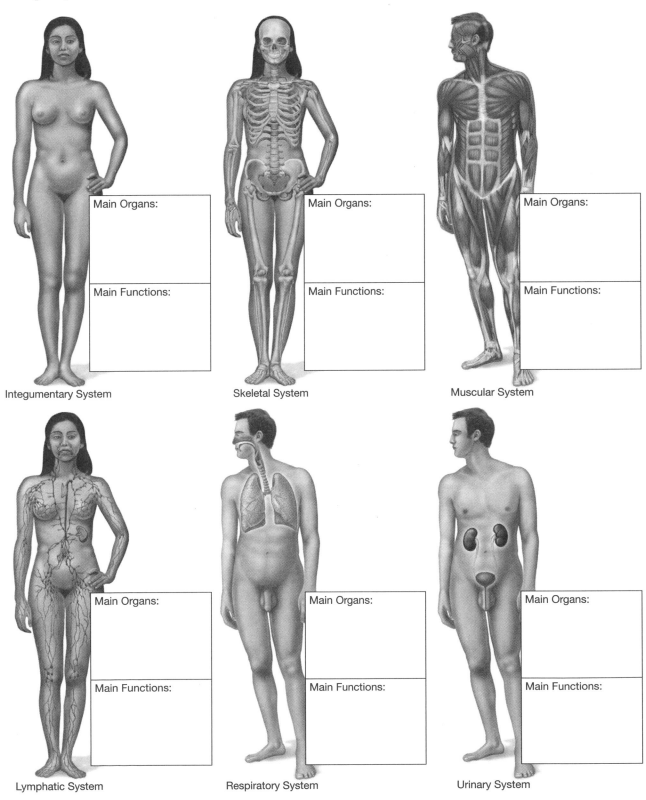

Main Organs:	Main Organs:	Main Organs:
Main Functions:	Main Functions:	Main Functions:

Integumentary System   Skeletal System   Muscular System

Main Organs:	Main Organs:	Main Organs:
Main Functions:	Main Functions:	Main Functions:

Lymphatic System   Respiratory System   Urinary System

FIGURE 1.11 Organ systems of the body

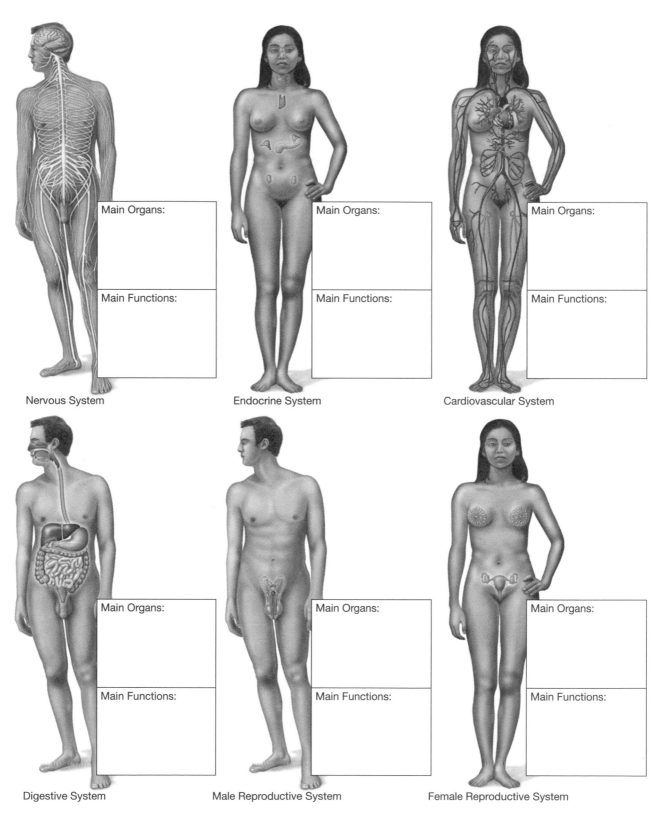

Main Organs:

Main Functions:

Nervous System

Main Organs:

Main Functions:

Endocrine System

Main Organs:

Main Functions:

Cardiovascular System

Main Organs:

Main Functions:

Digestive System

Main Organs:

Main Functions:

Male Reproductive System

Main Organs:

Main Functions:

Female Reproductive System

FIGURE **1.11** Organ systems of the body (cont.)

EXPLORING ANATOMY & PHYSIOLOGY IN THE LABORATORY

NAME _____

SECTION _____ DATE _____

Check Your Recall

1 Which of the following best describes anatomical position?

a. Body facing forward, toes pointing forward, palms facing backward
b. Body, toes, and palms facing backward
c. Body facing forward, arms at the sides, palms facing forward
d. Body facing backward and palms facing outward

2 Match the directional term with its correct definition.

a. Distal _____   A. Away from the surface/toward the body's interior

b. Lateral _____   B. Toward the back of the body

c. Anterior _____   C. Closer to the point of origin (e.g., of a limb)

d. Proximal _____   D. Away from the body's midline

e. Inferior _____   E. Toward the head

f. Deep _____   F. Farther from the point of origin (e.g., of a limb)

g. Superficial _____   G. Toward the body's midline

h. Posterior _____   H. Away from the head/toward the tail

i. Medial _____   I. Toward the front of the body

j. Superior _____   J. Toward the surface/skin

3 Which of the following is an *incorrect* use of a directional term?

a. The ankle is inferior to the knee.
b. The sternum is superior to the abdomen.
c. The bone is deep to the muscle.
d. The mouth is medial to the ears.

4 Label the following anatomical regions on **Figure 1.12**:

- Inguinal region
- Carpal region
- Leg
- Orbital region

- Otic region
- Forearm
- Upper limb
- Cervical region

- Digital region
- Brachial region
- Sternal region
- Lumbar region

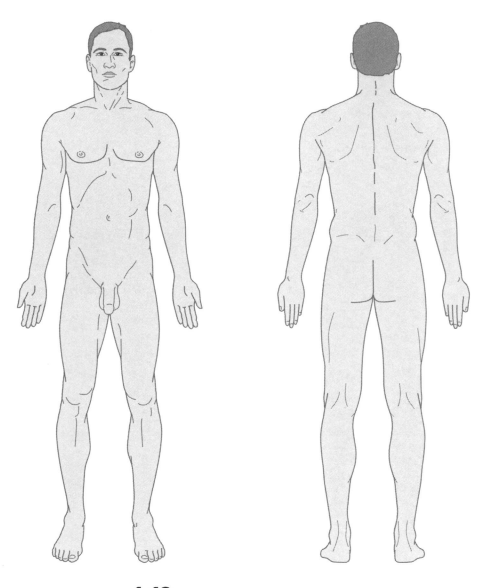

FIGURE 1.12 Anterior and posterior views of the body

5 Anatomical position and specific directional and regional terms are used in anatomy and physiology to:

a. standardize units of measure.
b. provide a standard that facilitates communication and decreases the chances for errors.
c. provide a standard that is used to develop drug delivery systems.
d. make students' lives difficult.

6 Label the following body cavities on Figure 1.13 and indicate with an asterisk (*) which cavities are surrounded by serous membranes.

- Dorsal cavity
- Cranial cavity
- Vertebral cavity
- Ventral cavity
- Thoracic cavity
- Pleural cavities

- Mediastinum
- Pericardial cavity
- Abdominopelvic cavity
- Abdominal cavity
- Pelvic cavity

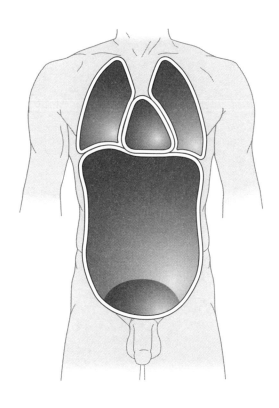
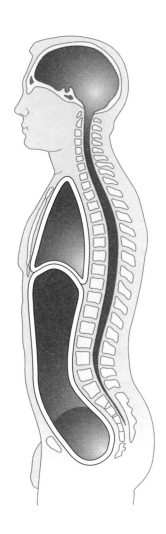

FIGURE 1.13 Anterior and lateral views of the body cavities

7 Fill in the blanks: A serous membrane secretes

_____, which _____ the organs in certain _____ body cavities.

8 Define the following planes of section:

a. Midsagittal plane _____

b. Parasagittal plane _____

c. Frontal plane _____

d. Transverse plane _____

e. Oblique plane _____

9 The following organs belong to the _____ system: esophagus, gallbladder, liver.
a. integumentary
b. reproductive
c. lymphatic
d. digestive

10 The following organs belong to the _____ system: pancreas, thyroid gland, adrenal glands.
a. endocrine
b. urinary
c. lymphatic
d. cardiovascular

Check Your Understanding | **Critical Thinking and Application Questions**

1 Figure 1.14 is not in anatomical position. List all of the deviations from anatomical position.

FIGURE **1.14** Figure not in anatomical position

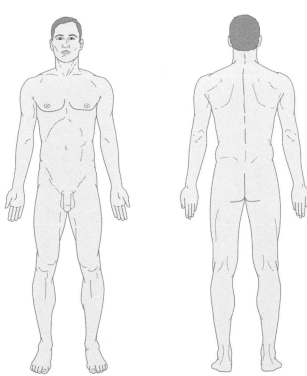

2 You are reading a surgeon's operative report. During the course of the surgery, she made several incisions. Your job is to read her operative report and determine where the incisions were made. Draw the incisions on **Figure 1.15**.

a. The first incision was made in the right anterior cervical region, 3 centimeters lateral to the trachea. The cut extended vertically, 2 centimeters inferior to the mental region, to 3 centimeters superior to the thoracic region.

b. The second incision was made in the left anterior axillary region and extended medially to the sternal region. At the sternal region the cut turned inferiorly to 4 centimeters superior to the umbilical region.

c. The third incision was made in the left posterior scapular region. The cut was extended medially to 2 centimeters lateral to the vertebral region, where it turned superiorly and progressed to one centimeter inferior to the cervical region.

FIGURE **1.15** Anterior and posterior views of the body

3 What may be some real-world applications of anatomical sections? (*Hint*: Think of the medical field.)

4 Which anatomical section(s) would provide a view of the internal anatomy of *both* kidneys?

5 The type of anatomy we are studying in this lab manual is called *systemic anatomy*, which means that we cover the organs related to a specific organ system. Some, however, choose to study anatomy from a regional point of view (e.g., the abdominal region or the thoracic region). Find at least two organ systems that contain organs in different regions of the body, and state where the organs are located in the body.

> *Example*: The nervous system has organs in the cranial cavity (the brain), the spinal cavity (the spinal cord), the thoracic and abdominopelvic cavities (spinal and cranial nerves), and the upper and lower limbs (spinal nerves).

Chemistry

OBJECTIVES

Once you have completed this unit, you should be able to:

- Demonstrate the proper interpretation of pH paper.

- Describe and apply the pH scale.

- Describe the purpose and effects of a buffer.

- Explain and demonstrate the purpose of an enzyme.

- Describe the differences in solubility of ionic, polar, and nonpolar covalent molecules in different solvents.

- Apply laboratory techniques to perform aspirin synthesis and DNA extraction.

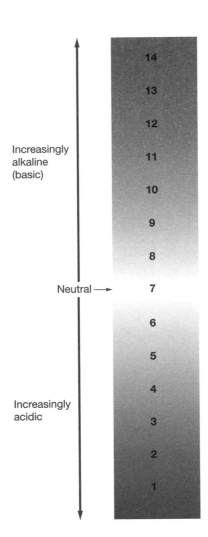

Increasingly
alkaline
(basic)

Neutral →

Increasingly
acidic

14

13

12

11

10

9

8

7

6

5

4

3

2

1

Pre-Lab Exercise 1
Key Terms

Table 2.1 lists the key terms with which you should be familiar before coming to lab.

TABLE 2.1	Key Terms

Term	Definition
pH	_____
	_____
Acid	_____
	_____
Base	_____
	_____
Buffer	_____
	_____
Chemical reaction	_____
	_____
Enzyme	_____
	_____
Polar covalent bond	_____
	_____
Nonpolar covalent bond	_____
	_____
Ionic bond	_____
	_____
Cation	_____
	_____
Anion	_____
	_____

Valence electron _____

Solute _____

Solvent _____

Solubility _____

Deoxyribonucleic acid (DNA) _____

✎ Pre-Lab Exercise 2

The pH Scale and –logarithms

The **pH scale** is a numerical scale from 0–14 that represents the hydrogen ion concentration of a solution. Solutions with a pH of 0 have the highest hydrogen ion concentration, and solutions with a pH of 14 have the lowest hydrogen ion concentration. This is because the pH isn't the actual hydrogen ion concentration; instead, it is a negative logarithm of a solution's hydrogen ion concentration. This actually isn't as difficult as it sounds. Grab your calculator and enter the following:

$-\log(0.01)$ =   _____

$-\log(0.001)$ =   _____

$-\log(0.0001)$ =   _____

$-\log(0.00001)$ =   _____

Notice the general trend here: *As the number gets smaller, its negative log gets bigger.*

Here's a quick example using hydrogen ions: Solution X has a hydrogen ion concentration of 0.05 M. Solution Y has a hydrogen ion concentration of 0.0002 M.

▶ Which solution has more hydrogen ions (*hint*: it's just the solution with the bigger number)? _____

▶ What is the –log of solution X? _____

▶ What is the –log of solution Y? _____

As you can see, the solution with the *higher* hydrogen ion concentration has the *lower* –log. This means that:

▶ Solution X is the more _____ (*acidic/basic*) solution and has a _____ (*lower/higher*) pH.

▶ Solution Y is the more _____ (*acidic/basic*) solution and has a _____ (*lower/higher*) pH.

Pre-Lab Exercise 3
Chemical Bonding

One of the more challenging concepts to grasp in chemistry is **chemical bonding,** a topic that pops up again and again in A&P. We revisit it in cytology, the nervous system, blood, respiration, and digestion—to name just a few.

First let's do some basics. Use your textbook and this unit to answer the following questions.

1. Do metals donate or accept electrons? What do they become after they donate/accept electrons?

2. Do nonmetals donate or accept electrons? What do they become after they donate/accept electrons?

3. Electron shells:

   How many electrons go in the first shell? _____

   How many electrons go in the second shell? _____

   How many electrons go in the third shell? _____

4. How many electrons does each element require in its valence shell to become stable? _____

Okay, now let's apply all of this to an example.

Example 1: Ionic Bond

Let's take a look at sodium (Na) and chlorine (Cl). Draw what I am describing and you will see it better.

   Sodium has how many protons? _____

   Sodium has how many electrons? _____

   How many electrons will go in the first shell? _____

   How many in the second shell? _____

   How many in the third? _____

Now draw this out on the diagram in **Figure 2.1,** and take a look at it, in particular the third (valence) shell.

   We know that Na requires eight electrons in its valence shell to

become stable. But how many does it have? _____

   So, to fill this shell, will it be easier for sodium to steal seven more electrons from another atom, or will it be easier for sodium to give up that one electron and get rid of that third shell? Sodium is simply going

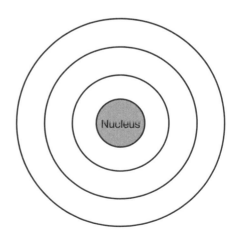

FIGURE 2.1 An orbital model of an atom

to *give away* that last electron. This means that it will lose an electron (negative charge) but will keep the same number of protons (positive charges). What will sodium's overall charge be now? _____

Now let's look at chlorine:

Chlorine has how many protons? _____

Therefore, we know that chlorine has how many electrons? _____

How many electrons will go in the first shell? _____

How many in the second shell? _____

How many in the third? _____

Draw this in Figure 2.2:

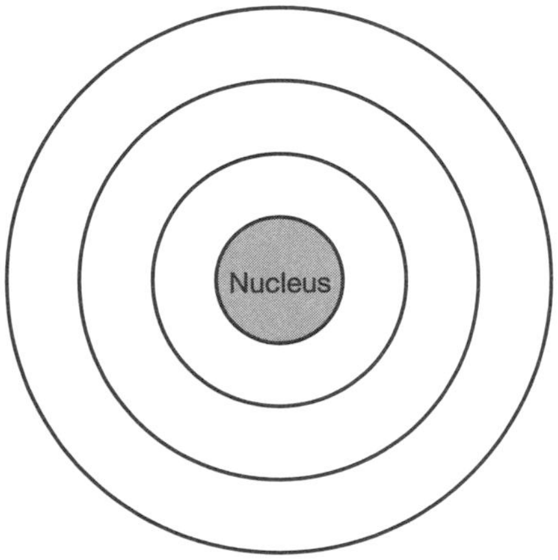

FIGURE **2.2** An orbital model of an atom

We know that chlorine requires eight electrons in its valence shell to become stable. Will it be easier for chlorine to steal one electron from another atom, or will it be easier for chlorine to give away seven electrons? It will be *much* easier for chlorine to take an electron from another atom than to try to pawn off seven electrons. This means that chlorine will gain an electron (negative charge) but will keep the same number of protons (positive charges). What charge will chlorine (now called *chloride*) have now? _____

Let's tie this in with what we learned about metals and nonmetals earlier.

We said that metals _____ electrons to become _____.

Is sodium a metal or a nonmetal? _____

Looking at our example above, did sodium donate or accept an electron? _____

We also said that nonmetals _____ electrons to become _____.

Is chlorine a metal or a nonmetal? _____

Looking at our example above, did chlorine donate or accept an electron? _____

Why do we care? Well, when sodium donates an electron to become a positively charged **cation**, and chlorine accepts an electron to become a negatively charged **anion**, this creates an attraction. (Think of it like "opposites attract.") This attraction leads the two ions to *bond* to one another, forming an **ionic bond**. See how simple that is?

Example 2: Polar Covalent Bond

Now let's tackle another difficult bonding concept, the **polar covalent bond**. The classic example of a polar covalent bond is represented by two bonds in water (Figure 2.3):

Hydrogen (H) has how many protons? _____

Hydrogen has how many electrons? _____

How many electrons will go in the first shell? _____

How many electrons would hydrogen require to fill up that first shell? _____

Oxygen (O) has how many protons? _____

Therefore, we know that oxygen has how many electrons? _____

How many electrons will go in the first shell? _____

How many electrons will go in the second shell? _____

How many electrons would oxygen require to fill up that second shell? _____

In this case, both hydrogen and oxygen are nonmetals, and neither element will give up electrons. So, instead of one element taking the electrons, they share them. Because oxygen requires two electrons to fill that valence shell, and hydrogen has only one electron to share, oxygen has to form a partnership with two hydrogen atoms, forming H_2O. But oxygen has a higher **electronegativity**, which means that it attracts electrons more strongly than hydrogen does. This means that the electrons will spend more time around the oxygen and less around the hydrogen. Thus, a polar covalent bond is born. This means that:

Oxygen will have a partial _____ charge, and the two hydrogens will have a partial _____ charge, which you can see in **Figure 2.3**.

A **nonpolar covalent bond** involves the same principles, except that the atoms share the electrons equally. As a result, there are no partial positive or negative charges because the electrons don't spend any more time around one or more of the atoms.

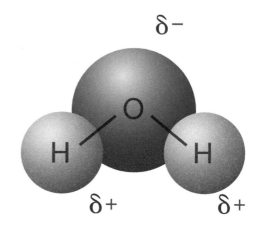

FIGURE **2.3** A water molecule

Exercises

Including a chemistry unit with an Anatomy and Physiology course may seem odd, but consider for a moment the simplest level of organization—the chemical level. Our cells, tissues, and organs, as well as our extracellular environments, are all composed of chemicals that undergo countless chemical reactions every second. So, to be able to understand our anatomy and physiology, we first must understand the most basic structures in our bodies—chemicals.

In the following exercises you will learn chemistry principles as you measure the pH of various solutions, demonstrate the effects of buffers on the pH of solutions, examine the effects of enzymes on chemical reactions, and determine different chemicals' solubilities. Then you will witness chemistry in action by synthesizing aspirin and extracting DNA. It's not every day that you get to say that you made aspirin and extracted DNA!

Safety Note: Many of the chemicals you will use in this lab are poisons and can burn your skin, eyes, and mucous membranes. Take care in handling all chemicals. Do not handle any chemicals without safety glasses, gloves, and a lab coat.

Exercise 1

pH, Acids, and Bases

MATERIALS NEEDED

- 3 glass test tubes
- Test tube rack
- pH paper
- Samples of various acids and bases
- 0.1M Hydrochloric acid
- Samples of various antacid tablets

The **pH** is a measure of the concentration of hydrogen ions present in a solution. As the hydrogen ion concentration increases, the solution becomes more **acidic**. As the hydrogen ion concentration decreases, the solution becomes more **alkaline, or basic**. Notice that the **pH scale**, shown in **Figure 2.4**, ranges from 0 (the most acidic) to 14 (the most basic). As you learned in the Pre-Lab Exercises, the reason for this is that the pH is actually a negative logarithm, so the lower the pH, the higher the hydrogen ion concentration. A pH of 7 is considered **neutral**, which is neither acidic nor basic.

Acids and bases are some of the most important chemicals in the human body: They are found in the stomach, in the blood and other extracellular body fluids, in the cytosol, and in the urine, and are released by cells of the immune system. The normal pH for the blood is about 7.35–7.45. The body regulates this pH tightly, as swings of even 0.2 points in either direction can cause serious disruptions in many of our physiologic processes.

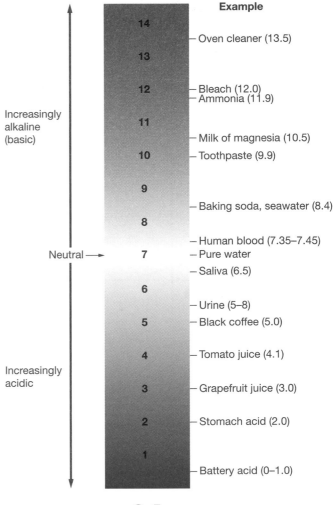

FIGURE **2.4** The pH scale

 ## Procedure Reading the pH

A simple way to measure pH is to use pH paper. To test the pH with pH paper, drop one or two drops of the sample solution on the paper and compare the color change with the colors on the side of the pH paper container. The pH is read as the number that corresponds to the color the paper turned. When you have completed the activity, answer Check Your Understanding question 1 (p. 47).

Note: Safety glasses and gloves are required.

---

1 Obtain two samples each of known acids and bases, and record their molecular formulae in Table 2.2 (the molecular formula should be on the side of each bottle).

2 Measure the pH of each acid and base using pH paper and record their pH values in Table 2.2.

3 Obtain two randomly selected unknown samples and measure their pH using pH paper.

4 Record the pH values in Table 2.2, and determine if this substance is an acid, a base, or is neutral.

TABLE 2.2	Samples of Acids and Bases	
Samples	**Molecular Formula**	**pH**
Acid #1		
Acid #2		
Base #1		
Base #2		
Unknown Samples	**pH**	**Acid, Base, or Neutral?**
Unknown #1		
Unknown #2		

Procedure pH Applications

Now let's apply the pH scale to physiologic systems. The stomach contains concentrated hydrochloric acid, and the pH of the stomach contents ranges between 1 and 3. **Antacids** are a group of medications that neutralize stomach acid to treat a variety of conditions, including *gastroesophageal reflux*, commonly known as *heartburn*. Antacids generally consist of a metal bound ionically to a base. In this activity you will compare the effectiveness of three widely available antacids in neutralizing concentrated hydrochloric acid.

Note: Safety glasses and gloves are required.

---

1 Obtain three glass test tubes and label them 1, 2, and 3.

2 Obtain a bottle of 0.1M hydrochloric acid (HCl), and record its pH in the blank below.

pH of 0.1M HCl: _____

3 In each tube, place 2 ml of the 0.1M HCl.

4 Add one-fourth of one crushed Tums® tablet to tube 1.

5 Add one-fourth of one crushed Rolaids® tablet to tube 2.

6 Add one-fourth of one crushed Alka-Seltzer® tablet to tube 3.

7 Allow the tubes to sit undisturbed for 3 minutes.

8 Measure the pH of the contents of each tube and record the values in Table 2.3.

9 What effect did the antacids have on the acid?

TABLE 2.3	Effectiveness of Three Antacids	
Antacid	**Active Ingredient**	**pH**
Tums®		
Rolaids®		
Alka-Seltzer®		

10 Based upon your observations, which antacid is the most effective at neutralizing the acid?

 # Exercise 2

Buffers

MATERIALS NEEDED

- Well plates
- Stirring rods
- 0.1M Hydrochloric acid
- Buffered solution
- Distilled water
- 0.1M NaOH
- pH paper

A **buffer** is a chemical that resists changes in pH, and a solution to which a buffer has been added is called a **buffered solution**. When acid is added to a buffered solution, the buffer binds the added hydrogen ions and removes them from the solution. Similarly, when a base is added to a buffered solution, the buffer releases hydrogen ions into the solution. Both effects minimize the pH changes that otherwise would occur in the solution if the buffer were not present.

 ## Procedure Testing Buffered Solutions

In this experiment you will examine the effects of adding an acid or a base to buffered solutions and non-buffered solutions. You will use distilled water as your non-buffered solution, and your instructor will choose an appropriate buffered solution for you to use. When you have completed the activity, answer Check Your Understanding question 2 (p. 47).

Note: Safety glasses and gloves are required.

---

1 Obtain a well plate and number four wells 1, 2, 3, and 4.

2 Fill wells 1 and 2 about half-full of distilled water. Measure the pH of the distilled water, and record that value in Table 2.4 ("pH before").

3 Fill wells 3 and 4 about half-full of buffered solution. Measure the pH of the buffered solution and record that value in Table 2.4 ("pH before").

4 Add two drops of 0.1M HCl to wells 1 and 3. Stir the solutions with a stirring rod (or a toothpick) and measure the pH of each well. Record the pH in Table 2.4 ("pH after").

TABLE 2.4	Buffered and Non-buffered Solutions		
Well	Contents	pH Before	pH After
1	Water and HCl		
2	Water and NaOH		
3	Buffer and HCl		
4	Buffer and NaOH		

5 Add two drops of 0.2M NaOH (a base) to wells 2 and 4. Stir the solutions and measure the pH of each well. Record the pH in Table 2.4 ("pH after").

6 Interpret your results. What effect did the buffer have on the pH changes that you saw?

 Exercise 3

Enzymes and Chemical Reactions

MATERIALS NEEDED

- 9 glass test tubes
- Test tube rack
- Lipase
- Boiled lipase
- Bile salts
- Vegetable oil
- Ice water bath
- Warm-water bath (set to 37° C)
- Distilled water
- Phenol red

Often when atoms or molecules interact, chemical bonds are formed, broken, or rearranged, or electrons are transferred between molecules. These interactions are called **chemical reactions**. Most chemical reactions can proceed spontaneously, but they often take an extremely long time.

One factor that can alter the rate at which a reaction takes place is *temperature*. Generally, when temperatures increase (up to a point), molecules move faster and they collide and react at a faster rate. The opposite happens when temperatures are decreased.

Another factor that can affect reaction rate is a substance called a **catalyst**, which is added to a reaction to increase its rate. Catalysts are not consumed in the reaction and may be reused after the reaction has completed.

In the body, biological catalysts called **enzymes** speed up essentially all of the chemical reactions. Nearly all enzymes in the body are proteins that work by binding specifically to the reacting components and by reducing the amount of energy required for a reaction to proceed (called the *activation energy*). Note here that an enzyme must bind to the reactants for it to work. This means that if the enzyme is damaged and loses its shape, it will not be able to function. Enzymes may be damaged by the same processes that damage all proteins, including extreme heat and extreme pH swings. An enzyme that has lost its shape as a result of such damage is said to be **denatured**.

 ## Procedure Test Enzymatic Activity

In the following procedure, you will be comparing the ability of three solutions—lipase, boiled lipase, and water—to digest vegetable oil at three different temperatures. **Lipase** is an enzyme found in the human digestive tract that speeds up the digestion of dietary fats. Note that you also must add another component, called **bile**, to your mixture. Bile is not an enzyme itself, but it does increase the ability of lipase to facilitate fat digestion.

You will check for the presence of digestion using an indicator called **phenol red**. Phenol red appears pink at an alkaline (basic) pH, changes to an orange-red color at a neutral pH, and changes to a yellow color when the pH becomes acidic.

If fat has been digested, fatty acids will be released that will decrease the pH of the contents of your tube and turn them orange-red or yellow. You may interpret your results in the following way:

▶ Pink color = the pH is basic and no (or limited) fat digestion occurred

▶ Red-orange color = the pH is neutral and some fat digestion occurred

▶ Yellow color = the pH is acidic and significant fat digestion occurred

---

Note: Safety glasses and gloves are required.

---

1 Obtain nine glass test tubes and label them 1 through 9 with a marker.

2 Add 3 ml of vegetable oil to each test tube.

3 Add 8–10 drops of the pH indicator phenol red to each test tube. The oil now should appear pink. If it does not, add drops of 0.1M NaOH (a base) to each tube until the indicator turns pink.

4 Add the following ingredients to each test tube:

▶ Tubes 1, 4, and 7: 1 ml lipase, 1 ml bile

▶ Tubes 2, 5, and 8: 1 ml boiled lipase, 1 ml bile

▶ Tubes 3, 6, and 9: 2 ml water

5 Place tubes 1, 2, and 3 in an ice bath and incubate them for 30 minutes.

6 Leave tubes 4, 5, and 6 in your test tube rack to incubate at room temperature for 30 minutes.

7 Place tubes 7, 8, and 9 in a warm-water bath set to 37° C and incubate them for 30 minutes.

8 After 30 minutes have passed, remove the tubes from the ice bath and warm-water bath and place them in your test tube rack.

9 Record the color of each tube in Table 2.5 and interpret your results.

TABLE **2.5**	**Enzymatic Activity of Lipase**		
Tube Number	**Color**	**pH (Acidic, Neutral, or Alkaline)**	**Amount of Digestion that Occurred**
1			
2			
3			
4			
5			
6			
7			
8			
9			

10 Answer the following questions about your results:

a. What effect does temperature have on enzyme activity?

b. Were your results with the boiled lipase different from the results with non-boiled lipase? Why?

c. Predict what color change you would see in tubes 3, 6, and 9 if you were to leave them in a warm-water bath for several days.

Exercise 4

Chemical Solubilities

MATERIALS NEEDED

- 10 glass test tubes
- 10 rubber stoppers
- Test tube rack
- Distilled water
- Paint thinner
- Table salt
- Sucrose
- Iodine crystals
- Unknown solids

In the Pre-Lab Exercises you learned that three main types of bonds are present between atoms:

1. **Ionic bonds.** An ionic bond forms when one or more electrons is transferred from a metal to a nonmetal. In this transfer, the metal becomes a positive cation and the nonmetal becomes a negative anion. These positive and negative charges attract and result in the ionic bond.

2. **Covalent bonds.** Covalent bonds form when two or more atoms share electrons. This results in more stable atoms with eight electrons in their valence shells (or two electrons for atoms with five or fewer protons). The two types of covalent bonds are:

   a. **Polar covalent bonds.** A polar covalent bond results from the unequal sharing of electrons between atoms with unequal electronegativities. Essentially, one atom is "greedier" (more electronegative) than the other, and the electrons spend more time around the "greedier" atom. This creates what is known as a **dipole**—a molecule with a partial positive pole and a partial negative pole ("dipole" literally means "two poles").

   b. **Nonpolar covalent bonds.** The electrons are shared equally between the atoms involved in a nonpolar covalent bond because the two atoms have nearly the same electronegativities. As a result, they have no partial positive or negative charges.

Chemicals are not limited to having only ionic, polar covalent, or nonpolar covalent bonds. Certain chemicals are **amphipathic**, meaning that they have both polar and nonpolar parts. Many important biological molecules, including phospholipids and bile salts, are amphipathic.

How to Determine the Types of Bonds in a Molecule

Following are some simple rules for determining whether a compound is a polar covalent, a nonpolar covalent, or an ionic compound.

▶ Any metal bound to a nonmetal is an ionic compound.

▶ All compounds that contain only C and H are nonpolar covalent.

▶ Any molecule containing only identical atoms (e.g., H_2) is a nonpolar covalent molecule.

▶ A compound containing a significant amount (about ¼ or more of the total atoms) of O, N, or P is a polar covalent compound.

▶ Any compound containing mostly C and H is overall a nonpolar covalent compound. Note that many nonpolar covalent compounds do contain elements such as O or N, but these atoms are in the minority.

Determining the types of bonds present in a molecule is important because molecules will generally only mix well with other molecules that have similar bonds. In general, the following rule can be used to determine whether one molecule will mix with another molecule:

Like interacts with like.

This means that chemicals with charged ends (polar covalent and ionic molecules) prefer to interact with chemicals that also have charged ends (other polar covalent and ionic molecules). Chemicals without charged ends (nonpolar covalent molecules) prefer to interact with chemicals that also have no charged ends (other nonpolar covalent molecules). This is something you have seen in action if you have ever tried to combine oil (nonpolar covalent compound) and water (polar covalent compound). In their natural conditions, oil and water don't mix because of their types of chemical bonds.

The degree to which a solid substance, also called a **solute**, will mix with a liquid substance, also called a **solvent**, is known as its **solubility**. A solute that dissolves completely in a solvent is said to be highly *soluble* in that solvent. Conversely, a solute that does not dissolve in a solvent is said to be *insoluble* in that solvent. The rules for which solutes are soluble in which solvents are simple:

▶ Polar covalent and ionic solutes are soluble in most polar covalent solvents.

▶ Nonpolar covalent solutes are soluble in most nonpolar covalent solvents.

Procedure Test Chemical Solubilities

You now will determine whether solutes are polar covalent, nonpolar covalent, or ionic based upon the type of solvent in which they dissolve. In this exercise, you will have two solvents—one that is polar covalent and one that is nonpolar covalent. You will attempt to dissolve three known samples of solute and two unknown samples of solute in these solvents, after which you will use these solubilities to determine the types of bonds that may be present in your unknown samples.

After you test them, your instructor may wish to give you the molecular formulae of your unknown samples so you may determine the types of molecules more precisely. When you have completed the activity, answer Check Your Understanding question 3 (p. 47).

Note: Safety glasses and gloves are required.

---

1 Obtain 10 glass test tubes, a test tube rack, and 10 rubber stoppers. Number the test tubes 1 through 10.

2 Add 2 ml of distilled water to tubes 1–5. Add 2 ml of paint thinner to tubes 6–10.

    a. Which solvent is the polar solvent?

    b. Which solvent is the nonpolar solvent?

3 Add a pinch of table salt to tubes 1 and 6. What is the molecular formula for salt?

4 Add a pinch of sucrose to tubes 2 and 7. What is the molecular formula for sucrose?

5 Add a pinch of iodine crystals to tubes 3 and 8. What is the molecular formula for iodine?

6 Add a pinch of unknown substance A to tubes 4 and 9.

7 Add a pinch of unknown substance B to tubes 5 and 10.

8 Place a stopper firmly in each of the tubes. Shake each tube for several seconds until you are certain that no further change will take place in the solute.

9 Examine each tube and determine if the solute dissolved or remained separate. Record your results in Table 2.6.

TABLE **2.6** Solubilities		
Tube	**Dissolved in Polar Solvent?**	**Dissolved in Nonpolar Solvent?**
1 (salt + water)		
2 (sucrose + water)		
3 (iodine + water)		
4 (unknown A + water)		
5 (unknown B + water)		
6 (salt + paint thinner)		
7 (sucrose + paint thinner)		
8 (iodine + paint thinner)		
9 (unknown A + paint thinner)		
10 (unknown B + paint thinner)		

10 Based on your results, answer the following questions:

a. What type of chemical bonds are present in salt?

b. What type of chemical bonds are present in sucrose?

c. What type of chemical bonds are present in iodine?

d. What type of chemical bonds are present in unknown A?

e. What type of chemical bonds are present in unknown B?

Exercise 5

Aspirin Synthesis

MATERIALS NEEDED

- Well plates
- Willow bark
- Ethanol
- Glacial acetic acid
- Ferric chloride
- Pipettes

In this exercise you will use the principles you learned in the previous exercises to make one of the most commonly used over-the-counter anti-inflammatory drugs—**aspirin**. Aspirin is derived from the chemical **salicylic acid**, found naturally in willow bark. The "all natural" salicylic acid, however, contains a chemical group called a *phenol*, which makes it highly irritating to the stomach.

Fortunately, we can remove the phenol group and replace it with another chemical group called an *acetyl* group, which is much less irritating to the stomach. This simple substitution changes the drug from salicylic acid to **acetylsalicylic acid**, the chemical name for aspirin.

 Procedure Synthesize Aspirin

The first step in aspirin synthesis is to isolate salicylic acid from its source, willow bark. This is done by soaking the willow bark in ethanol. Salicylic acid is soluble in the slightly polar ethanol, which pulls it out of the willow bark and into solution. In the second step, the phenol group is removed and replaced with an acetyl group. This is accomplished by adding strong acetic acid (which is strong vinegar) to the ethanol-salicylic acid solution. When you have completed the activity, answer Check Your Understanding question 4 (p. 48).

Note: Safety glasses and gloves are required.

---

1 Obtain a well plate, some willow bark, and ethanol.

2 Fill the large well with willow bark, and cover the bark with ethanol. Label two smaller wells as 1 and 2.

3 Soak the bark in ethanol for 15 minutes.

4 After 15 minutes, use a pipette to remove several drops of the ethanol-salicylic acid solution, and place equal amounts into well 1 and well 2.

5 Add two drops of iron chloride (also called ferric chloride) into well 1. Iron chloride reacts with phenol groups to turn a dark purple-brown color. If no phenol groups are present, the solution remains orange.

Color of solution in well 1: _____

Does the solution in well 1 contain salicylic acid? How do you know?

6 Add 15 drops of glacial acetic acid (vinegar) to the solution in well 2, and stir for 15 seconds. You have now synthesized aspirin!

7 Into well 2, add two drops of iron chloride.

Color of solution in well 3: _____

Was your synthesis of aspirin successful? How can you tell?

Exercise 6

DNA Extraction

MATERIALS NEEDED

- 50 ml beaker
- Test tube
- Graduated cylinder
- Pea mixture
- Isopropyl alcohol
- Enzymes (meat tenderizer)
- Liquid detergent
- Wooden applicator stick

The chemical **deoxyribonucleic acid,** or **DNA,** is located in the nucleus of nearly all cells (mature red blood cells are an exception). DNA contains the instructions for building every protein in an organism. It is the small differences in DNA from one organism to another that account for the vast diversity we see on this planet. DNA extraction and analysis have become commonplace in the search for diagnoses of genetic diseases, our evolutionary history, and many other topics in science and medicine. DNA analysis requires specialized equipment, but DNA extraction is surprisingly simple and can be performed using things you have around the house.

 Procedure DNA Extraction

In this exercise, we will be extracting DNA from dried green peas. Although this obviously is not animal DNA, keep in mind that it is composed of precisely the same nucleic acid bases as animal DNA, and that the plant cell membrane structure is identical to that of animal cells. The first part of the extraction involves removing the DNA from the cell and the nucleus using a mild detergent. Then you will use enzymes to degrade the proteins around the DNA and free up the nucleic acids. Finally, you will take advantage of DNA's weak polarity to extract it from strongly polar water into weakly polar isopropyl alcohol. When you have completed the activity, answer Check Your Understanding questions 5 and 6 (p. 48).

Note: Safety glasses and gloves are required.

---

1 Obtain about 10 ml of the blended pea mixture (which contains only dried peas and water), and pour it into a 50 ml beaker.

2 Add 1.6 ml of liquid detergent to the pea soup. Stir gently with a glass stirring rod and let the mixture sit undisturbed for 5–10 minutes.

3 After 5–10 minutes, fill about one-third of a test tube with the pea-soap mixture.

4 Add a pinch of meat tenderizer (which contains enzymes) to the test tube, and stir with a glass stirring rod *very gently* (rough stirring will break up the DNA, making it harder to see).

5 Tilt the test tube, and slowly add isopropyl alcohol so the alcohol forms a layer on the top of the pea mixture. The amount of alcohol you add should be approximately equal to the volume of the pea mixture.

6 Watch as the DNA (it looks like long, stringy stuff) rises to the alcohol layer.

7 Use an applicator stick to fish out the DNA and set it on a paper towel. Voila! You have just extracted DNA!

UNIT 2 REVIEW

NAME _____

SECTION _____ DATE _____

1 Fill in the blanks: A solution with a pH less than 7 is considered _____, a solution with a pH of 7 is

considered _____, and a solution with a pH greater than 7 is considered _____.

2 The pH is a measure of a solution's _____.
   a. hydrogen concentration
   b. oxygen concentration
   c. cation concentration
   d. hydrogen ion concentration

3 A buffer is a chemical that
   a. keeps the pH of a solution neutral.
   b. resists a change in pH.
   c. adds hydrogen ions to a solution when the pH decreases.
   d. removes hydrogen ions from a solution when the pH increases.

4 A chemical reaction takes place when:
   a. chemical bonds are formed.
   b. chemical bonds are broken.
   c. chemical bonds are rearranged.
   d. electrons are transferred between molecules.
   e. All of the above.

5 Fill in the blanks: Increasing the temperature generally _____ the rate of reactions, and decreasing

the temperature generally _____ the rate of reactions.

6 Which of the following is *not* a property of enzymes?
   a. They increase the rate of a reaction.
   b. They are not consumed in the reaction.
   c. They change the products of a reaction.
   d. They are biological catalysts.

7 An enzyme that has lost its shape
   a. is denatured and no longer functional.
   b. is denatured but still functional.
   c. is not different to a regular enzyme as shape is not important.
   d. can be restored by extreme temperatures or pH swings.

8 Predict whether each of the following has ionic, polar covalent, or nonpolar covalent bonds.

O_2 _____

$MgBr_2$ _____

$C_6H_{12}O_6$ _____

CH_4 _____

I_2 _____

NH_3 _____

9 Fill in the blanks: The degree to which a solid substance, also called a _____, will mix with a liquid substance, also called a _____, is known as its _____.

10 Which of the following molecules would you expect to dissolve in water?
a. Polar covalent molecules
b. Ionic molecules
c. Nonpolar covalent molecules
d. Both a and b are correct
e. Both a and c are correct

Check Your Understanding | Critical Thinking and Application Questions

1 Your lab partner argues with you that if you add acid to a solution, the hydrogen ion concentration should increase and, therefore, the pH should increase. Is he correct? What do you tell him?

2 What is the normal pH range for human blood? Which buffer systems in the body maintain the blood pH within this range?

Use your text to determine what happens if the buffer systems are overwhelmed and

   a. the blood becomes excessively basic (a condition called alkalosis).

   b. the blood becomes excessively acidic (called acidosis).

3 The blood is largely composed of water. Which types of biological molecules can the blood transport safely throughout the body? Which types of biological molecules can the blood not transport safely throughout the body without modification? Why?

4 The popularity of herbal medical remedies has surged recently. Most of these remedies claim that, because they are derived from plants, they are "all natural" and "drug-free." Given the aspirin synthesis you just performed, what do you think of such claims? Furthermore, which preparation is less irritating to the stomach: the "all natural" willow bark or the synthetic aspirin?

5 Detergents have a structure very similar to the phospholipids in the plasma membrane. Explain why you used detergents in the first step of your DNA extraction. (*Hint*: Think about the rule that "like dissolves like.")

6 The enzymes in meat tenderizer are called *proteases*. What reactions do proteases catalyze? Why must we have proteases in this extraction? (*Hint:* Think about the structure of DNA. What is in it besides nucleic acids?)

Introduction to the Microscope

OBJECTIVES

Once you have completed this unit, you should be able to:

- Identify the major parts of the microscope.

- Define the magnification of high, medium, and low power.

- Demonstrate proper use of the microscope.

- Practice focusing on low, medium, and high power.

- Define depth of focus.

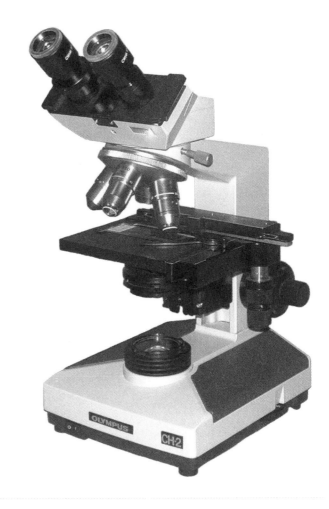

Exercises

Working with the microscope and slides seems to be one of the least favorite tasks of anatomy and physiology students. But with a bit of help, a fair amount of patience, and a lot of practice, the use of microscopes becomes progressively easier with each unit (and, yes, we do look at slides in almost every unit!).

Exercise 1

Introduction to the Microscope

MATERIALS NEEDED

> Light microscopes with three objective lenses

The microscopes that you will use in this lab are called **light microscopes** (Figure 3.1). This type of microscope shines light through the specimens to illuminate them, and the light is refracted through objective lenses to magnify the image. Light microscopes have the following components:

- **Ocular lens.** The ocular lens is the lens through which you look to examine the slide. The microscope may have one ocular lens (a **monocular** microscope) or two ocular lenses (a **binocular** microscope). Many ocular lenses have pointers that can be moved by rotating the black **eyepiece.**

- **Objective lenses.** These are lenses with various powers of magnification. Most microscopes have low (4×), medium (10×), and high (40×) power objective lenses. The objective lenses are attached to the **nosepiece,** which allows the operator to switch between objectives. Please note that certain microscopes have a higher power objective (100×), called the **oil immersion lens,** which requires that a drop of oil be placed between the slide and objective lens. Your instructor may have an oil immersion microscope set up for demonstration purposes.

- **Stage.** The stage is the surface on which the slide sits. It typically has stage clips to hold the slide in place. The stage on many microscopes is moveable using the mechanical stage adjustment knob. Others require you to move the slide manually.

- **Arm.** The arm supports the body of the microscope and typically houses the adjustment knobs.

- **Coarse adjustment knob.** This large knob on the side of the arm moves the stage up and down to change the distance of the stage from the objective lenses. It allows gross focusing of the image.

- **Fine adjustment knob.** This smaller knob allows fine-tuning of the image's focus.

- **Lamp.** The lamp, also called the *illuminator*, provides the light source. It rests on the *base* of the microscope.

- **Iris diaphragm.** This adjustable wheel on the underside of the stage controls the amount of light that is allowed to pass through the slide.

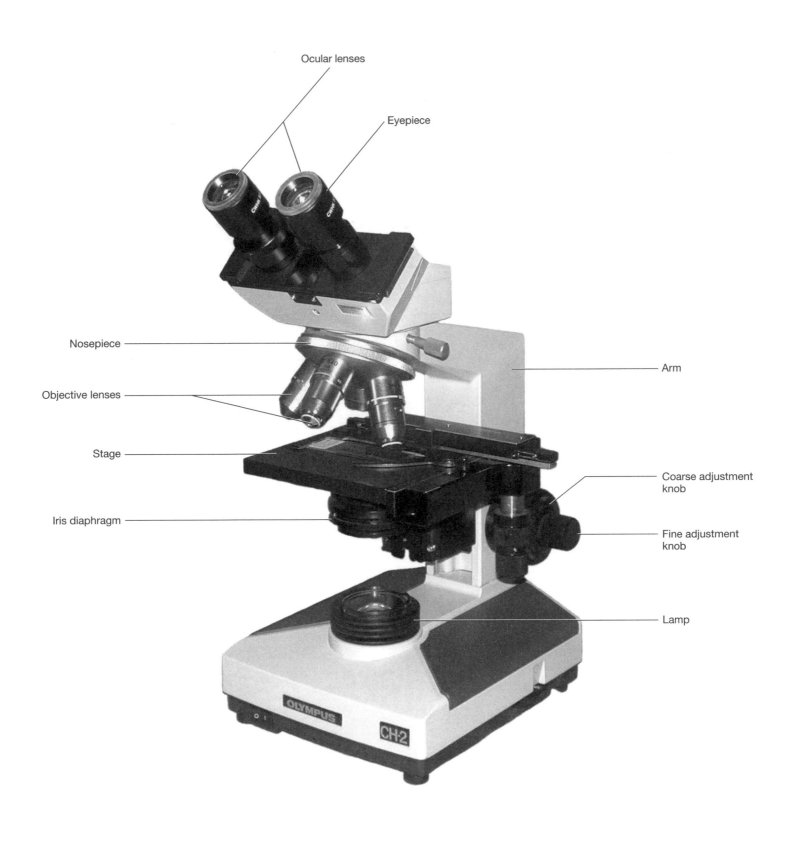

Ocular lenses

Eyepiece

Nosepiece

Arm

Objective lenses

Stage

Coarse adjustment knob

Iris diaphragm

Fine adjustment knob

Lamp

OLYMPUS

CH-2

FIGURE **3.1** A compound light microscope

Procedure Magnification

Light is refracted through two lenses to obtain magnification—the ocular lens and the objective lens. Magnification of the ocular lens is always 10× (magnified 10 times). Magnification of the objectives lens varies, but typically is 4× for low power, 10× for medium power, and 40× for high power. (To verify that this is the case for your microscope, look at the side of the objective lens, which usually is labeled with its magnification.). Remember that oil immersion provides even greater magnification at 100×. The total magnification is obtained by multiplying the power of the ocular lens by the power of the objective lens.

Fill in Table 3.1 to determine total magnification at each power:

TABLE 3.1			
Power	**Magnification of Ocular Lens**	**Magnification of Objective Lens**	**Total Magnification**
Low			
Medium			
High			

 ## Exercise 2

Focusing and Using the Microscope

MATERIALS NEEDED

- Light microscope with three objective lenses
- Lens paper
- Introductory slides (three colored threads and the letter "e")

As I'm certain your instructor will point out, microscopes are expensive! Care must be taken to ensure that the microscopes stay in good working condition. Taking proper care of microscopes makes the histology sections of your labs run more smoothly and also ensures that you stay on your lab instructor's good side! Bearing that in mind, following are some general guidelines for handling the microscopes:

▶ Use two hands to support the microscope when carrying it—one hand to hold the arm and the other hand to support the base.

▶ Gather up the cord so it does not dangle off the lab table after you have plugged in the microscope. This will help to prevent people from tripping over loose cords.

▶ Clean the lenses with lens paper only. Do not use paper towels or cloth to clean a lens, as this will scratch its surface.

▶ Make sure the lowest power objective is in place before you begin.

▶ Get the image in focus on low power, then switch to higher power and adjust with the fine adjustment knob. Be careful not to use the coarse adjustment knob with the high power objective in place, as you could break the slide and damage the lens.

If you follow these general guidelines, you can rest assured that the microscopes (and your grade) will not suffer any harm.

Procedure Focusing the Microscope

Now that we know how to handle the microscope properly, let's practice using it.

1 Obtain a slide of the letter "e."

2 Examine the letter "e" slide before placing it on the stage. How is the "e" oriented on the slide? Is it right side up, upside down, backward, etc.?

3 Ensure that the nosepiece is switched to low power, place the slide on the stage, and secure it with the stage clips.

4 Use the coarse adjustment knob to bring the slide into focus slowly. Once it is grossly in focus, use the fine adjustment knob to sharpen the focus. How is the "e" oriented in the field of view? Is it different from the way it was when you examined it in item 2?

5 Move the nosepiece to medium power. You should only have to adjust the focus with the fine adjustment knob; no adjustment of the coarse focus should be necessary.

6 Once you have examined the slide on medium power, move the nosepiece to high power. Again, focus only with the fine adjustment knob.

Wasn't that easy?

 Procedure Depth of Focus

At times you will look at a slide and see something brown or black and kind of neat-looking with interesting swirls and specks. What is this fascinating discovery you've made? It's dirt on top of the slide. This happens because people tend to focus the objective on the first thing they can make out, which usually is the top of the coverslip on the slide, which has a tendency to be dirty.

These "dirt discoveries" can be avoided by appreciating what is known as **depth of focus**. Also called the *depth of field*, the depth of focus is the thickness of a specimen that is in sharp focus. Thicker specimens will require you to focus up and down to look at all levels of the specimen. This takes practice and skill. Let's get some practice doing this.

1 Obtain a slide with three colored threads. The threads are located on the slide at varying depths, and you will have to focus on each thread individually.

2 Examine the slide prior to putting it on the stage.

3 Ensure that the nosepiece is switched to low power, place the slide on the stage, and secure it with the stage clips.

4 Use the coarse adjustment knob to get the slide into focus on low power.

5 Switch to medium power, and use the fine adjustment knob to sharpen the focus. Which thread(s) is(are) in focus?

6 Move the objective up and down slowly with the coarse adjustment knob, focusing on each individual thread. Figure out which color thread is on the bottom, in the middle, and on the top, and write the color below:

Bottom   _____

Middle   _____

Top   _____

HINTS & TIPS

How to Approach Microscopy

Okay, so you can focus on newsprint and threads, but what about cellular structures and tissue sections? Well, those certainly are more difficult, but if you keep the following hints in mind, the task becomes much simpler:

▶ **Always start on low power.** You are *supposed* to start on low power anyway, to avoid damaging the objective lenses. Sometimes, though, students forget and jump straight to medium or high power. This risks damaging the lenses and also makes it harder on you. Bear in mind that most slides will have more than one histological or cellular structure on each slide. Starting on low power allows you to scroll through a large area of the slide and then focus in on the desired part of the section.

▶ **Beware of too much light.** It is easy to wash out the specimen with too much light. If you are having difficulty making out details, first adjust the focus with the fine adjustment knob. If this doesn't help, use the iris diaphragm to reduce the amount of light illuminating the specimen. This will increase the contrast and allow you to observe more details. It also helps to reduce headaches and eyestrain.

▶ **Keep both eyes open.** It is tempting to close one eye when looking through a monocular microscope. Admittedly, keeping both eyes open isn't easy at first, but it helps to reduce eyestrain and headaches.

▶ **Compare your specimen to the photos in your lab manual.** Although the slides you are examining will not necessarily be identical to the photos in this book, they should be similar in appearance. Generally speaking, if you are looking at something that is vastly different from what is in this book, you probably should move the slide around a bit or change to a different power objective to find the correct tissue or cell type on the slide.

▶ **Remember that the slides aren't perfect.** Not all slides will clearly demonstrate what you need to see. Some aren't stained adequately or properly. Some are sectioned at a funny angle. Some don't contain all of the tissue or cell types you need to see. What should you do about this? See the next hint for the answer.

▶ **Look at more than one slide of each specimen.** This will help you in the face of sub-par slides and also will assist you overall in gaining a better understanding of the specimens you are examining.

▶ **Draw what you see.** Students tend to resist drawing, often claiming artistic incompetence as an excuse. But even the most basic picture is helpful for two reasons. First, it allows you to engage more parts of your brain in the learning process. The more areas of your brain that are engaged, the better are the chances that you will retain the information. Also, drawing is helpful in that you actually have to look at the specimen long enough to draw it!

▶ **Have patience!** It really does get easier. Don't get frustrated, and don't give up. By the end of the semester, you may come to appreciate the microscope and the fascinating world it reveals!

NAME _____

SECTION _____ DATE _____

 Check Your Recall

1 Match the following terms with the correct definition from the column.

a. Ocular lens   _____   A. The surface on which the slide sits

b. Iris diaphragm   _____   B. The microscope's light source

c. Arm   _____   C. Allows fine tuning of the focus

d. Objective lens   _____   D. The lens through which you look to view the slide

e. Coarse adjustment knob _____   E. Supports the body of the microscope

f. Stage   _____   F. Controls the amount of light that passes through the slide

g. Fine adjustment knob   _____   G. Moves the stage up and down; provides gross focusing

h. Lamp   _____   H. Lenses of various powers of magnification

2 Magnification of the ocular lens is always _____.

3 The lens that provides $100\times$ magnification is the
a. low power lens.
b. medium power lens.
c. high power lens.
d. oil immersion lens.

4 Before you begin, you should make sure that
a. the objective lens is switched to the highest-power magnification.
b. the lamp is turned to its full brightness.
c. the ocular lens is switched to the highest-power magnification.
d. the objective lens is switched to the lowest-power magnification.

5 How should you carry the microscope?

6 What is the depth of focus?

7 What should you do if your slide looks dramatically different from the slide in your lab manual?

8 If you are having trouble making out details on your slide, what can you do?
   a. Adjust the focus with the fine adjustment knob.
   b. Reduce the amount of light with the iris diaphragm.
   c. Both a and b.
   d. Neither a nor b.

9 What should you use to clean the lenses?

10 What are the steps you should use to focus the image on the slide?

Cytology

OBJECTIVES

*Once you have completed this unit,
you should be able to:*

- Identify parts of the cell and organelles.

- Describe the process of diffusion.

- Describe the effects of hypotonic, isotonic, and hypertonic environments on cells.

- Prepare and observe a sample of cells.

- Identify the stages of the cell cycle and mitosis.

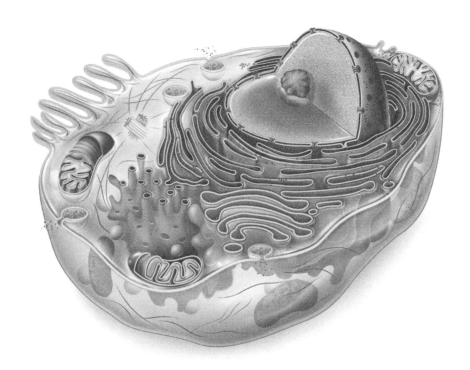

Pre-Lab Exercise 1
Key Terms

Table 4.1 lists the key terms with which you should be familiar before coming to lab.

TABLE 4.1	Key Terms

Term   **Definition**

Cell Structures and Organelles

Plasma membrane _____

_____

Cytoplasm _____

_____

Nucleus _____

_____

Mitochondrion _____

_____

Ribosome _____

_____

Peroxisome _____

_____

Smooth endoplasmic reticulum _____

_____

Rough endoplasmic reticulum _____

_____

Golgi apparatus (or complex) _____

_____

Lysosome _____

_____

Centrosome _____

_____

Cilia _____

_____

_____

Flagella _____

_____

Membrane Transport

Diffusion _____

_____

Osmosis _____

_____

Tonicity _____

_____

Cell Cycle and Mitosis

Cell cycle _____

_____

Interphase _____

_____

Mitosis _____

_____

Pre-Lab Exercise 2
The Plasma Membrane

Label and color the parts of the plasma membrane depicted in **Figure 4.1** with the terms from Exercise 1. Use your text and Exercise 1 in this unit for reference.

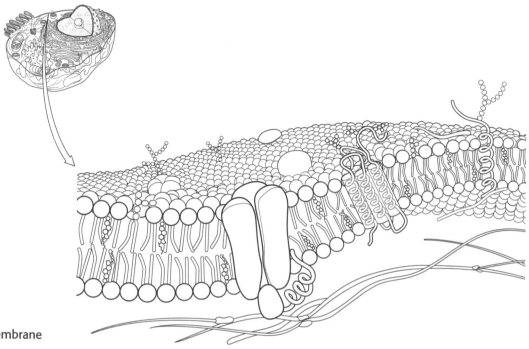

FIGURE **4.1**
The plasma membrane

Pre-Lab Exercise 3
The Parts of the Cell

Label and color the parts of the cell depicted in **Figure 4.2** with the terms from Exercise 1 (note that you may not see all structures in this diagram). Use your text and Exercise 1 in this unit for reference.

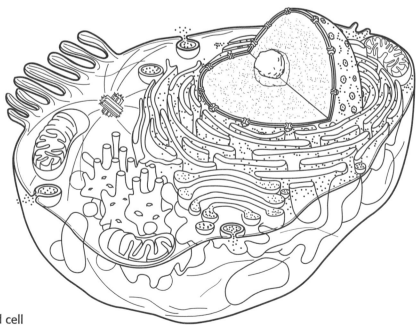

FIGURE **4.2** A generalized cell

Pre-Lab Exercise 4
The Cell Cycle

Use your text and lab manual to answer the following questions pertaining to the cell cycle and mitosis.

1. Describe the following stages of the cell cycle:

   a. G1:

   b. S:

   c. G2:

   d. M:

2. Describe the events that are occurring in the cell in each of the following phases of mitosis in Table 4.2.

TABLE **4.2**	Stages of Mitosis	
Stage of Mitosis	**Events Taking Place in the Cell**	**Cell Appearance**
Prophase		
Metaphase		
Anaphase		
Telophase		

Exercises

A basic principle that we will revisit repeatedly in our study of anatomy and physiology is:

Form follows function.

This principle may be stated in a variety of ways, and is alternatively called the "principle of complementarity of structure and function." Essentially, what this means is that the anatomy (the *form*) is always suited for the structure's physiology (the *function*). This is obvious at the organ level: Imagine if the heart were solid rather than composed of hollow chambers, or if the femur were pencil-thin rather than the thickest bone in the body. These organs wouldn't be able to carry out their functions of pumping blood and supporting the weight of the body very well, would they? But this principle is applicable even at the cellular level, which we will see in this unit.

We begin with an examination of the cell, its structures, and the relationship of structure and function. Then we turn our attention to two critical processes at the cellular level: *diffusion* and *osmosis*. In the final exercise we explore the *cell cycle* and the process of cell division.

Note that this is the first lab in which you will use a *Model Inventory*—something you will use throughout this lab manual. In this inventory you will list the anatomical models or diagrams that you use in lab (if the model is not named, make up a descriptive name for it) and record in a table the structures you are able to locate on each model. This is particularly helpful for study purposes, as it allows you to return to the proper models to locate specific structures.

 Exercise 1

Organelles and Cell Structures

MATERIALS NEEDED

- Cell models and diagrams
- Five colors of modeling clay
- Fetal pigs or other preserved small mammal
- Wooden applicator sticks
- Methylene blue dye
- Glass slide (blank)
- Distilled water
- Coverslip
- Light microscope
- Cell slides (with red blood cells, skeletal muscle, and sperm cells)
- Colored pencils

Most cells in the body are composed of three basic parts—the plasma membrane, the cytoplasm, and the nucleus.

1. **Plasma membrane.** The plasma membrane is the outer boundary of the cell (Figure 4.3). It is composed of a **phospholipid bilayer** with multiple components interspersed throughout, including **integral proteins, peripheral proteins, cholesterol, glycoproteins,** and **glycolipids**. It is a dynamic, fluid structure that acts as a selectively permeable barrier. In parts of the body where rapid absorption is necessary, the plasma membrane is folded into projections called **microvilli**, which increase its surface area.

2. **Cytoplasm.** The cytoplasm is the material inside the cell. It consists of three parts: **cytosol,** the **cytoskeleton,** and **organelles** (Figure 4.4). Cytosol is the fluid portion of the cytoplasm and contains water, solutes, enzymes, and other proteins. The cytoskeleton is a collection of protein filaments including **microtubules, actin filaments,** and **intermediate filaments**. Together, these filaments support the cell, function in cell movement, and move substances within the cell.

In addition, microtubules form the core of motile extensions from the cell called **cilia** and **flagella**. Cilia are small, hairlike extensions that beat rhythmically together to propel substances past the cell. Flagella are single extensions that propel the cell itself (sperm cells are the only flagellated cells in the human body).

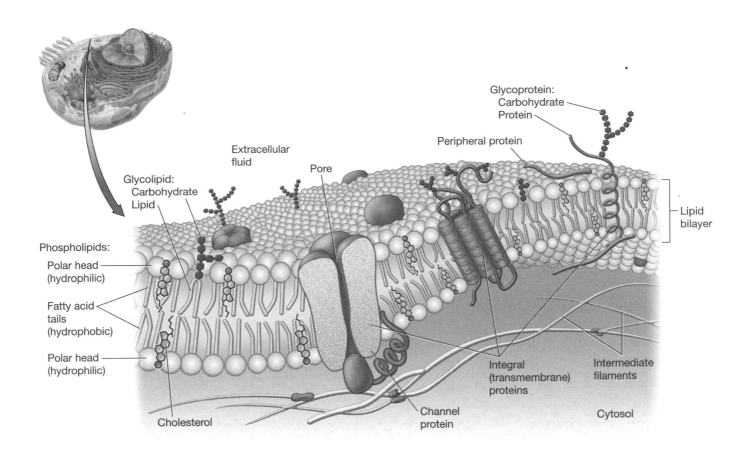

FIGURE **4.3** The plasma membrane

Organelles are specialized cellular compartments that carry out a variety of functions. The organelles we cover in this unit include the following:

a. **Ribosomes.** The small, granular ribosomes are not surrounded by a membrane and are composed of two subunits. Some ribosomes float freely in the cytosol, whereas others are bound to the membrane of another organelle or the nucleus. Ribosomes are the sites of protein synthesis in the cell.

b. **Peroxisomes.** The small, vesicular organelles known as peroxisomes contain enzymes that detoxify chemicals produced by cellular reactions, metabolize fatty acids, and synthesize certain phospholipids.

c. **Mitochondria.** The bean-shaped mitochondria are double-membrane-bounded organelles that produce the bulk of the cell's ATP (energy).

d. **Endoplasmic reticulum.** The series of membrane-enclosed sacs known as the endoplasmic reticulum may be of two types: **rough endoplasmic reticulum,** or **RER,** which has ribosomes on its surface, and **smooth endoplasmic reticulum,** or **SER,** which lacks ribosomes. RER functions in protein synthesis and modifies proteins that the ribosomes have made. SER has multiple functions including lipid synthesis and detoxification reactions.

e. **Golgi apparatus.** The Golgi apparatus is a stack of flattened sacs near the RER. Its membrane-enclosed sacs receive vesicles from the RER and other places in the cell and process, modify, and sort the products within the vesicles.

f. **Lysosomes.** Lysosomes are sacs filled with digestive enzymes that digest particles brought into the cell, old and worn-out organelles, and even the cell itself.

g. **Centrioles.** Centrioles are paired organelles composed primarily of microtubules that are located in the central area of the cell called the **centrosome.** They appear to be microtubule organizing centers and are important in facilitating the assembly and disassembly of microtubules.

3. **Nucleus.** The third component in nearly all cells, the large nucleus, is the cell's biosynthetic center. The nucleus is surrounded by a double membrane called the **nuclear envelope,** which contains holes called **nuclear pores.** Within the nucleus we find **chromatin,** a ball-like mass of tightly coiled DNA and proteins, RNA, and a dark-staining region called the **nucleolus.** The nucleolus contains a type of RNA called **ribosomal RNA** and is the "birthplace" of ribosomes.

Note that the cell shown in **Figure 4.4** is a **generalized cell** that contains each organelle. Most cells in the body don't look like this and instead are specialized so their forms follow their functions. For example, the cells of the liver contain a large amount of smooth endoplasmic reticulum, and immune cells (phagocytes) house a large number of lysosomes.

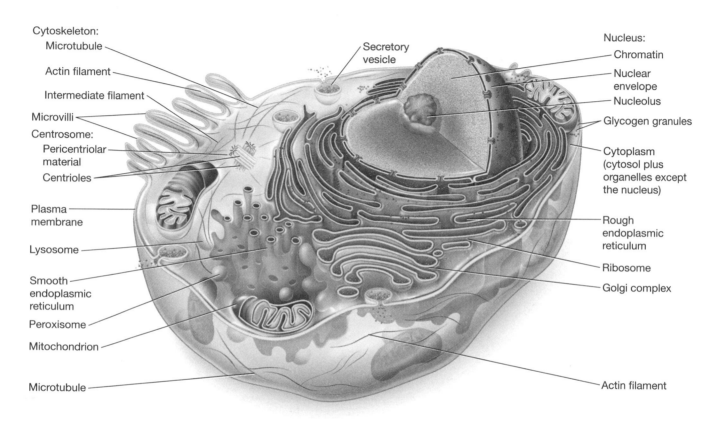

Sectional view

FIGURE 4.4 A generalized cell

 Procedure Model Inventory

Identify the following structures of the cell on models and diagrams, using your textbook and this unit for reference. As you examine the anatomical models and diagrams, record on the model inventory in Table 4.3 the name of the model and the structures that you were able to identify. When you have completed this activity, answer Check Your Understanding questions 1 and 2 (p. 79).

1. Plasma membrane
   a. Phospholipid bilayer
   b. Integral proteins
   c. Peripheral proteins
   d. Carbohydrates
2. Nucleus
   a. Nuclear membrane
   b. Nuclear pores
   c. Chromatin and chromosomes
   d. Nucleolus
3. Cytoplasm
4. Ribosomes
5. Smooth endoplasmic reticulum
6. Rough endoplasmic reticulum
7. Golgi apparatus
8. Lysosomes
9. Peroxisomes
10. Mitochondria
11. Centrioles
12. Microtubules, actin filaments, and intermediate filaments
13. Vesicle
14. Microvilli
15. Cilia
16. Flagella

TABLE 4.3	Cellular Structures Model Inventory
Model/Diagram	**Structures Identified**

Procedure Building a Plasma Membrane

In this exercise, you will build a model of the plasma membrane with modeling clay. Refer to Figure 4.3 for assistance. Use the following color code to build your model:

- Phosphate heads: blue
- Fatty acid tails: yellow
- Integral proteins: red
- Peripheral proteins: green
- Carbohydrates: orange

Procedure Preparing a Cell Sample

Now let's attempt to identify some organelles and cell structures on an actual cell. You will obtain a sample of cells from the mouth (either your own or from a preserved small mammal) and then stain them with methylene blue so that the structures become visible.

Note: Safety glasses and gloves are required. *Methylene blue stains hands and clothes readily.*

1 Obtain a blank slide and a coverslip.

2 Clean the slide with lens paper.

3 Swab the inside of your cheek with a sterile wooden applicator stick. As an alternative, swab the inside cheek of a fetal pig or other preserved small mammal. Do not get large chunks of tissue on the swab or individual cells will not be visible.

4 Wipe the swab with the cheek cells on the blank slide. If you are using your own cheek cells, dispose of the swab in a biohazard bag.

5 Place one drop of methylene blue dye onto the slide. Wait one minute.

6 Rinse the dye off the slide with distilled water and pat dry. The blue dye should be barely visible on the slide. If you see large areas of blue, rinse the slide again or get a new sample of cheek cells.

7 Place a coverslip over the stained area and place the slide on the stage of a microscope. Focus the image grossly on low power, then switch to the high power objective lens to find individual cells (use oil immersion if available).

8 Draw an individual cell, and label all cell structures and organelles that you can identify.

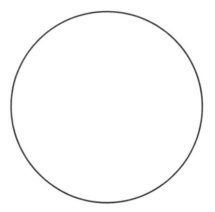

 Procedure Examining Cellular Diversity with Microscopy

The structure of different cell types can vary drastically. Cells differ not only in size and shape but also in the types and prevalence of organelles in the cell. In this activity you will examine prepared microscope slides of red blood cells, sperm cells, and skeletal muscle cells. Use the techniques you learned in Unit 3: Begin your observation on low power and advance to high power for each slide. Note that sperm cells can be difficult to find, so an oil-immersion lens is helpful to find the tiny cells. Draw, color, and label the cellular structures and organelles that you see on each slide. You may wish to look at Figures 5.2K (p. 94) and 5.3A (p. 99) for reference for red blood cells and skeletal muscle. When you have completed this activity, answer Check Your Understanding question 3 (p. 79).

1. Red blood cell

2. Sperm cell

3. Skeletal muscle cell

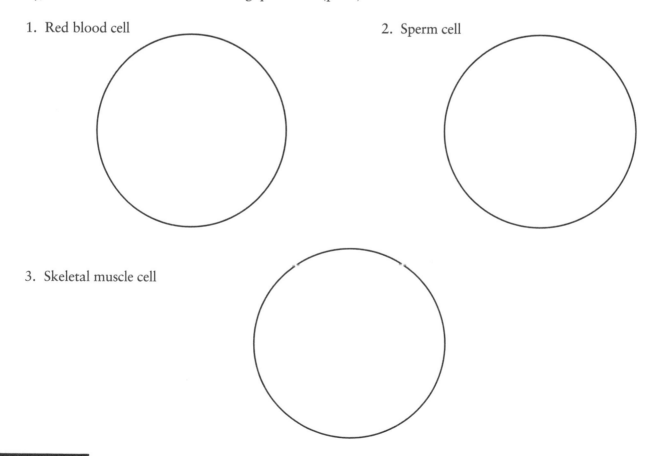

Exercise 2

Diffusion

MATERIALS NEEDED

- 2-100 ml beakers
- Food coloring
- Hot water
- Ice water
- 4 agar plates
- 0.1M potassium permanganate
- 0.1M methylene blue
- 1.0M methylene blue
- 0.1M congo red
- Ruler

Diffusion is defined as the movement of solute from a high concentration to a low concentration (**Figure 4.5**). Diffusion is a **passive process**—one that requires no net input of energy by a cell—because the energy for diffusion comes from a **concentration gradient**. A concentration gradient is defined as having different concentrations of the same substance at two different points. A concentration gradient is illustrated in **Figure 4.5** in the panels "Time 1" and "Time 2."

The rate at which diffusion takes place depends upon several factors, including the steepness of the concentration gradient, the temperature, and the size of the particles. Generally, smaller particle size, steeper concentration gradients, and higher temperatures will increase the rate of diffusion.

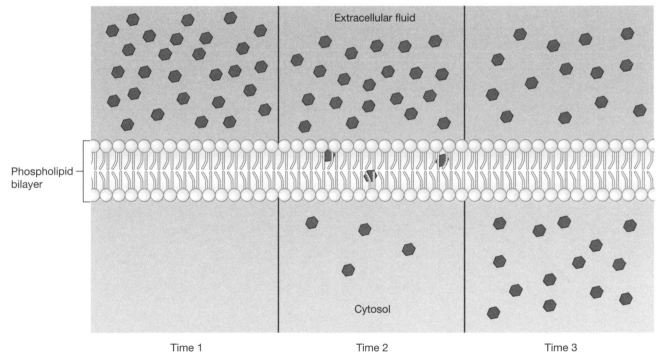

Time 1   Time 2   Time 3

FIGURE 4.5 Diffusion across a plasma membrane

Procedure Measuring Rates of Diffusion

Diffusion is a process that occurs all around us and therefore is easy to witness in action. In this experiment you will examine the effects of temperature, concentration, and particle size on the rate of diffusion by placing dyes in water and in agar. When you have completed the activities, answer Check Your Understanding question 4 (p. 79).

Note: Safety glasses, gloves, and lab coats are required. Use caution with the following dyes, as they will stain clothes and irritate and stain skin.

Part 1

1 Obtain two 100 ml glass beakers. Label one beaker "cold" and the other "hot." Fill the "cold" beaker with ice water (take care not to get ice in the beaker, though), and fill the "hot" beaker with water that has been heated.

2 Add two drops of food coloring to each beaker.

3 Observe the beakers from the side and measure with a ruler the distance that the dye spreads every minute for 5 minutes. Record your results in Table 4.4.

TABLE 4.4	Diffusion Results for Food Coloring in Water	
Time	Distance of Dye Diffusion: Cold Water	Distance of Dye Diffusion: Hot Water
1 minute		
2 minutes		
3 minutes		
4 minutes		
5 minutes		

4 Interpret your results. What effect does temperature have on the rate of diffusion?

Part 2

1 Obtain 4 agar plates with 1 cm wells scooped out of the middle. Label the plates 1, 2, 3, and 4.

2 Fill the well of plate 1 with 0.1M potassium permanganate (molecular weight = 158 g/mol).

3 Fill the well of plate 2 with 0.1M methylene blue (molecular weight = 374 g/mol).

4 Fill the well of plate 3 with 1.0M methylene blue (molecular weight = 374 g/mol).

5 Fill the well of plate 4 with 0.1M congo red (molecular weight = 697 g/mol).

6 Measure with a ruler the distance that each dye has diffused in 10-minute intervals for 50 minutes. Record your results in Table 4.5.

TABLE 4.5	Diffusion Results for Dyes in Agar			
Time	Distance of Diffusion: Plate 1, 0.1 Potassium permanganate	Distance of Diffusion: Plate 2, 0.1M Methylene blue	Distance of Diffusion: Plate 3, 1.0M Methylene blue	Distance of Diffusion: Plate 4, 0.1M Congo red
10 minutes				
20 minutes				
30 minutes				
40 minutes				
50 minutes				

7 Interpret your results:

a. Which dye diffused the fastest?

b. Which dye diffused the slowest?

c. What effect does particle size have on the rate of diffusion?

d. What effect does concentration have on the rate of diffusion?

Exercise 3

Osmosis and Tonicity

MATERIALS NEEDED

- Animal or human blood cells
- 3 glass slides (blank)
- Coverslips
- Dropper
- Wooden applicator stick
- 5% dextrose in water solution
- Deionized water
- 25% NaCl in water solution
- Lens paper/paper towel
- Light microscope with 40X objective

Whereas diffusion refers to the movement of solute, a separate passive process called **osmosis** refers to the movement of *solvent* (**Figure 4.6**). Specifically, osmosis is the movement of solvent (usually water) from a solution with a lower solute concentration to a solution with a higher solute concentration through a selectively permeable membrane. Notice in **Figure 4.6** that the solvent (water) moves to the more concentrated solution and dilutes it until the concentrations are approximately equal on each side because the solute particles cannot pass through the membrane. At this point, net movement ceases because the concentration gradient is extinguished.

In discussing biological solutions, it is useful to describe the concentration of a solution outside a cell compared to the concentration of the cytosol. This is called **tonicity**, and it determines how water crosses the plasma membrane. The three variations of tonicity are:

 ▶ **Hypotonic.** A hypotonic solution has a lower solute concentration than the cytosol. A cell placed in a hypotonic solution will gain water by osmosis and may swell and burst.

▶ **Isotonic.** An isotonic solution has the same solute concentration as the cytosol. There is no concentration gradient to drive osmosis, so there is no net movement of water into or out of a cell in an isotonic solution.

▶ **Hypertonic.** A hypertonic solution has a higher solute concentration than the cytosol. A cell in a hypertonic environment will lose water by osmosis and shrivel or **crenate** because water will be drawn toward the more concentrated solution.

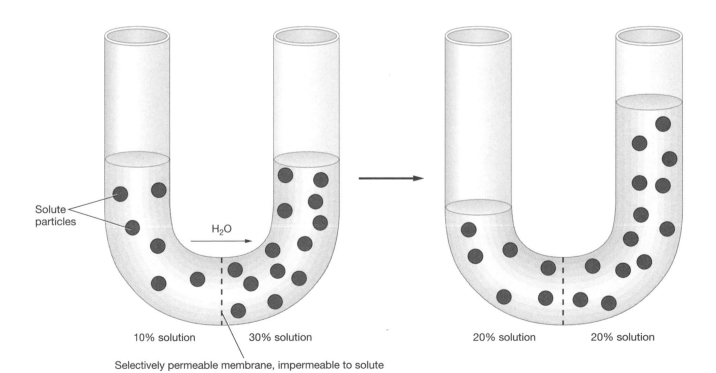

FIGURE 4.6 Osmosis

Procedure Watching Osmosis in Action

The following experiment will allow you to watch osmosis in action by placing cells—in this case, red blood cells—in solutions of different tonicity. When you have completed the activities, answer Check Your Understanding questions 5 and 6 (p. 80).

Note: Safety glasses and gloves are required.

---

1 Obtain three blank slides and number them 1, 2, and 3.

2 Place one small drop of animal blood on each slide with a dropper. Take care to keep your droplet fairly small, otherwise you won't be able to see individual cells.

3 Gently spread the droplet around the center of the slide with a wooden applicator stick and place a coverslip on it.

4 On slide 1, place a drop of 5% dextrose solution on one side of the coverslip. On the other side of the coverslip, hold a piece of lens paper or a paper towel. The lens paper will draw the fluid under the coverslip.

5 Observe the cells under the microscope on high power.

6 Repeat the procedure by placing 25% NaCl solution on slide 2 and distilled water on slide 3.

7 Draw and describe what you see with each slide.

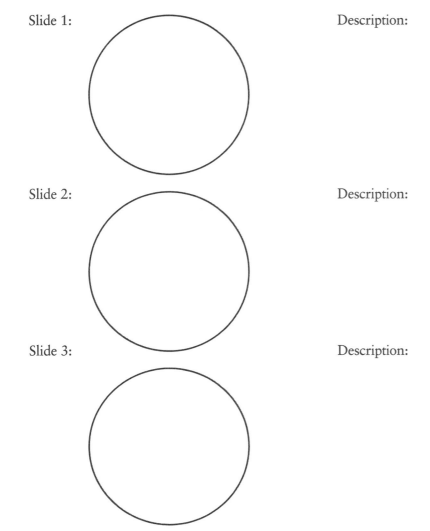

Slide 1:   Description:

Slide 2:   Description:

Slide 3:   Description:

8 Interpret your results:

a. Which solution was hypotonic? Explain your reasoning.

b. Which solution was isotonic? Explain your reasoning.

c. Which solution was hypertonic? Explain your reasoning.

 Exercise 4

Mitosis and the Cell Cycle

MATERIALS NEEDED

- Cell models or diagrams
- Mitosis models
- Mitosis slides
- Light microscope
- Colored pencils

Most cells go through a continual cycle of growth and replication called the **cell cycle**. The cell cycle consists of four phases:

1. **G1,** or the initial growth phase,
2. **S phase,** during which the DNA is replicated,
3. **G2,** the second growth phase, and
4. **M phase** or **mitosis,** during which the cell divides its organelles, cytosol, and replicated DNA among two identical **daughter cells.**

Each daughter cell has the same exact genetic and structural characteristics as the original cell. The portions of the cycle from G1–G2, when the cell is not dividing, are collectively called **interphase.**

Mitosis proceeds in the four general stages shown in Figure 4.7.

1. **Prophase.** During prophase, the nuclear membrane starts to degenerate and the DNA condenses so individual **chromosomes** are visible. Also during this stage, we see the formation of a structure called the **mitotic spindle.**

2. **Metaphase.** In metaphase we see the chromosomes line up along the central portion of the cell. Microtubules called **spindle fibers** emanate from the mitotic spindle and attach to the center of each pair of chromosomes.

3. **Anaphase.** During anaphase we see the chromosomes start to move toward opposite poles of the cell as the spindle fibers shorten. In addition, a process called **cytokinesis** begins, during which the cytoplasm is divided up among the two cells.

4. **Telophase.** In the final phase of mitosis, a divot forms between the two cells, called a **cleavage furrow**, which will pinch the cell into two identical daughter cells. In addition, during this stage the nuclear membranes begin to reassemble, the mitotic spindle becomes less visible, and cytokinesis completes.

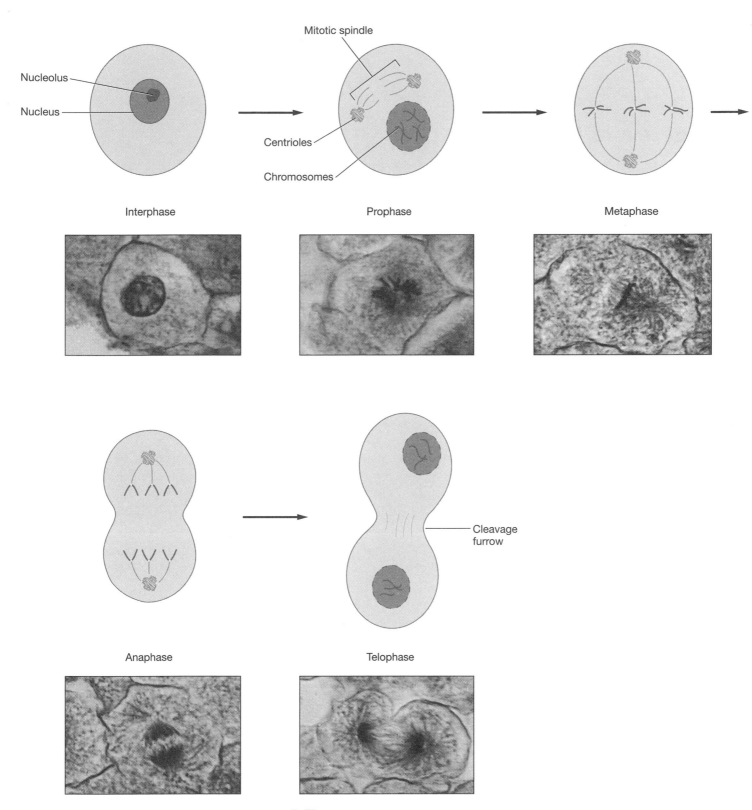

FIGURE **4.7** Interphase and the stages of mitosis

Procedure Identify Structures of Cell Division

Identify the following structures associated with mitosis on cell models and diagrams.

1. Mitotic spindle
   a. Centrioles
   b. Microtubules (spindle fibers)

2. Nucleus
   a. Nucleolus
   b. Nuclear membrane

3. Chromosomes
   a. Chromatids
   b. Centromere

 Procedure Model Mitosis

Arrange models of the cell cycle and mitosis in the proper order. As an alternative, build a set of cell cycle models with modeling clay and arrange them in the proper order of the cycle.

1. Interphase
2. Mitosis
   a. Prophase
   b. Metaphase
   c. Anaphase
   d. Telophase

 Procedure Microscopy of the Cell Cycle

Examine the five phases of the cell cycle on prepared whitefish mitosis slides using the highest-power objective. Note that every stage of the cell cycle may not be visible on one single slide, so you may have to use more than one slide. Note also that most of the cells you see will be in interphase.

   Draw and describe what the cell looks like during each phase of the cell cycle, and label your drawing with as many of the structures of cell division (see above) as you can see in each cell. Use your Pre-Lab exercises and **Figure 4.7** for reference. When you have completed the activities, answer Check Your Understanding question 7 (p. 80).

1. Interphase

   Description:

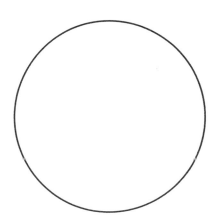

2. Mitosis

   a. Prophase

      Description:

   b. Metaphase

      Description:

   c. Anaphase

      Description:

   d. Telophase

      Description:

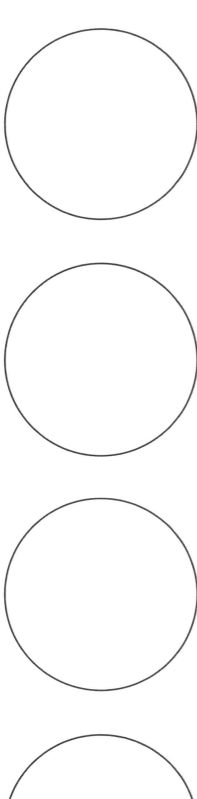

NAME _____

SECTION _____ DATE _____

 Check Your Recall

1 Label the following parts of the cell on **Figure 4.8**.

- Plasma membrane
- Nucleus
- Nuclear membrane
- Nucleolus
- Ribosomes
- Smooth endoplasmic reticulum
- Rough endoplasmic reticulum
- Golgi apparatus
- Lysosome
- Peroxisome
- Mitochondria
- Microvilli

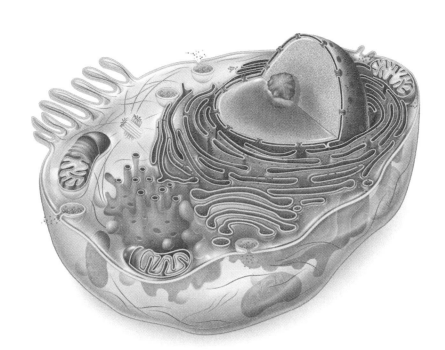

FIGURE **4.8** A generalized cell

2 Which of the following is not a basic component of most cells?

a. Microvilli
b. Plasma membrane
c. Nucleus
d. Cytoplasm

3 Matching: Match the following organelles and cell structures with the correct definitions.

Plasma membrane	_____	A.	Biosynthetic center of the cell
Smooth ER	_____	B.	Produce(s) the bulk of the cell's ATP
Mitochondria	_____	C.	Contain(s) digestive enzymes
Ribosomes	_____	D.	Stack of flattened sacs that modify and sort proteins
Rough ER	_____	E.	Composed of a phospholipid bilayer
Nucleus	_____	F.	Series of membrane-enclosed sacs with ribosomes on the surface
Nucleolus	_____	G.	Cytoskeletal filament found in cilia and flagella
Lysosome	_____	H.	Series of membrane-enclosed sacs that detoxify substances and synthesize lipids
Microtubule	_____	I.	The cell's "ribosome factory"
Golgi apparatus	_____	J.	Granular organelles that are the sites of protein synthesis

4 Fill in the blank: Diffusion is a _____ process where a solute moves from a _____ concentration to a _____ concentration.

5 Which of the following factors influences the rate at which diffusion takes place?
   a. Size of the particles
   b. Temperature
   c. Steepness of the concentration gradient
   d. All of the above

6 Fill in the blanks: Osmosis is the movement of _____ from a solution with a _____ solute concentration to a solution with a _____ solute concentration.

7 How do isotonic, hypertonic, and hypotonic solutions differ?

8 Label the stages of mitosis and the cell cycle on **Figure 4.9**.

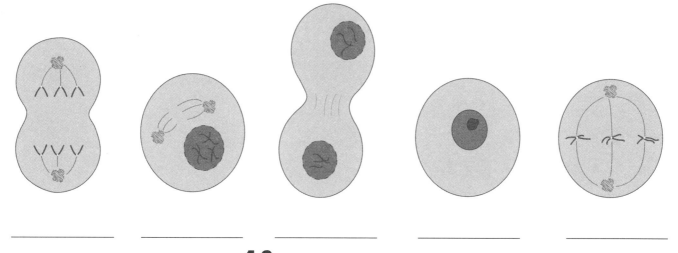

_____   _____   _____   _____   _____

FIGURE **4.9** The stages of the cell cycle and mitosis

9 Which of the following is *not* a phase of mitosis?
   a. Prophase
   b. Telophase
   c. Interphase
   d. Anaphase

10 What is the function of the mitotic spindle and spindle fibers during mitosis?

Check Your Understanding | **Critical Thinking and Application Questions**

1 Mitochondria contain their own DNA, which encodes 13 proteins. Defects on the mitochondrial DNA can be passed down maternally, leading to a group of diseases called *mitochondrial cytopathies*. Explain which cell populations you think would be most affected by this disease.

2 Another group of diseases that affect a specific organelle are the *lysosomal storage diseases*. What potential problems could be caused by defective lysosomes?

3 Which cell type that you observed lacks a nucleus? What functions would this cell be unable to carry out?

4 Two drugs are injected into the same thigh muscle. Drug A is body temperature (37° C) and has a molecular weight of 150 g/mol. Drug B has been in the refrigerator (5° C) and has a molecular weight of 600 g/mol. Predict which drug will diffuse into the blood more rapidly. Explain your reasoning.

5 Isotonic saline and 5% dextrose in water are solutions that are considered isotonic to human blood. What effect on red blood cells would you expect if a patient were given these fluids intravenously? A solution of 10% dextrose in water is hypertonic to human blood. What would happen if you were to infuse your patient with this solution?

6 Gatorade and other sports drinks are actually hypotonic solutions. How do these drinks rehydrate their user's cells?

7 List examples of locations in the body where cell populations undergo rapid mitosis. Why would you expect frequent cell division in these locations?

Histology

OBJECTIVES

Once you have completed this unit, you should be able to:

- Identify epithelial tissues by number of layers, cell shape, and specializations.

- Identify and describe connective tissues.

- Identify and describe muscle and nervous tissues.

- Relate tissue structure to tissue function and describe how organs are formed from two or more tissue types.

- Give examples of organs where each tissue type is found.

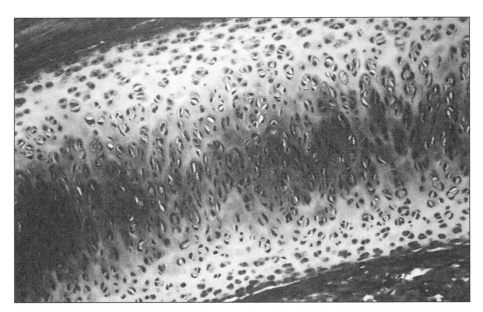

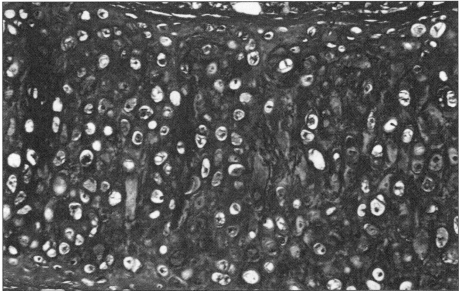

Pre-Lab Exercises

Complete the following exercises prior to coming to lab, using your textbook and lab manual for reference.

Pre-Lab Exercise 1

Key Terms

Table 5.1 lists the key terms with which you should be familiar before coming to lab.

TABLE 5.1	Key Terms

Term	Definition
Epithelial Tissue	
Simple epithelial tissue	_____
	_____
Stratified epithelial tissue	_____
	_____
Pseudostratified epithelial tissue	_____
	_____
Squamous cell	_____
	_____
Cuboidal cell	_____
	_____
Columnar cell	_____
	_____
Basal lamina	_____
	_____
Connective Tissue	
Ground substance	_____
	_____
Collagen fiber	_____
	_____
Elastic fiber	_____
	_____

Reticular fiber _____

_____

Loose connective tissue _____

_____

Dense connective tissue _____

_____

Cartilage _____

_____

Bone _____

_____

Blood _____

_____

Muscle Tissue

Striated _____

_____

Skeletal muscle tissue _____

_____

Cardiac muscle tissue _____

_____

Smooth muscle tissue _____

_____

Nervous Tissue

Neuron _____

_____

Neuroglial cell _____

_____

Exercises

The histology labs can be some of the more intimidating and frustrating labs for beginning anatomy and physiology students. The subjects are somewhat abstract and unfamiliar and require use of a complicated tool—the microscope. The best way to approach this subject is to be systematic and let your lab manual walk you through it step-by-step. If you get confused, don't despair. With the help of this book, your lab instructor, and a little patience, you can do it! Before you begin, you may wish to review the "Hints and Tips" box on page 54.

The exercises in this unit introduce you to the four basic types of tissue: **epithelial tissue**, **connective tissue**, **muscle tissue**, and **nervous tissue**. All four of these tissue types have two main components:

1. cells, which are unique for each tissue type, and

2. the **extracellular matrix** (**ECM**), which is produced largely by the tissue's cells.

ECM generally consists of a gelatinous substance called **ground substance** and numerous different **protein fibers** that provide a tissue with distensibility and tension resistance. Let's now begin our exploration of this fascinating level of organization.

 ## Exercise 1

Epithelial Tissue

MATERIALS NEEDED

- Epithelial tissues slides
- Light microscope
- Colored pencils

Epithelial tissues are our covering and lining tissues. They are found covering body surfaces, lining body passageways, lining body cavities, and forming glands. Epithelia predominantly contain cells called **epithelial cells**, and their ECM is limited mostly to the space underneath the cells in a layer called the **basal lamina**. The basal lamina adheres to another layer of ECM produced by the tissues deep to the epithelium. Together, these two structures are called the **basement membrane**.

Epithelial tissues are all **avascular**; they have no blood vessels to supply them directly and rely on oxygen and nutrients diffusing up from deeper tissues. For this reason, epithelial tissues can be only a certain number of cell layers in thickness. If they are too thick, oxygen and nutrients will not reach the more superficial cells and they will die.

The many types of epithelia are classified according to the number of cell layers and the shape of the epithelial cells. The classes include the following (see **Figure 5.1**):

1. **Simple epithelia** have only one layer of cells and include:

   a. **Simple squamous epithelium**. This type of epithelium, shown in **Figure 5.1A**, consists of a single layer of flat cells with a flattened nucleus. We often find simple squamous epithelium in places where substances have to cross the epithelium quickly, such as the air sacs of the lungs.

   b. **Simple cuboidal epithelium**. Note in **Figure 5.1B** that the cells of simple cuboidal epithelium are about as wide as they are tall, with a round, central nucleus. Simple cuboidal epithelium is found lining glands such as the thyroid gland, certain respiratory passages, and certain tubes in the kidneys.

   c. **Simple columnar epithelium**. The cells of simple columnar epithelium, shown in **Figure 5.1C**, are taller than they are wide, with round nuclei located near the base of the cell. These cells line certain respiratory passages, much of the digestive tract, and the genitourinary tract. The surfaces of simple columnar epithelial cells often contain cilia or are folded into microvilli.

2. **Stratified epithelia** have two or more layers of cells and include the following:

   a. **Stratified squamous epithelium**. This type of epithelium consists of many layers of flattened cells. There are two variants of simple stratified epithelium. The first, shown in **Figure 5.1D**, is **stratified squamous keratinized epithelium**, which consists of epithelial cells called **keratinocytes** that produce the hard protein keratin. Note in

Figure 5.1D that the more superficial cells are dead and are filled with keratin. This is because they are too far away from the blood supply in the deeper tissues and, as a result, they harden and die. Stratified squamous keratininzed epithelial cells are well-structured to resist mechanical stresses and, accordingly, they are found in the outer layer of the skin, where they protect the deeper tissues.

      Stratified squamous nonkeratinized epithelium (Figure 5.1E) contains no keratin and is located in places that are subject to lesser degrees of mechanical stress, such as the oral cavity, the pharynx (throat), the anus, and the vagina. This type of epithelium is not as thick and as a result the superficial cells are alive and much different in appearance than those of keratinized epithelium.

b. **Stratified cuboidal epithelium** and **stratified columnar epithelium**. Both of these types of epithelium are rare in the human body (**Figure 5.1F** shows stratified cuboidal epithelium only), and are found lining the ducts of certain glands.

    Not all epithelial tissues fit this classification scheme. One such epithelial tissue is called **transitional epithelium** (**Figure 5.1G**). This type of epithelial tissue is stratified but is not classified by its shape because its cells can change shape. Typically, the apical or surface cells are dome-shaped, but when the tissue is stretched, they flatten and are squamous in appearance. Transitional epithelium is found lining the urinary bladder and ureters.

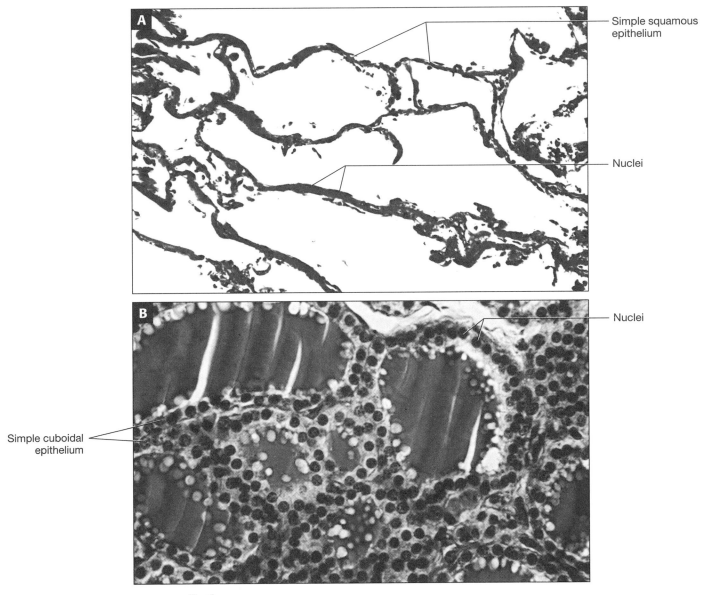

FIGURE 5.1 Epithelial tissues: (A) simple squamous epithelium from the lungs; (B) simple cuboidal epithelium from the thyroid gland

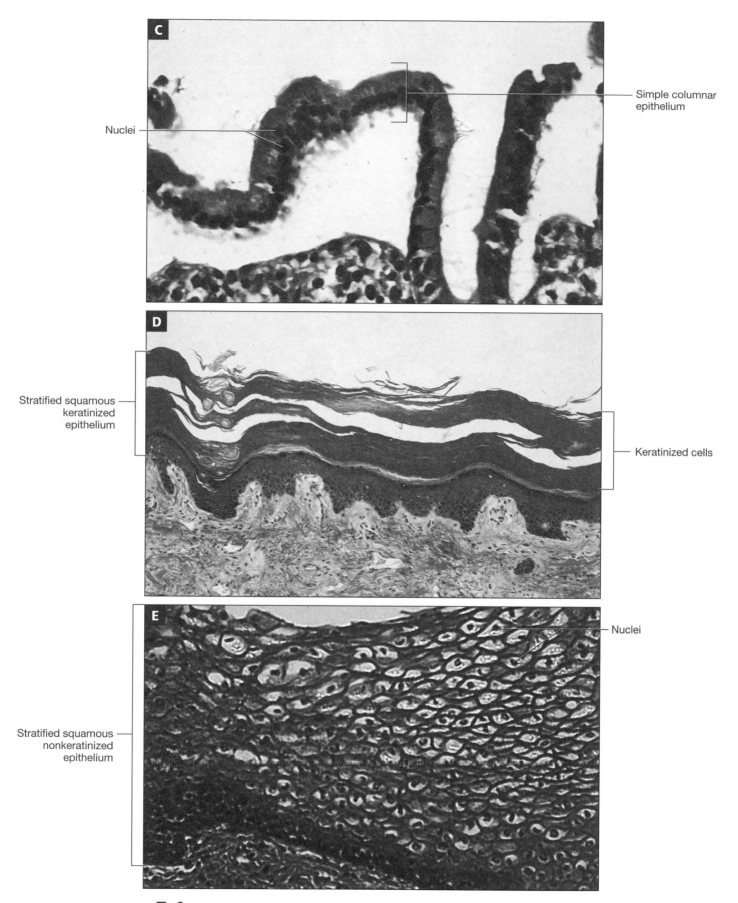

FIGURE 5.1 Epithelial tissues (cont.): (C) simple columnar epithelium from the digestive tract; (D) stratified squamous keratinized epithelium from the skin; (E) stratified squamous nonkeratinized epithelium from the esophagus

EXPLORING ANATOMY & PHYSIOLOGY IN THE LABORATORY

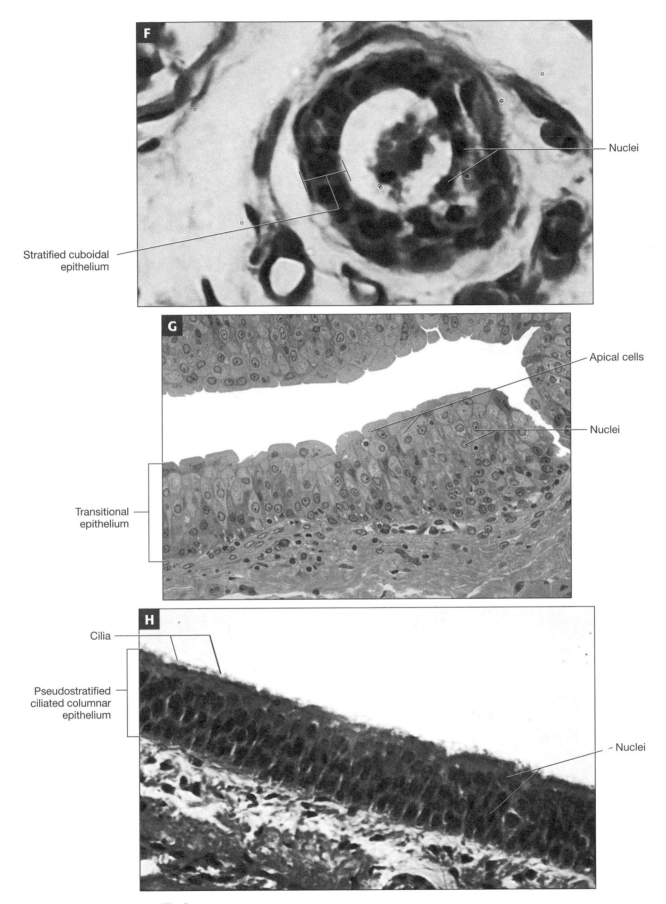

Nuclei

Stratified cuboidal
epithelium

Apical cells

Nuclei

Transitional
epithelium

Cilia

Pseudostratified
ciliated columnar
epithelium

Nuclei

FIGURE 5.1 Epithelial tissues (cont.): (F) stratified cuboidal epithelium from a sweat gland;
(G) transitional epithelium from the urinary bladder; (H) pseudostratified ciliated columnar epithelium from the trachea

Another type of epithelial tissue that does not fit neatly into this classification system is **pseudostratified ciliated columnar epithelium** (Figure 5.1H). This epithelium has the appearance of having many layers but actually has only one layer of cells (*pseudo* = false). Note that this type of epithelium usually has cilia and that the cell shape is always columnar. It is found lining the nasal cavity and much of the respiratory tract.

HINTS & TIPS

How to Approach Epithelium

Before you start, remember how to approach any slide: First examine the slide with the naked eye, then begin on low power and scan the slide, and advance progressively to higher power to see details. Use Figure 5.1 as a guide—if your slide looks nothing like the tissue in Figure 5.1, scroll around the slide and keep looking. But do not rely completely on Figure 5.1, as the slides that your lab uses may be prepared with different stains or from different tissues. A few simple things will help you distinguish epithelial tissues from other tissues, including:

1. Epithelial tissues are all avascular so you won't see any blood vessels in epithelium.

2. Epithelial tissues are typically on the outer edge of the slide. Keep in mind that most slides have several tissues in each section. To find the epithelial tissue, scroll to one end of the slide or the other.

3. Epithelial tissues consist mostly of cells. If you aren't sure if something is a cell, look for a nucleus. If you can see a nucleus, you typically will be able to see the plasma membrane surrounding the cell as well, which will help you define borders between cells.

Procedure Microscopy of Epithelial Tissues

Examine prepared slides of the following epithelial tissues. Use colored pencils to draw what you see under the microscope, and label your drawings with the terms from **Figure 5.1**. Then (a) describe what you see, and (b) give examples of locations in the body where this tissue is found. When you have completed the microscope slides, answer Check Your Understanding question 1 (p. 109).

1. Simple squamous epithelium

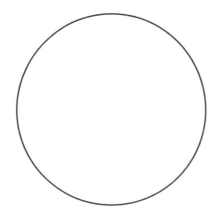

a. _____

b. _____

2. Simple cuboidal epithelium

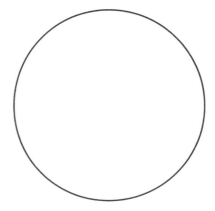

a. _____

b. _____

3. Simple columnar epithelium

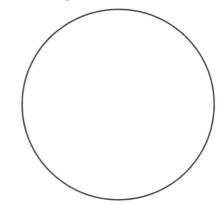

a. _____

b. _____

4. Stratified squamous keratinized epithelium

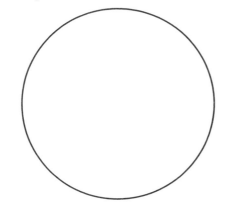

a. _____

b. _____

5. Stratified squamous nonkeratinized epithelium

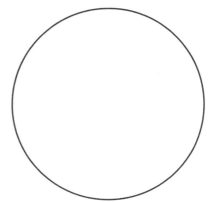

a. _____

b. _____

6. Stratified cuboidal epithelium

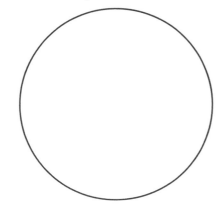

a. _____

b. _____

7. Transitional epithelium

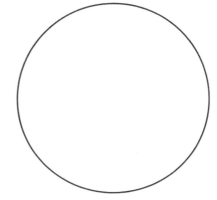

a. _____

b. _____

8. Pseudostratified ciliated columnar epithelium

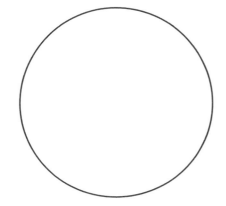

a. _____

b. _____

Exercise 2

Connective Tissue

MATERIALS NEEDED

- Connective tissue slides
- Light microscope
- Colored pencils

Connective tissues are found throughout the body. They have a variety of functions, most of which serve to *connect*, as their name implies (blood is an exception). All connective tissues stem from a common embryonic tissue called **mesenchyme.** Connective tissues are distinguished easily from epithelial tissues by the prominence of their extracellular matrices. Typically, connective tissues contain few cells and have an extensive ECM consisting of ground substance and three main types of protein fibers:

1. **collagen fibers,** composed of the thick protein **collagen** that give a tissue tensile strength,

2. **elastic fibers,** composed of the protein **elastin,** which makes a tissue distensible, and

3. thin **reticular fibers** that interweave to form networks that support blood vessels, nerves, and other structures.

The four general types of connective tissue (CT) are as follows (**Figure 5.2**):

1. **Connective tissue proper.** Connective tissue proper, the most widely distributed class of connective tissue in the body, consists of scattered cells called **fibroblasts** that secrete an extensive ECM filled with many types of protein fibers. This tissue is highly vascular, with an extensive blood supply. The subclasses of CT proper include the following:

   a. **Loose (areolar) CT.** You can see in **Figure 5.2A** that the primary element in loose CT is ground substance, which gives it a "loose" appearance on a slide. All three types of protein fibers are scattered in loose CT ground substance. Loose CT is found as part of the basement membrane and in the walls of hollow organs.

   b. **Dense regular collagenous CT.** The difference between loose and dense CT is obvious in **Figure 5.2B**. Dense regular collagenous CT consists primarily of collagen fibers arranged in parallel bundles with little ground substance and few cells. It is exceptionally strong and makes up structures that require tensile strength in a single plane, such as tendons and ligaments.

   c. **Dense irregular collagenous CT.** Like dense regular collagenous CT, dense *irregular* collagenous CT consists of bundles of collagen fibers. Note in **Figure 5.2C**, however, that these collagen bundles are arranged in an irregular, haphazard fashion without a consistent pattern. Dense irregular collagenous CT also is quite strong and is located in places that require tensile strength in multiple planes, such as the dermis and joint and organ capsules.

   d. **Reticular CT.** As you can guess by the name, reticular CT consists of many reticular fibers produced by cells called **reticular cells** (**Figure 5.2D**). It is located in the spleen and lymph nodes, where the fine reticular fibers interweave to form "nets" that trap pathogens and foreign cells. Reticular CT also is located around blood vessels and nerves, where it forms supportive networks.

   e. **Adipose tissue.** Notice in **Figure 5.2E** that adipose tissue (fat tissue) has a much different appearance than the other types of CT proper. It consists of mostly cells, with little visible ECM, and the huge cells are called **adipocytes.** Each adipocyte contains a large lipid droplet that occupies most of its cytoplasm. The nucleus and other organelles are barely visible because they are pushed to the periphery of the cell against the plasma membrane. Adipose tissue is distributed widely throughout the body under the skin and around organs.

   f. **Dense regular elastic CT.** This tissue type contains elastic fibers arranged in parallel bundles (**Figure 5.2F**). We find dense regular elastic CT lining large blood vessels and in certain ligaments.

2. **Cartilage.** Cartilage is a tough but flexible tissue that is resistant to tension, twisting, and compressive forces. It consists of cells called **chondrocytes** located in cavities called **lacunae** embedded in the ECM. Among the connective tissues, cartilage is notable for being avascular. Each of the three types of cartilage has a different ECM composition.

   a. **Hyaline cartilage.** Notice in **Figure 5.2G** that hyaline cartilage contains mostly chondrocytes scattered in ground substance with few visible protein fibers. This lack of protein fibers gives hyaline cartilage a smooth, glassy appearance and makes it an ideal tissue to cover the ends of bones where they form joints with another bone. The smooth texture of hyaline cartilage provides a nearly frictionless surface on which bones can articulate. Hyaline cartilage also is found connecting the ribs to the sternum, in the nose, and forming the framework for certain respiratory passages.

b. **Fibrocartilage**. As you can see in **Figure 5.2H**, fibrocartilage is named appropriately, as it is full of protein fibers (mostly collagen). This makes fibrocartilage tough and extremely strong but not at all smooth (think of the surface of fibrocartilage like a flannel sheet, with the cotton fibers representing the protein fibers). For this reason, fibrocartilage does not cover the ends of bones, but it does reinforce ligaments and form *articular discs*, tough structures that improve the fit of two bones. In addition, fibrocartilage is found in joints where hyaline cartilage has been damaged.

c. **Elastic cartilage**. The final type of cartilage, elastic cartilage, is shown in **Figure 5.2I**. It is filled with elastic fibers that give it the property of distensibility. Elastic cartilage is found in the ear and in the epiglottis.

3. **Bone**. Bone tissue, also called **osseous tissue**, consists of bone cells called **osteocytes** that are encased in an ECM that contains collagen fibers and calcium hydroxyapatite crystals. Note in **Figure 5.2J** that the ECM is arranged in concentric layers called **lamellae,** with the osteocytes sandwiched between. This structure makes bone the hardest tissue in the body and the most resistant to mechanical stresses.

4. **Blood**. Blood (**Figure 5.2K**) is unique in that it is the only connective tissue that doesn't actually connect anything physically. It consists of cells called **erythrocytes** (red blood cells) and **leukocytes** (white blood cells), and an ECM called **plasma**.

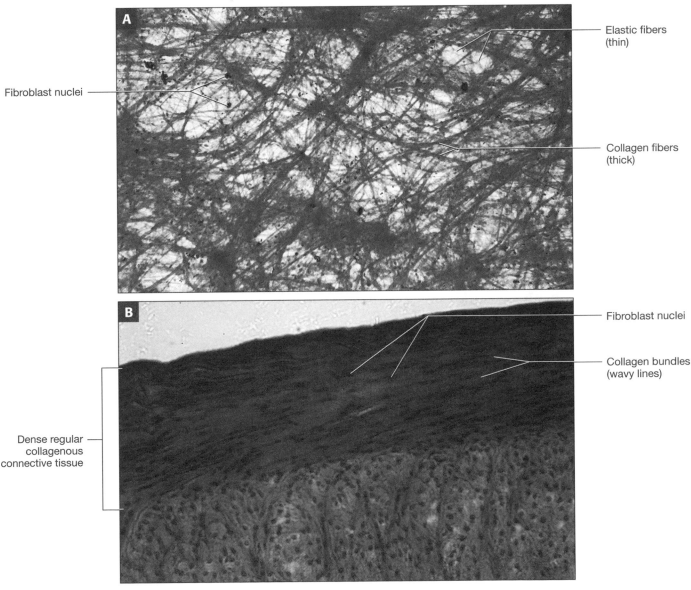

FIGURE 5.2 Connective tissues: (A) loose (areolar) CT; (B) dense regular collagenous CT from a tendon

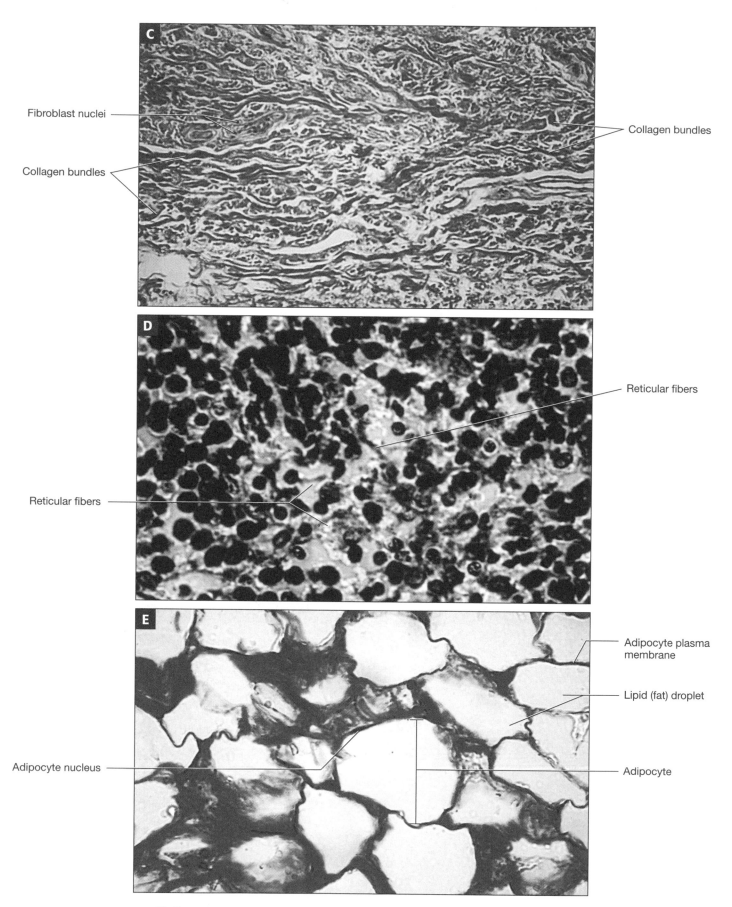

Fibroblast nuclei

Collagen bundles

Collagen bundles

Reticular fibers

Reticular fibers

Adipocyte plasma membrane

Lipid (fat) droplet

Adipocyte nucleus

Adipocyte

FIGURE **5.2** Connective tissues (cont.): (C) dense irregular collagenous CT from the dermis; (D) reticular CT from the spleen; (E) adipose tissue

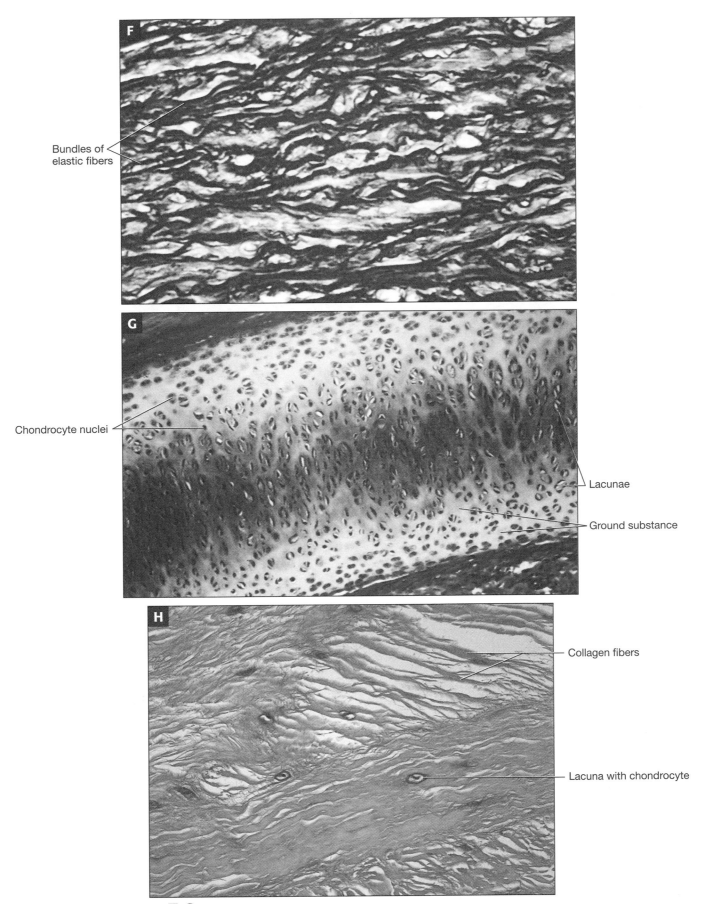

Bundles of
elastic fibers

Chondrocyte nuclei

Lacunae

Ground substance

Collagen fibers

Lacuna with chondrocyte

FIGURE **5.2** Connective tissues (cont.): (F) dense regular elastic CT from the aorta;
(G) hyaline cartilage from a joint; (H) fibrocartilage from an articular disc

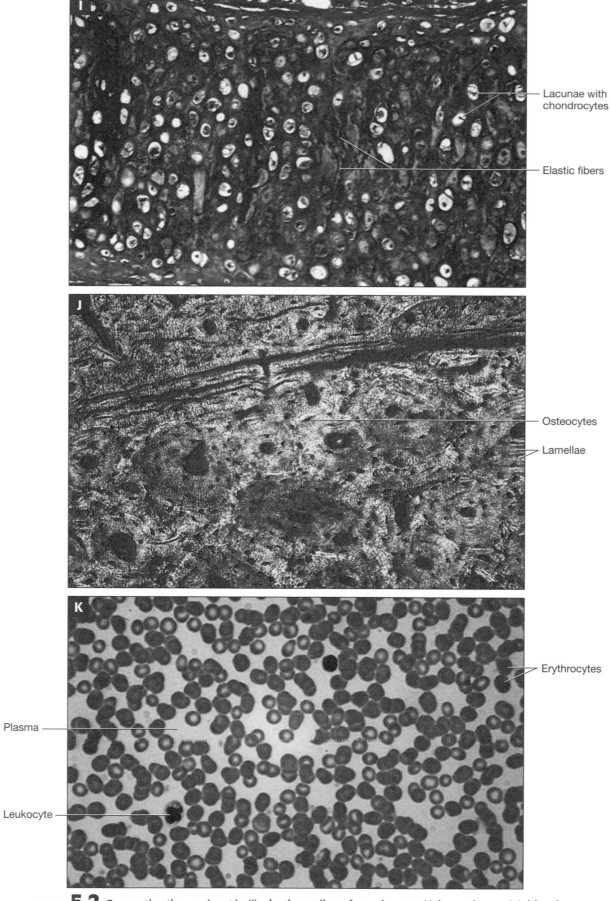

Lacunae with chondrocytes

Elastic fibers

Osteocytes

Lamellae

Erythrocytes

Plasma

Leukocyte

FIGURE 5.2 Connective tissues (cont.): (I) elastic cartilage from the ear; (J) bone tissue; (K) blood

HINTS & TIPS

How to Approach Connective Tissue

Students typically find connective tissue to be the most difficult of the tissues to identify. Don't let this scare you away. Simply take care to approach these slides systematically in the same manner as the epithelial slides, and use Figure 5.2 as a guide. The following points will help you to identify the various connective tissues and differentiate them from other tissue types:

1. The cells typically are not densely packed together and you usually will see a large amount of space between connective tissue cells. Remember to look for the nucleus and plasma membrane to discern the borders of a cell.

2. Generally connective tissue, particularly connective tissue proper, contains a lot of ground substance and protein fibers in the extracellular matrix. One exception is adipose tissue, which consists of densely packed adipocytes filled with a large lipid droplet.

3. Many types of CT can be distinguished by the types of fibers they contain:
   a. Reticular fibers are the thinnest fibers, which typically stain brown or black. Look for reticular fibers in loose and reticular CT.
   b. Collagen fibers are thick fibers that often stain pink. Look for collagen fibers in fibrocartilage, dense regular collagenous CT, dense irregular collagenous CT, and loose CT.
   c. Elastic fibers are fairly thick and wavy fibers. Their color ranges from purple-black to blue, depending on the stain used. Look for them in elastic tissue, elastic cartilage, and loose CT.

4. Cartilage is easy to discern from CT proper by looking at the shape of the cells. Fibroblasts are generally small and flat, whereas chondrocytes are much larger and round. In addition, chondrocytes sit in hollow cavities called lacunae.

5. Blood and bone are perhaps the two easiest tissues you will examine in this lab. They should look much like they do in Figure 5.2, and no other tissues resemble them.

Procedure Microscopy of Connective Tissue Proper

View prepared slides of each type of connective tissue proper. Use colored pencils to draw pictures of what you see under the microscope, and label your drawings with the terms from **Figure 5.2**. Then (a) describe what you see, and (b) give examples of locations in the body where this tissue is found.

1. Mesenchyme (embryonic CT)

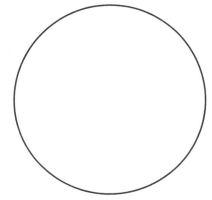

a. _____

b. _____

2. Loose (areolar) CT

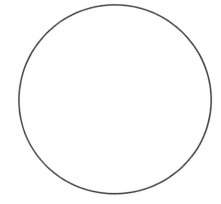

a. _____

b. _____

3. Reticular CT

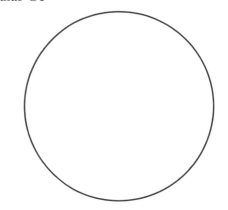

a. _____

b. _____

4. Adipose tissue

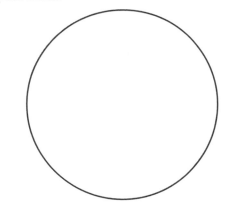

a. _____

b. _____

5. Dense regular elastic CT

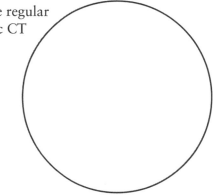

a. _____

b. _____

6. Dense regular collagenous CT

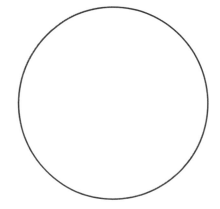

a. _____

b. _____

7. Dense irregular collagenous CT

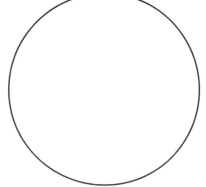

a. _____

b. _____

Procedure Microscopy of Cartilage

View prepared slides of the three types of cartilage. Use colored pencils to draw pictures of what you see under the microscope, and label your drawings with the terms from **Figure 5.2**. Then (a) describe what you see, and (b) give examples of locations in the body where this tissue is found.

8. Hyaline cartilage

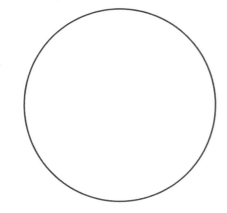

a. _____

b. _____

9. Fibrocartilage

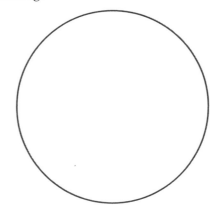

a. _____

b. _____

10. Elastic cartilage:

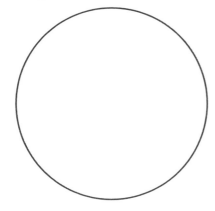

a. _____

b. _____

Procedure **Microscopy of Bone and Blood**

View prepared slides of bone and blood. Use colored pencils to draw pictures of what you see under the microscope, and label your drawings with the terms in **Figure 5.2**. Then (a) describe what you see, and (b) give examples of locations in the body where this tissue is found. When you have completed all of the connective tissue slides, answer Check Your Understanding questions 2 and 3 (p. 109–110).

11. Bone

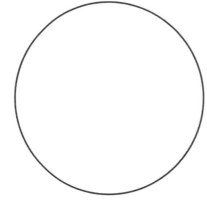

a. _____

b. _____

12. Blood

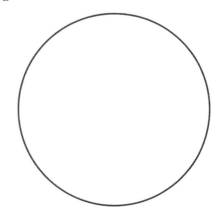

a. _____

b. _____

 Exercise 3

Muscle Tissue

MATERIALS NEEDED

- Muscle tissue slides
- Light microscope
- Colored pencils

Muscle tissue is located in skeletal muscles, in the walls of hollow organs, in the heart, and in other locations such as the iris of the eye. It consists of muscle cells, sometimes called **myocytes** or **muscle fibers,** and a small amount of ECM called the **endomysium.** Notice in Figures 5.3A–C that myocytes aren't shaped like the cells that you are accustomed to seeing. For this reason, muscle tissue is easy to discern from the other tissue types.

There are three types of muscle tissue:

1. **Skeletal muscle tissue.** The myocytes of skeletal muscle tissue are long, tubular, and **striated** (striped) in appearance (**Figure 5.3A**). The striations result from the arrangement of proteins within the muscle fiber called **myofilaments.** Skeletal muscle fibers are formed from the fusion of cells called **myoblasts,** and for this reason have multiple nuclei.

2. **Cardiac muscle tissue.** The myocytes of cardiac muscle tissue, located in the heart, are short, fat, striated, and tend to be branching (**Figure 5.3B**). Adjacent myocytes are linked by specialized junctions called **intercalated discs,** which contain desmosomes and gap junctions. Cardiac myocytes typically have only one nucleus, but some may have two or more.

3. **Smooth muscle tissue.** The myocytes of smooth muscle tissue are flat with one nucleus in the center of the spindle (**Figure 5.3C**). The arrangement of myofilaments within smooth muscle fibers differs from that of skeletal and cardiac muscle fibers, and as a result, these cells lack noticeable striations (hence the name *smooth* muscle). It lines all hollow organs and is found in the skin, the eye, and surrounding many glands.

We revisit each type of muscle tissue in several units in this book.

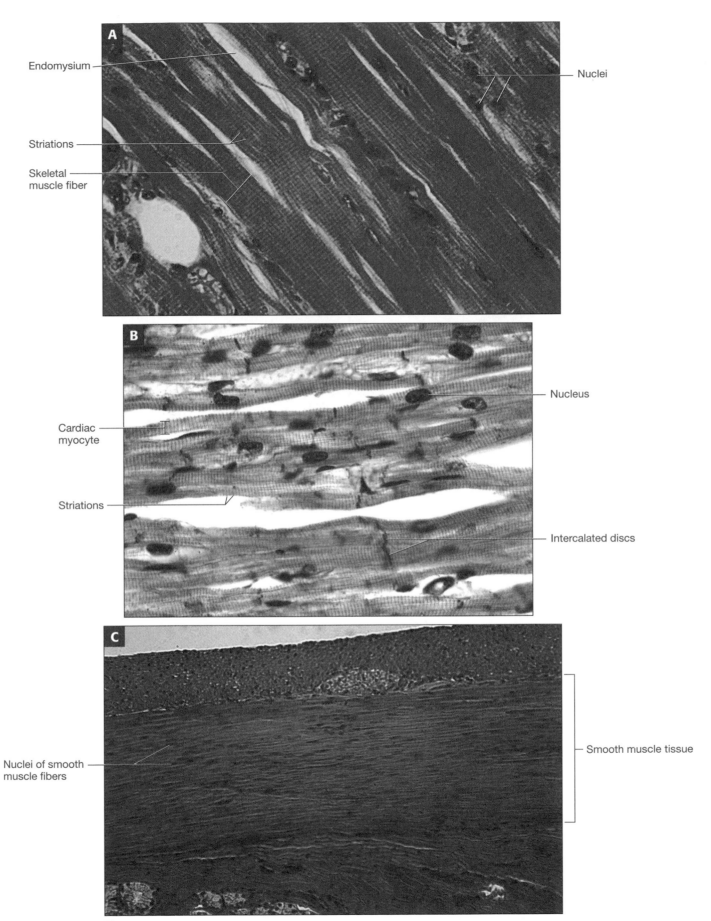

Endomysium

Nuclei

Striations

Skeletal muscle fiber

Cardiac myocyte

Nucleus

Striations

Intercalated discs

Nuclei of smooth muscle fibers

Smooth muscle tissue

FIGURE 5.3 Muscle tissue: (A) skeletal muscle tissue; (B) cardiac muscle tissue; (C) smooth muscle tissue

Procedure Microscopy of Muscle Tissue

View prepared slides of skeletal, smooth, and cardiac muscle tissue. Use colored pencils to draw what you see under the microscope, and label your drawing with the terms from **Figure 5.3**. Record your observations of each slide in Table 5.2. When you have completed the microscope slides, answer Check Your Understanding question 4 (p. 110).

1. Skeletal muscle:

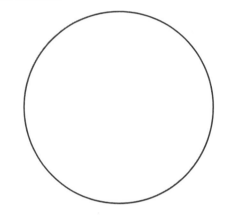

   a. _____

   b. _____

2. Smooth muscle:

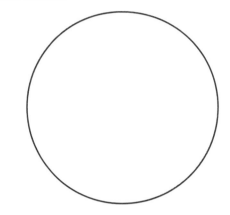

   a. _____

   b. _____

3. Cardiac muscle:

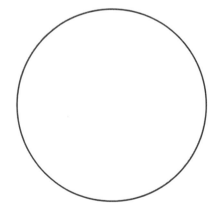

   a. _____

   b. _____

TABLE **5.2**	Characteristics of Muscle Tissues			
Muscle Tissue Type	**Striated or Nonstriated**	**One or Multiple Nuclei**	**Size and Shape of Cells**	**Special Features**
Cardiac muscle				
Skeletal muscle				
Smooth muscle				

 Exercise 4

Nervous Tissue

MATERIALS NEEDED

- Nervous tissue slide
- Light microscope
- Colored pencils

Nervous tissue (Figure 5.4) is the primary component of the brain, the spinal cord, and the peripheral nerves. It consists of a unique ECM and two main cell types:

1. **Neurons.** The neurons are responsible for sending and receiving messages within the nervous system. On your slide they are the larger of the two cell types. The large, central portion of the neuron is called the **cell body**. Within the cell body we find the nucleus and most of the neuron's organelles including clusters of rough endoplasmic reticulum called **Nissl bodies**. Most neurons contain two types of long "arms" extending from the cell body—the **dendrites**, which receive messages from other neurons, and the **axon**, which sends messages to other neurons, muscle cells, or gland cells.

2. **Neuroglial cells.** The smaller and more numerous cells around the neurons are the neuroglial cells. The six different types of neuroglial cells vary significantly in shape and appearance. Neuroglial cells in general perform functions that support the neurons in some way.

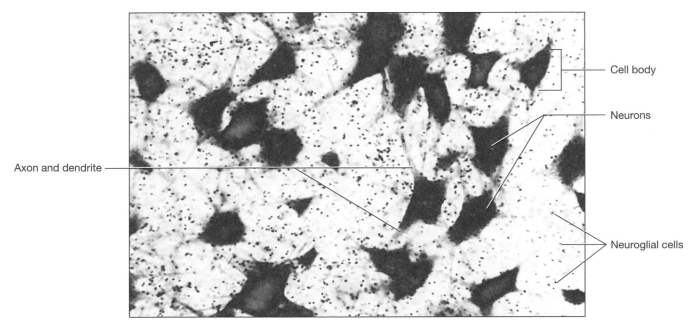

FIGURE 5.4 Nervous tissue

Procedure Microscopy of Nervous Tissue

View a prepared slide of nervous tissue (the slide might be called a "motor neuron smear"). Use colored pencils to draw a picture of what you see under the microscope, and label your drawing with the terms from **Figure 5.4**. Then (a) describe what you see, and (b) give examples of locations in the body where this tissue is found.

Nervous tissue

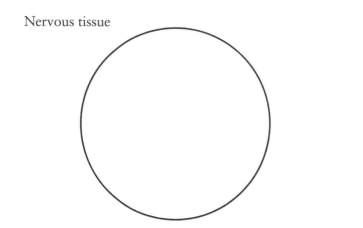

a. _____

b. _____

 # Exercise 5

Organology

MATERIALS NEEDED

▶ 5 colors of modeling clay

All organs consist of two or more tissues that must work together to enable the organ to function properly. The study of the tissues that make up the body's organs is called **organology**. Most organs are made of layers of tissues stacked upon one another and "glued" together by proteins and other molecules in the ground substance. This exercise introduces you to organology, a topic that we will explore repeatedly in the remainder of this lab manual.

 ## Procedure Determine Major Tissue Types of Organs

Use your textbook to determine the main tissue types that comprise each of the following organs, and record this information in Table 5.3. Be specific about which type of muscle, epithelial tissue, and connective tissues are found in the organ.

TABLE **5.3**	Organs and their Component Tissues
Organ	**Major Tissue Types**
Urinary bladder	
Blood vessel	
Stomach	
Skin	
Lymph node	
Tendon (and tendon sheath)	
Knee joint and capsule	
Heart	
Trachea	
Esophagus	
Auricle (external ear)	
Brain (and brain coverings)	

Procedure Build an Organ

For this exercise, choose one or more of the organs from Table 5.3, and build it with modeling clay. Use the following color code when building your organ:

Yellow = epithelial tissue (Be sure to use more than one layer if the tissue is stratified!)

Blue = connective tissue proper

Orange = cartilage

Red = muscle tissue

Green = nervous tissue (Don't forget that nearly every organ in the body has nerves that supply it. To place the nerves properly, find out which tissue layer of the organ is innervated.)

NAME _____

SECTION _____ DATE _____

Check Your Recall

1 Identify each of the following tissues in **Figure 5.5**.

a. _____

b. _____

c. _____

d. _____

e. _____

f. _____

g. _____

h. _____

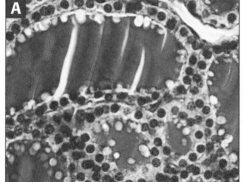

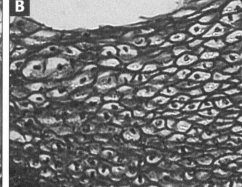

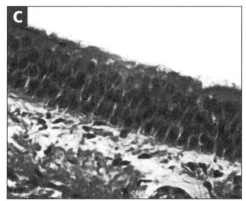

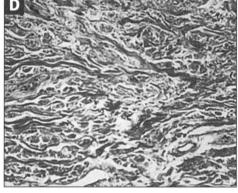

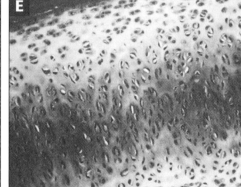

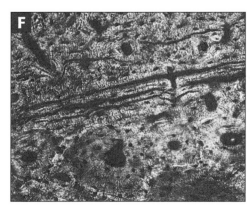

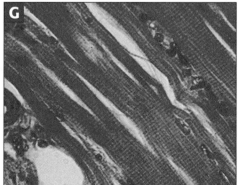

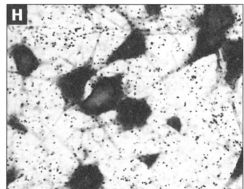

FIGURE 5.5 Unknown tissues for question 1

2 List the four basic tissue types:

_____   _____

_____   _____

3 Fill in the blanks: All tissues consist of two main components: _____ and _____, which consists of a gelatinous substance called _____ and numerous different _____.

4 Which of the following statements about epithelial tissue is *false*?
   a. Epithelial tissues are avascular.
   b. Epithelial tissues consist of few cells and an extensive ECM.
   c. The ECM of epithelial tissues is located in the basal lamina.
   d. Epithelial tissues are our covering and lining tissues.

5 How do simple and striated epithelial tissues differ?

6 Which of the following statements about connective tissue is *false*?
   a. All connective tissues stem from a common embryonic tissue called mesenchyme.
   b. Connective tissues may contain three types of protein fibers: collagen, elastic, and reticular fibers.
   c. Most connective tissues are highly vascular, with the exception of cartilage.
   d. Most connective tissues consist largely of cells with little ECM.

7 How do loose and dense connective tissues differ?

8 Which of the following statements about muscle tissue is *true*?
   a. Skeletal muscle and cardiac muscle tissues have no striations.
   b. Smooth muscle tissue is found in the heart.
   c. The cells of skeletal muscle tissue are long, tubular, and multinucleated.
   d. Smooth muscle cells are joined by intercalated discs.

9 Fill in the blanks: Nervous tissue is composed of _____ and _____.

10 Matching: Match the tissue type with its location in the body.

Simple squamous epithelium   _____   A. Part of the basement membrane, walls of hollow organs

Transitional epithelium   _____   B. Spleen and lymph nodes

Loose CT   _____   C. Air sacs of the lungs

Cardiac muscle   _____   D. Oral cavity, pharynx, vagina, anus

Nervous tissue   _____   E. Lining hollow organs, in the skin, in the eye, and surrounding many glands

Stratified squamous epithelium   _____

Hyaline cartilage   _____   F. Brain, spinal cord, peripheral nerves

    G. Urinary bladder

Smooth muscle   _____

    H. Tendons and ligaments

Dense regular collagenous CT   _____

    I. Joints, connecting the ribs to the sternum, nose

Reticular CT   _____

    J. In the heart

Check Your Understanding | Critical Thinking and Application Questions

1 Explain how the structure of each of the following epithelial tissues follows its function:

a. Stratified squamous keratinized epithelium:

b. Simple squamous epithelium:

c. Pseudostratified ciliated columnar epithelium:

2 Explain how the structure of each of the following connective tissues follows its function:

a. Dense regular collagenous CT:

b. Hyaline cartilage:

c. Bone:

3 The formation of fibrocartilage is a common response to injury of hyaline cartilage. Do you think that fibrocartilage would provide an articular surface (i.e., the cartilage in joints) that is as smooth as the original hyaline cartilage? Why or why not?

4 When muscle tissue dies, it usually is replaced with dense irregular collagenous connective tissue. How do these tissues differ in structure? Will the muscle be able to function normally? Why or why not?

Integumentary System

OBJECTIVES

Once you have completed this unit, you should be able to:

- Identify structures of the integumentary system.

- Map the distribution of touch receptors on different areas of the body.

- Perform and interpret finger-printing procedures.

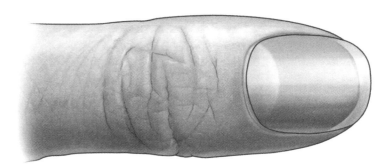

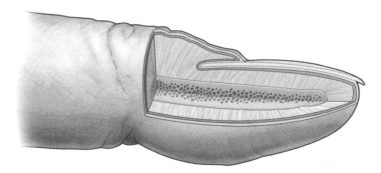

Pre-Lab Exercise 1
Key Terms

Table 6.1 lists the key terms with which you should be familiar before coming to lab.

TABLE 6.1	Key Terms

Term	Definition
Epidermal Structures	
Epidermis	
Keratinocyte	
Melanocyte	
Stratum corneum	
Stratum lucidum	
Stratum granulosum	
Stratum spinosum	
Stratum basale	
Structures of the Dermis	
Dermis	
Dermal papillae	

Pacinian (lamellated) corpuscle _____

_____

Meissner's (tactile) corpuscle _____

_____

Other Structures

Hypodermis _____ _____

_____

Sweat gland _____

_____

Sebaceous gland _____

_____

Hair follicle _____

_____

Nails _____

_____

Arrector pili muscle _____

_____

✎ Pre-Lab Exercise 2

Skin Anatomy

Label and color the structures of the skin depicted in Figure 6.1 with the terms from Exercise 1. Use your text and Exercise 1 in this unit for reference.

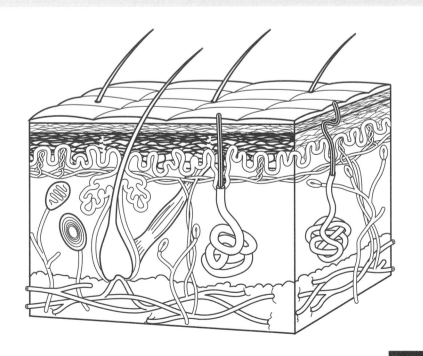

FIGURE 6.1 Skin section

Pre-Lab Exercise 3

Hair and Nail Anatomy

Label and color the structures of the hair and nail depicted in **Figure 6.2** with the terms from Exercise 1. Use your text and Exercise 1 in this unit for reference.

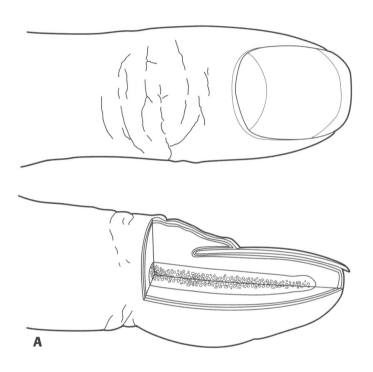

A

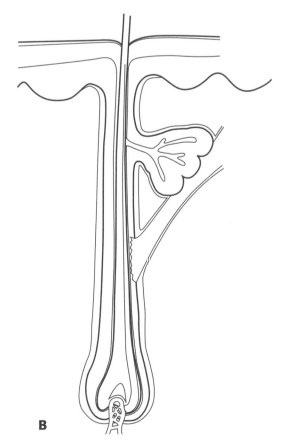

B

FIGURE **6.2** (A) Nail anatomy; (B) Hair structure

114

Although the skin is the largest organ in the body, most people don't realize that it is actually an organ. Like all organs, the skin is composed of several tissue types, including several layers of epithelial tissue, connective tissue, muscle tissue, and nervous tissue. In the following exercises you will examine the tissues of this organ, the other structures of the integumentary system, and apply your knowledge of this system to perform fingerprinting procedures.

Exercise 1

Skin Anatomy and Accessory Structures

MATERIALS NEEDED

▶ Skin models and diagrams

The **integumentary system** is composed of the skin (the **integument**) and its **accessory structures**: the **hair**, **glands**, and **nails**. The skin is composed of two layers, the **epidermis** and the **dermis** (**Figure 6.3**). The tissue beneath the dermis, sometimes called the **hypodermis** or the *subcutaneous tissue*, connects the skin to the underlying tissues and is not considered to be part of the integument.

The epidermis contains layers (or *strata*) of stratified squamous keratinized epithelium. From superficial to deep, the layers are as follows:

1. **Stratum corneum**. This superficial layer is composed of dead cells called **keratinocytes**. On microscopic examination, the cells of the stratum corneum bear little resemblance to living cells and have a dry, flaky appearance. These cells contain a hard protein called **keratin**, which protects the underlying layers of cells.

2. **Stratum lucidum**. This is a single layer of translucent, dead cells that is found only in the skin of the palms and the soles of the feet.

3. **Stratum granulosum**. The superficial cells of the stratum granulosum are dead, but the deeper cells are alive. This layer is named for the cells' cytoplasmic granules, which contain the protein keratin and an oily waterproofing substance.

4. **Stratum spinosum**. The first actively metabolizing cells are encountered in the stratum spinosum. The pigment **melanin** is found in this layer, which provides protection from UV light and also decreases production of vitamin D so the body does not overproduce it.

5. **Stratum basale**. The stratum basale is the deepest layer and contains a single layer of actively dividing cells. It is often combined with the stratum spinosum and called the **stratum germinativum**.

Why does the epidermis have so many dead cells? Recall that the epidermis is composed of epithelial tissue, and that epithelial tissue is avascular (has no blood supply). All epithelial tissues require oxygen and nutrients to be delivered from the deeper tissues. In the case of the epidermis, this deeper tissue is the dermis. Only the cells of the deeper parts of the stratum granulosum, the stratum

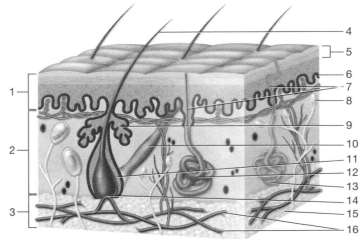

1. Epidermis	6. Stratum basale	11. Hair follicle
2. Dermis	7. Sweat duct	12. Apocrine sweat gland
3. Hypodermis	8. Sensory receptor	13. Eccrine sweat gland
4. Shaft of hair	9. Sebaceous gland	14. Bulb of hair
5. Stratum corneum	10. Arrector pili muscle	15. Adipose tissue
		16. Cutaneous blood vessels

FIGURE 6.3 Skin section

spinosum, and the stratum basale are close enough to the blood supply in the dermis to get adequate oxygen and nutrients for survival. As the cells migrate farther away from the blood supply, they begin to die.

Immediately deep to the stratum basale of the epidermis is the dermis. The dermis is composed of highly vascular connective tissue and contains two layers:

1. **Papillary layer.** The superficial papillary layer is composed of loose connective tissue. It contains fingerlike projections called the **dermal papillae,** which project into the epidermis. These dermal papillae contain touch receptors called **Meissner's corpuscles** and capillary loops that provide blood supply to the avascular epidermis. They also indent the overlying epidermis and create **epidermal ridges,** which produce the characteristic patterns of fingerprints.

2. **Reticular layer.** The thick reticular layer is composed of dense irregular collagenous connective tissue. It houses structures such as **sweat glands,** oil-producing **sebaceous glands,** blood vessels, and pressure receptors called **Pacinian corpuscles** (also known as *lamellated corpuscles*).

Both hair and nails are considered to be accessory structures of the integumentary system (**Figure 6.4**). A hair consists of two basic parts: (a) the long, slender **shaft,** composed of dead keratinized cells that projects from the skin's surface, and (b) the hair **root** or **bulb** that is embedded in the dermis (**Figure 6.4A**). The structure surrounding the hair bulb is an epithelial tissue-lined sheath known as the **hair follicle.**

Like hairs, nails are composed primarily of dead, keratinized cells (**Figure 6.4B**). A nail consists of a **nail plate** that is surrounded by folds of skin on all three sides, known as **nail folds.** The nail plate is formed by a group of dividing cells proximal to the nail fold called the **nail matrix.**

Other accessory structures of the integumentary system are its **sebaceous glands** and the **sweat glands,** both of which are exocrine glands that secrete their products onto the skin's surface. Sebaceous glands secrete sebum (oil) into a hair follicle, and sweat glands secrete sweat through a small pore.

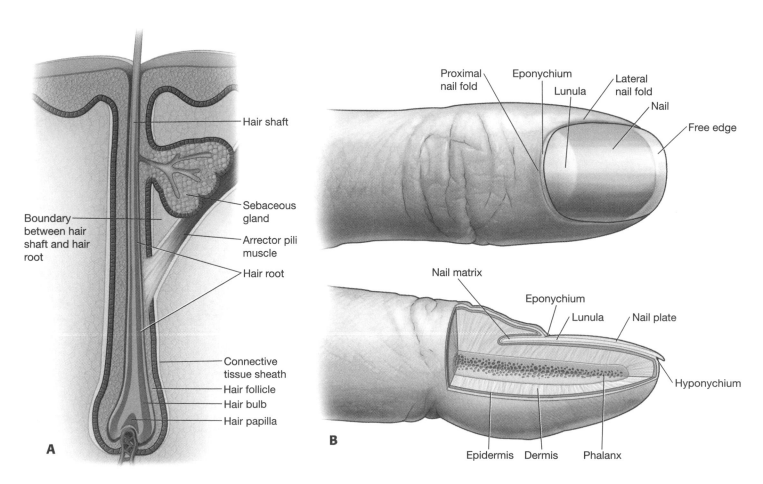

FIGURE 6.4 (A) Hair structure; (B) Nail anatomy

Procedure Model Inventory

Identify the following structures of the integumentary system on models and diagrams, using your textbook and this unit for reference. As you examine the anatomical models and diagrams, record on the model inventory in Table 6.2 the name of the model and the structures you were able to identify. When you have completed this activity, answer Check Your Understanding questions 1 and 2 (p. 125).

Structures of the Skin

1. Epidermal layers
   a. Stratum corneum
   b. Stratum lucidum
   c. Stratum granulosum
   d. Stratum spinosum
   e. Stratum basale
2. Dermal layers
   a. Papillary layer
   b. Reticular layer
3. Dermal papillae
4. Blood vessels
5. Nerves:
   a. Pacinian (lamellated) corpuscle
   b. Meissner (tactile) corpuscles
   c. Merkel discs

Other Structures

1. Sweat gland
2. Hair
   a. Hair follicle
   b. Hair shaft
   c. Arrector pili muscle
3. Nail
   a. Eponychium
   b. Nail matrix
   c. Nail fold
   d. Nail plate
   e. Lunula
4. Sebaceous gland
5. Hypodermis
   a. Subcutaneous fat
   b. Subcutaneous blood vessels

TABLE 6.2	Model Inventory for the Integumentary System
Model/Diagram	**Structures Identified**

Exercise 2

Histology of Integument

MATERIALS NEEDED

- Slide of thick skin
- Slide of thin skin
- Colored pencils
- Light microscope

In this exercise we will examine prepared slides of skin from different regions of the body so we can compare and contrast two types of skin: (a) **thick skin**, found on the palms and soles of the feet, and (b) **thin skin**, found virtually everywhere else.

Before moving on, you may wish to review the basics of microscopy from Unit 3. Remember to follow a step-by-step approach when examining the slides: Look at the slide with the naked eye first, then begin your examination on low power, and advance to higher power to see more details.

Procedure Microscopy of Thick Skin

Obtain a prepared slide of thick skin (which may be labeled "Palmar Skin"), and examine it with the naked eye to get oriented. Once you are oriented, place the slide on the stage of the microscope and scan it on low power. You should be able to see the epidermis with its superficial layers of dead cells, and the dermis with its pink clusters of collagen bundles that make up the dense irregular collagenous connective tissue. Advance to higher power to see the cells and associated structures in greater detail.

Use your colored pencils to draw what you see in the field of view (you will be able to see the most structures on low power). Label your drawing with the following terms, using **Figure 6.1** for reference. When you have completed your drawing, fill in Table 6.3.

1. Epidermis
   a. Stratum corneum
   b. Stratum lucidum
   c. Stratum granulosum
   d. Stratum spinosum
   e. Stratum basale

2. Dermis
   a. Dermal papillae
   b. Collagen bundles

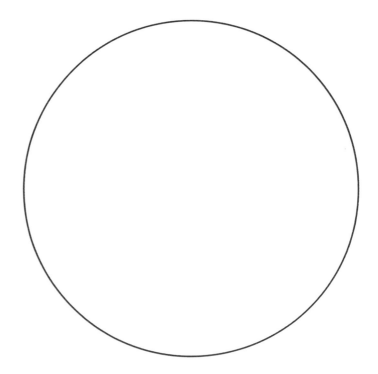

Procedure Microscopy of Thin Skin

Obtain a prepared slide of thin skin (which may be called "Scalp Skin"). As before, examine the slide with the naked eye, then scan the slide on low power, advancing to higher power as needed to see the structures more clearly.

Use your colored pencils to draw what you see in the field of view (you will be able to see the most structures on low power). Label your drawing using the terms below, using **Figure 6.1** for reference. When you have completed your drawing, fill in the remainder of Table 6.3.

1. Epidermis
   a. Stratum corneum
   b. Stratum granulosum
   c. Stratum spinosum
   d. Stratum basale

2. Dermis
   a. Dermal papillae
   b. Collagen bundles

3. Hair follicle

4. Sebaceous gland

5. Sweat gland

6. Arrector pili muscle

TABLE **6.3** Characteristics of Thick and Thin Skin		
Characteristic	**Thick Skin**	**Thin Skin**
Thickness of stratum corneum		
Hair follicles present?		
Sebaceous glands present?		
Stratum lucidum present?		
Arrector pili muscles present?		

Exercise 3

Touch Receptor Distribution

MATERIALS NEEDED

- Water-soluble marking pen
- Ruler
- Monofilament

The two types of sensory receptors in the skin that detect fine-touch are called **Meissner's corpuscles** and **Merkel discs**. The distribution of these receptors can be mapped using an instrument called a monofilament, which applies 10 grams of force as it is pressed on the skin (10 grams of force is generally accepted as the maximum amount of force required to activate the Meissner's corpuscles and Merkel discs). If a monofilament is not available, a similar instrument may be made using a stiff-bristle hair glued to a toothpick. When you have completed the activity, answer Check Your Understanding questions 3 and 4 (pp. 125–126).

 Procedure **Mapping Touch Receptor Distribution**

1 Use a centimeter ruler to measure a 2-centimeter square on your partner's palm. Mark this square with a water-soluble marking pen.

2 Have your partner close his or her eyes.

3 Apply the monofilament to one of the corners of the square. Press only until the monofilament bends slightly.

4 If your partner can perceive the sensation of the monofilament, mark this spot with a different-colored water-soluble marker.

5 Repeat this process for the remainder of the square, advancing the monofilament 1–2 millimeters each time. Be certain to apply the same amount of pressure with each application.

6 When you have completed the square, record the number of receptors that were found on the palm.

Number of receptors on palm: _____

7 Repeat this process on the posterior shoulder. Record the number of receptors that were found on this location.

Number of receptors on the posterior shoulder: _____

Exercise 4

Fingerprinting

MATERIALS NEEDED

- Blank glass slide (one per person)
- Inkpad
- Fingerprint card (one per person)
- Dusting powder
- Dusting brush
- Fingerprint lifting tape
- Latex gloves
- Sharpie® marker

Fingerprinting was not a widely accepted means of identification in the United States until 1902, when the New York Civil Service began collecting and using fingerprints. Prior to this, fingerprints had been studied widely and even were used to solve a murder case in Argentina in 1892.

Fingerprints can be broadly classified into three ridge flow patterns: loop, arch, and whorl (Figure 6.5). Once the ridge flow pattern has been identified, the characteristics of the ridges, also called *Galton points*, are examined. The three basic ridge characteristics are the ridge ending, the island (or the dot), and the bifurcation (Figure 6.6). In professional laboratories the fingerprint is analyzed further on the basis of minute details, or "points," in the fingerprint. Much of this is done currently with the help of computers.

Loop   Whorl   Arch

FIGURE **6.5** The three basic patterns of fingerprints

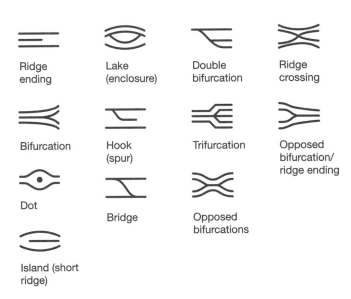

FIGURE **6.6** The Galton points

In this exercise you will be playing the role of detective, searching for a "thief" among your group members (which may end up being you!). Your group will determine the thief's identity by analyzing the ridge flow patterns and ridge characteristics of the fingerprints of your group members. When you have completed the activity, answer Check Your Understanding question 5 (p. 126).

Procedure Fingerprint Analysis

1 Assemble into groups of a minimum of four people.

2 Obtain one blank glass slide and handle it by the edges as you clean it off with a paper towel and water.

3 Once the slide is clean and dry, place two or three large fingerprints (use different fingers) somewhere on one side of the slide.

4 After placing one or more fingerprints on the slide, put on gloves and use the Sharpie at your table to write your initials on the upper righthand corner of the slide.

5 Bring the slides for the entire table to your instructor, and your instructor will pick one slide from the group, cover the initials in the corner, and designate that person as the "thief." The thief is one of your group members, so you are all suspects.

6 Fingerprint each member of the group (using the ink and cards provided):
   a. Obtain a blank fingerprinting card or a blank sheet of paper.
   b. Roll each finger individually on the inkpad from right to left. Be certain not to get excessive ink on your fingers.
   c. Roll each finger individually on the fingerprinting card slowly from left to right.

7 When all of the fingerprinting is completed, don a pair of latex gloves.

8 Use a dusting brush to lightly sweep some dusting powder onto the slide containing the thief's fingerprints. Use the dusting powder *sparingly*, as too much powder will obscure your prints. Brush away the excess dusting powder. (You can tap the slide on its side to help with this.)

9 Place a piece of fingerprint lifting tape on the fingerprints that have appeared. Pat the tape into place firmly, and remove it slowly. Place the tape on a blank sheet of white paper for comparison.

10 With the help of **Figures 6.5** and **6.6**, identify the thief by comparing the fingerprints lifted from the slide with the fingerprints taken from the suspects. Be careful not to tell your group mates which fingers you used. Remember—you don't want to be caught!

Who is the thief from your table?

How did you arrive at this conclusion?

EXPLORING ANATOMY & PHYSIOLOGY IN THE LABORATORY

UNIT
6
REVIEW

NAME _____

SECTION _____ DATE _____

Check Your Recall

1 Label the following parts of the skin on **Figure 6.7**.
- Epidermis
- Dermis
- Hypodermis
- Dermal papillae
- Sweat gland
- Sweat pore
- Sebaceous gland
- Arrector pili muscle
- Hair follicle
- Hair bulb (root)
- Hair shaft
- Blood vessels

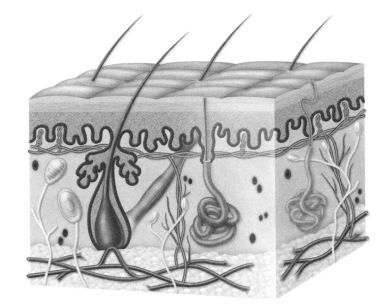

FIGURE **6.7** Skin section

2 The main cell type in skin is the
a. melanocyte.
b. reticulocyte.
c. monocyte.
d. keratinocyte.

3 Number the layers of the epidermis, with 1 being the most superficial layer and 5 being the deepest layer.

_____ Stratum lucidum

_____ Stratum basale

_____ Stratum corneum

_____ Stratum spinosum

_____ Stratum granulosum

4 Which layers of the epidermis contain living cells?
a. Stratum granulosum only
b. Stratum corneum, stratum granulosum, stratum lucidum
c. Stratum basale, stratum spinosum, stratum granulosum
d. All of the layers of the epidermis contain living cells.
e. None of the layers of the epidermis contain living cells.

5 From where do the cells of the epidermis obtain oxygen and nutrients?
   a. From blood vessels in the epidermis
   b. From blood vessels in the dermis
   c. Diffusion from the air
   d. From blood vessels in other epithelial tissues

6 Match the following terms with the correct description.

   a. Papillary layer _____   A. Secrete product through a pore

   b. Sebaceous gland _____   B. Pressure receptor in the dermis

   c. Pacinian corpuscle _____   C. Projections of the dermis that indent the epidermis

   d. Hair follicle _____   D. Superficial layer of the dermis; loose connective tissue

   e. Dermal papillae _____   E. Sheath of epithelial and connective tissue around a hair

   f. Hair shaft _____   F. Deep layer of the dermis; dense irregular collagenous connective tissue

   g. Reticular layer _____   G. Secretes sebum (oil)

   h. Sweat gland _____   H. Portion of the hair that projects from the skin's surface

7 Which of the following are characteristics of thick skin? (Circle all that apply.)
   a. Located over the palms and the soles of the feet
   b. Contains hair and arrector pili muscles
   c. Contains sweat glands
   d. Very thick stratum corneum
   e. Contains sebaceous glands
   f. Contains a stratum lucidum

8 The dividing cells of a nail are located in the
   a. nail fold.
   b. nail plate.
   c. nail bed.
   d. nail matrix.

9 The two types of receptors in the skin that detect fine touch are the
   a. Merkel discs.
   b. Pacinian corpuscles.
   c. Meissner's corpuscles.
   d. Both a and b are correct.
   e. Both a and c are correct.

10 Fingerprints are the result of
   a. dermal papillae that create epidermal ridges.
   b. projections from the hypodermis.
   c. epidermal papillae that create dermal ridges.
   d. dermal papillae that house sebaceous glands.

Check Your Understanding | Critical Thinking and Application Questions

1 Explain why a superficial skin scrape (such as a paper cut) doesn't bleed. Why don't you bleed when a hair is pulled out?

2 Shampoos and hair conditioners often claim to have nutrients and vitamins that your hair must have to grow and be healthy. Taking into account the composition of hair, do you think these vitamins and nutrients will be beneficial? Why or why not?

3 Would you have expected to have found more Meissner's corpuscles and Merkel discs on the palm or on the posterior shoulder? Why?

4 In which areas of the body would you expect to find the most Merkel discs and Meissner's corpuscles? Why?

5 Why do you have unique markings on your fingers, toes, palms, and soles but not on any other part of the body? Why do you leave behind fingerprints when you touch certain surfaces?

Introduction to the Skeletal System

OBJECTIVES

Once you have completed this unit, you should be able to:

- Identify the structures and components of osseous tissue.

- Explain the role of minerals and protein fibers in the function of osseous tissue.

- Classify bones according to their shape.

- Identify the parts of the long bone.

- Define and provide examples of various bone markings.

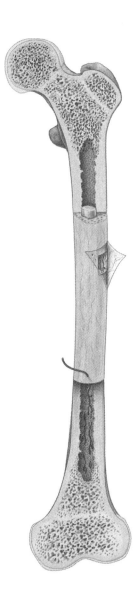

Pre-Lab Exercise 1
Key Terms

Table 7.1 lists the key terms with which you should be familiar before coming to lab.

TABLE 7.1	Key Terms

Term **Definition**

Types of Bone Tissue

Compact (cortical) bone _____

_____

Spongy (cancellous) bone _____

_____

Structures of the Osteon

Osteon _____

_____

Lamellae _____

_____

Central (Haversian) canal _____

_____

Perforating (Volkmann's) canal _____

_____

Lacunae _____

_____

Osteocytes _____

_____

Canaliculi _____

_____

Periosteum _____

_____

Endosteum _____

_____

Bone Shapes

Long bone _____

_____

Short bone _____

_____

Flat bone _____

_____

Irregular bone _____

_____

Structures of Long Bones

Diaphysis _____

_____

Epiphysis _____

_____

Epiphyseal plate/line _____

_____

Medullary cavity _____

_____

Bone marrow _____

_____

Pre-Lab Exercise 2
Microscopic Anatomy of Compact Bone

Label and color the microscopic anatomy of compact bone tissue depicted in **Figure 7.1** with the terms from Exercise 1. Use your text and Exercise 1 in this unit for reference.

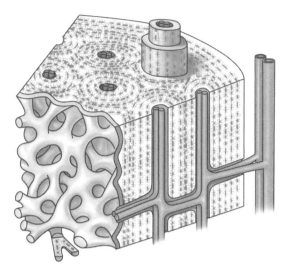

FIGURE 7.1 Microscopic anatomy of compact bone

Pre-Lab Exercise 3
Structure of a Long Bone

Label and color the diagram of a long bone (the femur) depicted in **Figure 7.2** with the terms from Exercise 4. Use your text and Exercise 4 in this unit for reference.

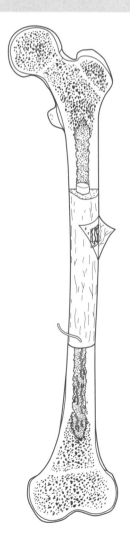

FIGURE 7.2
A long bone (the femur)

Exercises

The **skeletal system** consists of the bones and their cartilages. At first it might seem odd that a set of bones makes up an organ system, but remember that each bone is considered to be an organ. A bone consists of many tissue types, including osseous tissue, epithelial tissue, dense irregular collagenous connective tissue, and adipose tissue. The following exercises will introduce you to these complex organs and the histology of osseous tissue.

 ## Exercise 1

Histology of Osseous Tissue

MATERIALS NEEDED

- Osteon model
- Slide of compact bone
- Light microscope
- Colored pencils

The most superficial tissue of a bone is called the **periosteum**. The periosteum is composed of dense irregular collagenous connective tissue that is richly supplied with blood vessels. The innermost layer of the periosteum contains cells called **osteoblasts**, which secrete bone matrix and build new bone, and **osteoclasts**, which secrete enzymes that break down bone matrix. The periosteum is anchored to the bone by bundles of collagen called **perforating** or **Sharpey's fibers**.

Deep to the periosteum we find osseous tissue. The two general types of **osseous tissue** are

1. **compact** (or *cortical*) **bone**, and
2. **spongy** (or *cancellous*) **bone**.

Compact bone is hard, dense bone tissue found immediately deep to the periosteum. Its hardness comes from its structure, which consists of repeating, densely packed subunits called **osteons** (**Figure 7.3**). Osteons contain several features, including the following:

1. **Lamellae.** Lamellae are concentric rings of bone matrix. The lamellae give compact bone a great deal of strength, much in the way that a tree's rings make it a strong structure.

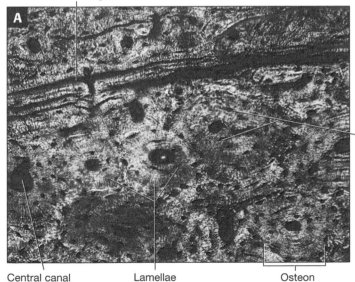

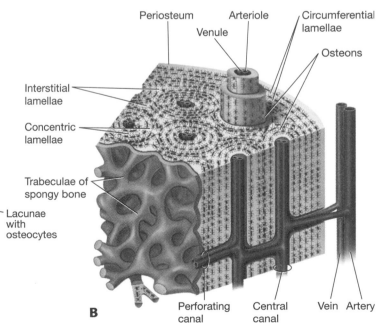

FIGURE 7.3 Compact bone: (A) photomicrograph; (B) illustration

2. **Central (Haversian) canal.** Running down the center of each osteon is a central or Haversian canal. Each central canal contains blood vessels and nerves and is lined with a connective tissue membrane called the **endosteum**. Like the periosteum, the endosteum has an inner layer of osteoblasts, which secrete bone matrix, and osteoclasts, which degrade bone.

3. **Lacunae.** Situated between the lamellae are small cavities called lacunae. Lacunae contain mature osteoblasts called **osteocytes**, which monitor and maintain the bone matrix. Neighboring lacunae are connected to each other by tiny canals called **canaliculi**.

4. **Perforating (Volkmann's) canals.** The perforating or Volkmann's canals lie perpendicular to the osteon and carry blood vessels into the bone from the periosteum. Like the central canals, perforating canals are lined by the endosteum.

Spongy bone is found on the inside of a bone deep to compact bone. As its name implies, it somewhat resembles a sponge and consists of a latticework-type structure with tiny bone spicules called **trabeculae** (Figure 7.4). The latticework structure of cancellous bone allows it to house another important tissue, the **bone marrow**. The two types of bone marrow are: (a) **red bone marrow**, which produces blood cells, and (b) **yellow bone marrow**, which is composed primarily of adipose tissue. As you can see in **Figure** 7.4, trabeculae are composed of lamellae but are not organized into osteons. For this reason, spongy bone lacks the hardness of compact bone.

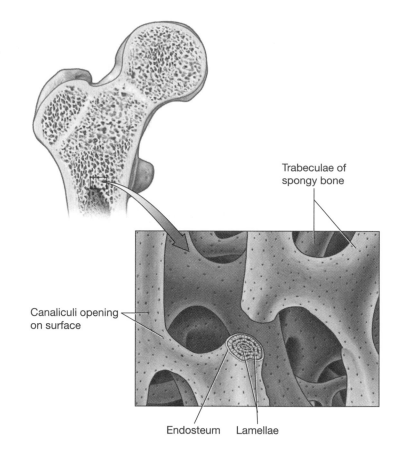

FIGURE 7.4 Microscopic anatomy of spongy bone tissue

Procedure Model Inventory

Identify the following structures of compact and spongy bone on models and diagrams, using your textbook and this unit for reference. As you examine the anatomical models and diagrams, record on the model inventory in Table 7.2 the name of the model and the structures you were able to identify. When you have completed the activity, answer Check Your Understanding questions 1 and 2 (p. 139).

Compact Bone Structures

▶ Osteon (Haversian system)
▶ Central (Haversian) canal
▶ Perforating (Volkmann's) canal
▶ Lamellae
▶ Lacunae
  — Osteocytes
▶ Canaliculi
▶ Periosteum
  — Sharpey's fibers
▶ Endosteum

Spongy Bone Structures

▶ Trabeculae (bone spicules)
▶ Red bone marrow
▶ Yellow bone marrow
▶ Endosteum
▶ Lacunae
▶ Osteocytes

TABLE **7.2**	Model Inventory for Osseous Tissue
Model	**Bone Structures Identified**

 Procedure Microscopy

View a prepared slide of compact bone. The structures on the slide should look similar to the osteon models you viewed in lab earlier and **Figure 7.3A**.

Use colored pencils to draw a picture of what you see under the microscope, and label your drawing with the compact bone structures from Exercise 1.

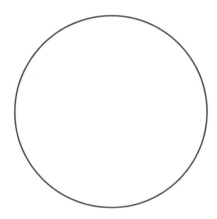

 Exercise 2

Chemical Components of Bone Tissue

MATERIALS NEEDED

▸ 1 piece of normal bone
▸ 1 piece of oven-heated bone
▸ 1 piece of bone soaked in nitric acid

Osseous tissue is a type of connective tissue composed of two primary chemical components.

1. The **organic component** consists mostly of protein fibers such as **collagen**. Collagen provides a supportive network that gives the bone tensile strength (the ability to withstand stretching and twisting forces).

2. The **inorganic component** consists mostly of calcium in the form of calcium hydroxyapatite crystals. The inorganic component provides the bone with compressional strength (the ability to withstand compressive forces).

As you will see in this exercise, both chemical components of bone are required for bone to remain strong and perform its functions.

 **Procedure**

In this exercise you will compare the effects of removing different chemical components of bone. One sample has been heated in an oven to destroy the organic components of the bone, and another sample has been treated with nitric acid to destroy the inorganic components of the bone. When you have completed the activity, answer Check Your Understanding questions 3 and 4 (pp. 139–140).

1 Obtain three pieces of bone:

    a. one piece of bone that has been baked in an oven for a minimum of 2 hours,

    b. one piece of bone that has been soaked in nitric acid, and

    c. one piece of untreated bone.

2 Place the three pieces of bone side by side. Describe their appearance. Do they look different from one another?

3 Compress or squeeze each sample using your fingers. What happens to:

    a. The heated bone?

    b. The bone treated with nitric acid?

    c. The untreated bone?

 Exercise 3

Classification of Bones by Shape

MATERIALS NEEDED

- Disarticulated bones
- Articulated skeleton

One way in which bones are classified is by their shape. Note in **Figure** 7.5 the four general shapes of bones:

1. **Long bones** are longer than they are wide, and include the bones of the upper and lower extremities excluding the ankle and wrist bones.

2. **Short bones** are about as long as they are wide. The bones of the wrist and the ankle are short bones.

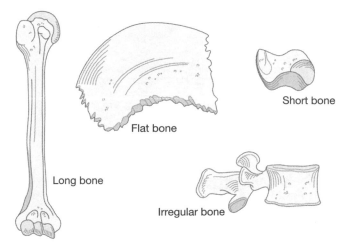

FIGURE **7.5** The four shapes of bones

3. **Flat bones** are shaped exactly as they're named—they are flat. Flat bones include the ribs, the sternum, the clavicle, certain skull bones, and the bones of the pelvis.

4. **Irregular bones** are those whose shape doesn't fit into any of the other classes. Irregular bones include the vertebrae and certain bones of the skull.

Procedure

Identifying Bone Shapes

Obtain a set of disarticulated bones and classify them according to their shape. Identify examples of long bones, short bones, flat bones, and irregular bones, and record them in Table 7.3. You may wish to use the figures in your text or Unit 8 (p. 141) to help you identify the disarticulated bones.

TABLE 7.3	Examples of Bone Shapes
Bone	**Shape**

Exercise 4

Anatomy of Long Bones

MATERIALS NEEDED

▶ Long bone, sectioned

All long bones share common structures and parts, illustrated in Figure 7.6 with the example of the femur. The **diaphysis** is the shaft of the long bone. As you can see in **Figure 7.6**, it consists of a thick collar of compact bone surrounding a hollow area called the **medullary cavity**. This collar of compact bone makes long bones quite strong and able to support the body's weight. The medullary cavity has sparse trabeculae and generally is filled with yellow bone marrow in living bone.

The ends of a long bone are called the **epiphyses**. Each epiphysis contains a shell of compact bone surrounding the inner spongy bone. The spongy bone within the epiphyses contains either red or yellow bone marrow. The end of each epiphysis is covered with hyaline cartilage (sometimes called **articular cartilage**), which allows two bones to articulate during movement with minimal friction. At certain epiphysis–diaphysis junctions you will note a thin, calcified line called the **epiphyseal line**. This structure is the remnant of the **epiphyseal plate**, a band of hyaline cartilage from which long bones grow in length. When longitudinal growth ceases, the chondrocytes of the epiphyseal plate die and are replaced by calcified bone tissue.

 Procedure Identification of Long Bone Structures

Identify the following structures of long bones on specimens and X-rays (if available). Check off each structure as you identify it. After you have completed the activity, answer Check Your Understanding question 5 (p. 140).

▶ Diaphysis
▶ Epiphysis
▶ Hyaline (articular) cartilage
▶ Medullary canal
▶ Epiphyseal plate (may be visible only on X-ray)
▶ Epiphyseal line
▶ Compact bone
▶ Spongy bone
▶ Yellow bone marrow
▶ Red bone marrow

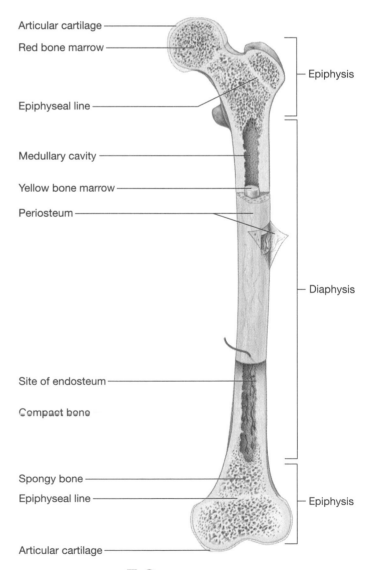

FIGURE **7.6** A long bone: the femur

NAME _____

SECTION _____ DATE _____

 Check Your Recall

1 Label the following parts of compact and spongy bone on Figure 7.7.

• Osteon

• Lamellae

• Central canal

• Lacunae

• Perforating canal

• Periosteum

• Endosteum

• Blood vessels

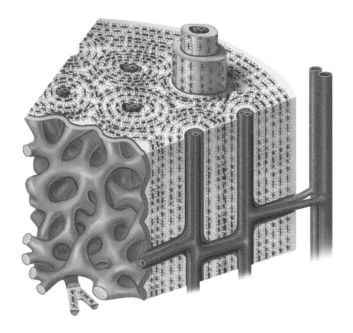

FIGURE 7.7 Microscopic anatomy of compact bone tissue

2 Mark the following statements as true or false. If the statement is false, correct it to make it a true statement.

a. The periosteum contains osteocytes and osteoclasts.

b. Osteoblasts secrete bone matrix.

c. Compact bone is the hard, outer bone.

d. Spongy bone is composed of spicules called osteons.

e. Spongy bone houses red and yellow bone marrow.

f. Perforating fibers anchor the endosteum to the superficial surface of a bone.

3 The organic component of bone consists of _____ and functions to _____.

a. collagen; provide compressional strength

b. calcium hydroxyapatite; provide compressional strength

c. collagen; provide tensile strength

d. calcium hydroxyapatite; provide tensile strength

4 The inorganic component of bone consists of _____ and functions to _____.

    a. Collagen; provide compressional strength

    b. Calcium hydroxyapatite; provide compressional strength

    c. Collagen; provide tensile strength

    d. Calcium hydroxyapatite; provide tensile strength

5 Which of the following statements about long bones is/are false? Circle all that apply.

    a. A long bone is longer than it is wide.

    b. They include the bones of the upper and lower extremities excepting the finger and toes.

    c. They include the bones of the upper extremities excepting the bones of the ankle and wrist.

    d. A long bone is named for its length.

6 Examples of flat bones include

    a. the clavicle.

    b. bones of the pelvis.

    c. ankle bones.

    d. Both a and b are correct.

    e. Both a and c are correct.

7 Short bones are

    a. short in length.

    b. about as long as they are wide.

    c. irregular in shape.

    d. flat.

8 Label the following parts of a long bone on **Figure 7.8**.

    • Diaphysis

    • Medullary cavity

    • Epiphysis

    • Epiphyseal line

    • Red bone marrow

    • Yellow bone marrow

    • Periosteum

    • Articular cartilage

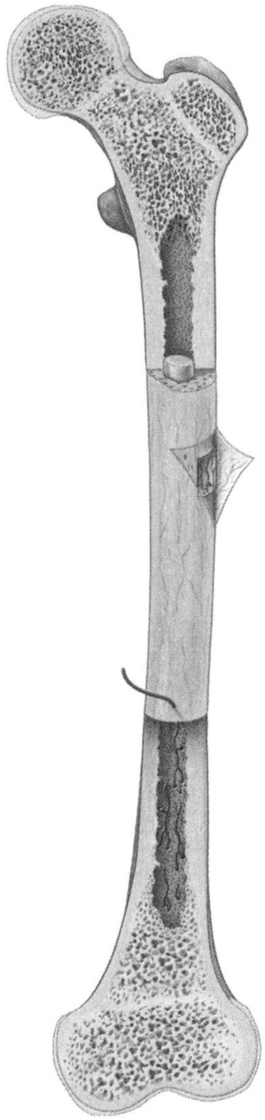

9 The epiphyseal plate is

    a. the structure from which long bones grow in length.

    b. a remnant of the structure from which long bones grow in length.

    c. composed of osseous tissue.

    d. found lining the surface of the epiphysis.

FIGURE 7.8 Structure of a long bone

10 The medullary cavity is found in the _____ and contains the _____.

    a. diaphysis; red bone marrow.

    b. epiphysis; red bone marrow.

    c. diaphysis; yellow bone marrow.

    d. epiphysis; yellow bone marrow.

Check Your Understanding | **Critical Thinking and Application Questions**

1 Explain how the structure of compact bone follows its function.

2 Explain how the structure of spongy bone follows its function.

3 *Osteogenesis imperfecta* is a congenital condition in which collagen synthesis is defective. Of the three samples of bone you tested in Exercise 3, which is most similar to the bones in osteogenesis imperfecta? What symptoms would you expect to find in this disease?

4 The diseases *rickets* and *osteomalacia* result from insufficient vitamin D intake, which decreases the amount of calcium available for synthesis of the inorganic component of bone. Of the three samples of bone you tested in Exercise 3, which is the most similar to the bones in rickets and osteomalacia? What symptoms would you expect to find in these diseases?

5 Explain how the structure of a long bone follows its function.

Skeletal System

OBJECTIVES

*Once you have completed this unit,
you should be able to:*

- Identify bones and markings of the axial skeleton.

- Identify bones and markings of the appendicular skeleton.

- Build a skeleton using disarticulated bones.

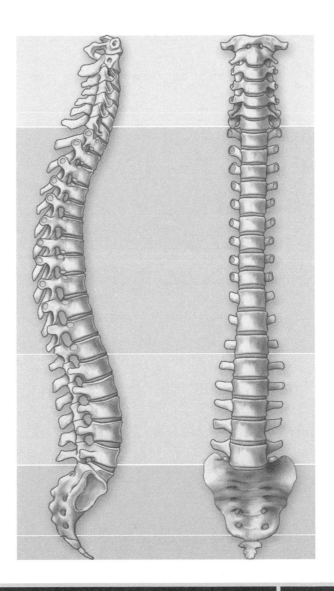

Pre-Lab Exercises

Complete the following exercises prior to coming to lab, using your textbook and lab manual for reference.

Pre-Lab Exercise 1

Key Terms

Table 8.1 lists the key terms with which you should be familiar before coming to lab.

TABLE 8.1 Key Terms

Term	**Definition**
Axial Skeleton	
Cranial bones	
Facial bones	
Suture	
Hyoid bone	
Vertebrae	
Ribs	
Sternum	
Appendicular Skeleton—Upper Limb	
Pectoral girdle	
Clavicle	

Scapula _____

_____

Humerus _____

_____

Radius _____

_____

Ulna _____

_____

Carpals _____

_____

Metacarpals _____

_____

Phalanges _____

_____

Appendicular Skeleton—Lower Limb

Pelvic girdle _____

_____

Femur _____

_____

Tibia _____

_____

Fibula _____

_____

Tarsals _____

_____

Metatarsals _____

_____

Pre-Lab Exercise 2
Bones of the Skull

Label and color the structures of the skull depicted in **Figure 8.1** with the terms from Exercise 1. Use your text and Exercise 1 in this unit for reference.

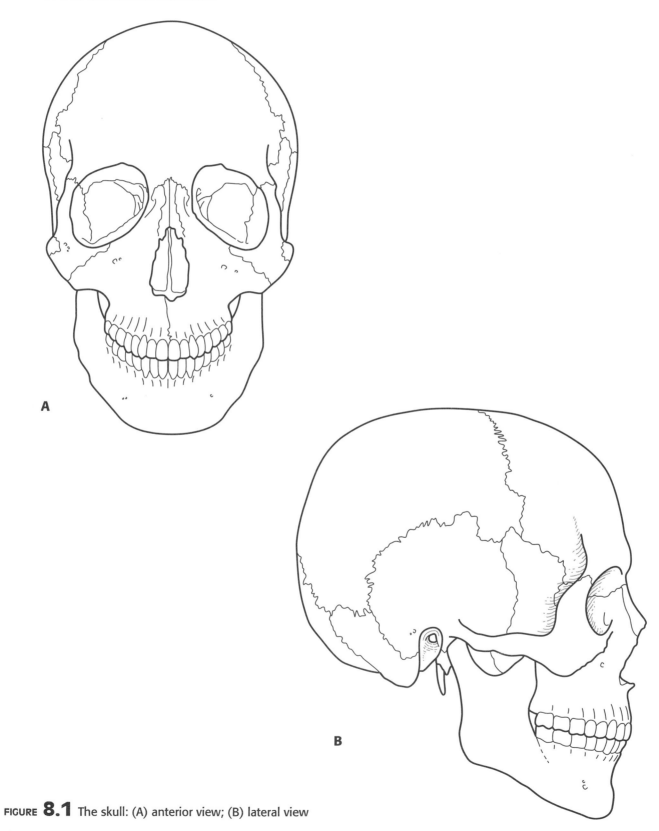

A

B

FIGURE **8.1** The skull: (A) anterior view; (B) lateral view

144

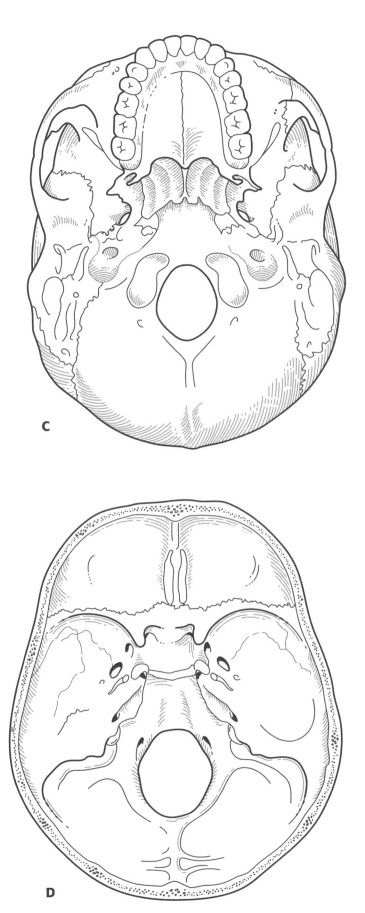

FIGURE 8.1 The skull (cont.): (C) inferior view (mandible removed); (D) internal view (calvaria removed)

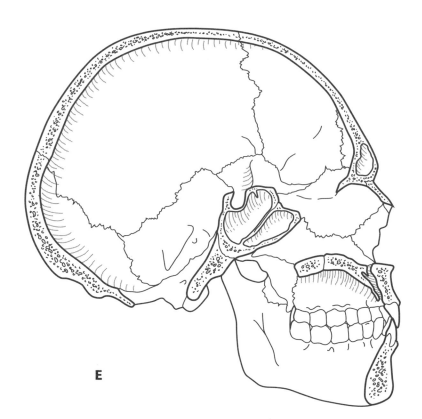

FIGURE **8.1** The skull (cont.):
(E) midsagittal section

E

✎ Pre-Lab Exercise 3

Remaining Bones
of the Axial Skeleton

Label and color the structures of the
axial skeleton depicted in **Figure 8.2**
with the terms from Exercise 2. Use
your text and Exercise 2 in this unit
for reference.

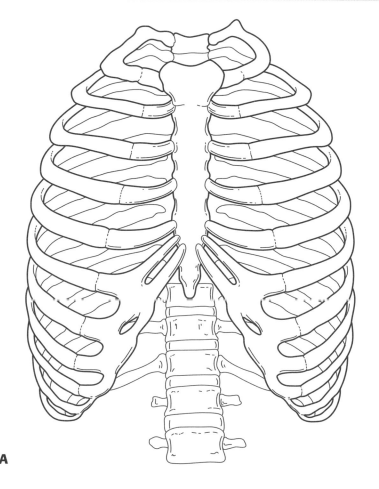

FIGURE **8.2** The axial skeleton:
(A) sternum and thoracic cage

A

EXPLORING ANATOMY & PHYSIOLOGY IN THE LABORATORY

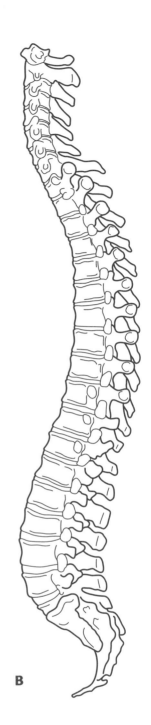

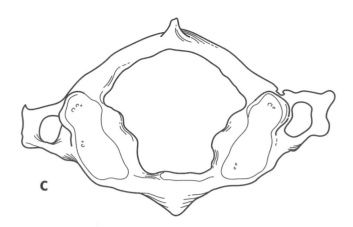

C

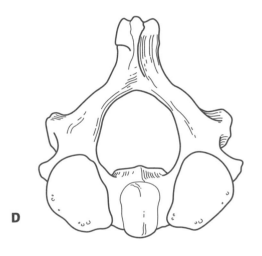

D

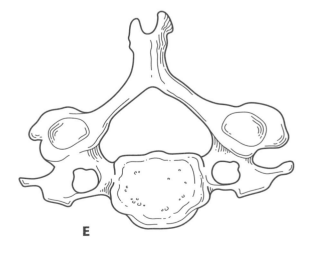

E

B

FIGURE **8.2** The axial skeleton (cont.):
(B) vertebral column;
(C) atlas (C1);
(D) axis (C2);
(E) typical cervical vertebra

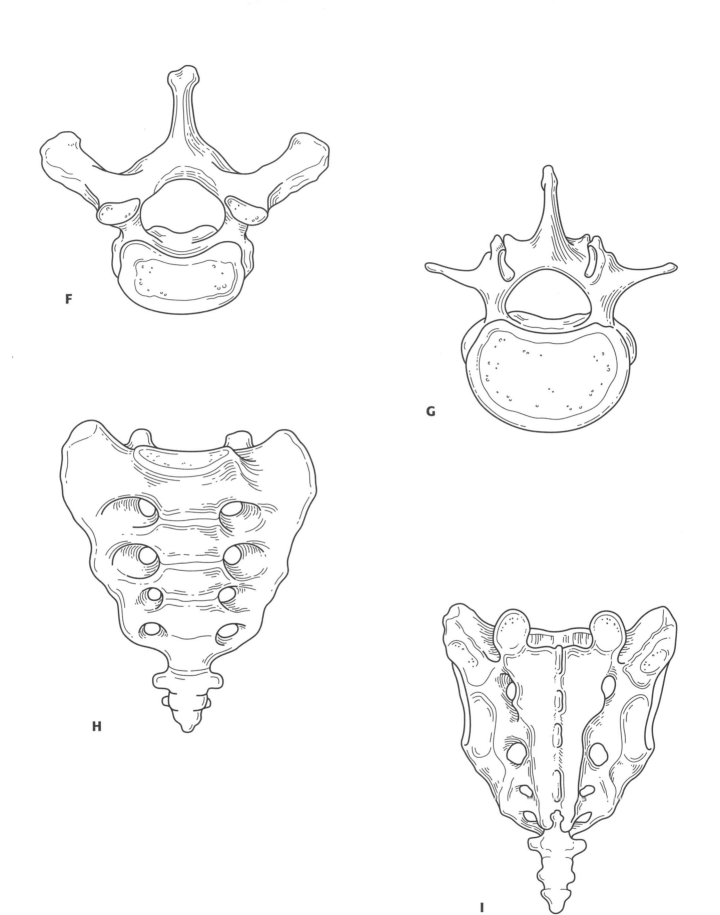

FIGURE 8.2 The axial skeleton (cont.): (F) typical thoracic vertebra; (G) typical lumbar vertebra; (H) sacrum, anterior view; (I) sacrum, posterior view

Whole Skeleton

Label and color the structures
of the skeleton depicted in
Figure 8.3 with the terms from
Exercises 1–3. Use your text
and Exercises 1–3 in this unit
for reference.

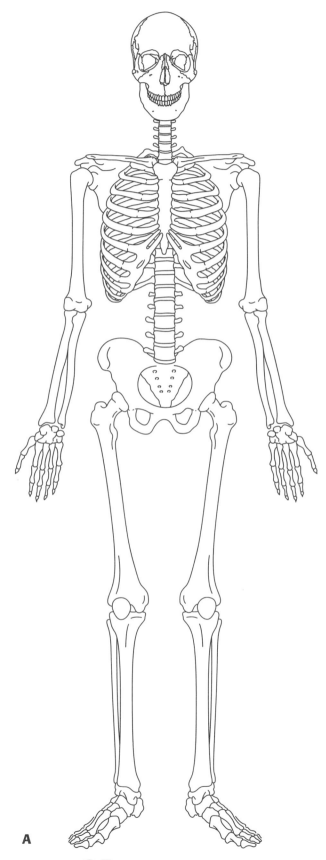

A

FIGURE **8.3** Skeleton: (A) Anterior view

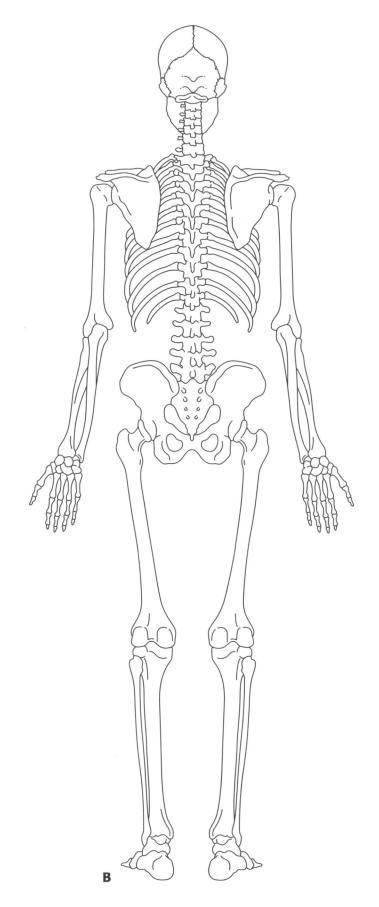

FIGURE **8.3** Skeleton (cont.): (B) posterior view

EXPLORING ANATOMY & PHYSIOLOGY IN THE LABORATORY

Exercises

The **skeletal system** consists of the bones, its associated cartilages, and the joints. Its two divisions are the **axial skeleton** and the **appendicular skeleton**. The axial skeleton is composed of the bones of the head, neck, and trunk—specifically, the cranial bones, the facial bones, the vertebral column, the hyoid bone, the sternum, and the ribs. The appendicular skeleton consists of the bones of the upper limbs, the lower limbs, the pectoral girdle (the bones forming the shoulder joint), and the pelvic girdle (the bones forming the pelvis and hip joint).

In this unit we explore the anatomy of the bones and bone markings of the skeletal system, which will serve as a foundation for later chapters. For example, the radial and ulnar arteries parallel the radius and the ulna, and the frontal, parietal, temporal, and occipital lobes of the brain are named for the cranial bones under which they are located.

 ## Exercise 1

The Skull

MATERIALS NEEDED

▍ Skulls, whole and sectioned

The skull is composed of two general types of bones—the **cranial bones** and the **facial bones** (**Figure 8.4A–E**). The cranial bones encase the brain and together form the **calvaria** (also known as the "skullcap"), which consists of several of the cranial bones joined at immovable joints called **sutures**. These bones also form the cranial base, which contains indentations that accommodate the brain, called the **anterior, middle,** and **posterior cranial fossae** (visible in **Figure 8.4D**). The eight cranial bones include the following:

1. **Frontal bone.** The frontal bone forms the anterior portion of the cranium. Internally it contains hollow spaces called the **frontal sinuses** (visible in **Figure 8.4E**). The frontal sinuses are part of a group of cavities that surround the nasal cavity, called the **paranasal sinuses**. Air from the nasal cavity enters the paranasal sinuses via small openings in the bones, and in the sinuses the air gets filtered, warmed, and humidified.

2. **Parietal bones.** The paired parietal bones form the superior and part of the lateral walls of the cranium. Note in **Figure 8.4A** and **8.4B** that they articulate with one another and many other cranial bones at sutures: They meet one another at the **sagittal suture**, they meet the frontal bone at the **coronal suture**, they meet the temporal bones at the **squamous sutures**, and they meet the occipital bone at the **lambdoid suture**.

3. **Temporal bones.** The paired temporal bones form the lateral walls of the cranium (best seen in **Figure 8.4B**). Each temporal bone has a complex shape with four general regions.

   a. The flat **squamous region** is the temporal bone's most lateral aspect. It contains a projection called the **zygomatic process**, which forms part of the cheekbone.

   b. The temporal bone's inferior region is called the **tympanic region**. It houses the **external acoustic meatus** and the needle-like **styloid process**.

   c. Posterior to the tympanic region is the **mastoid region**, which contains the large **mastoid process**.

   d. Internally, the temporal bone is shaped like a mountain ridge, and is accordingly called the **petrous region**.

4. **Occipital bone.** The posterior cranial bone is the occipital bone. Its most conspicuous feature is found in its base—a large hole called the **foramen magnum** through which the spinal cord passes.

5. **Sphenoid bone.** The butterfly-shaped sphenoid bone is posterior to the frontal bone on the interior part of the skull (best seen in **Figures 8.4C–E**). Its superior surface contains a saddle-like formation called the **sella turcica**, which normally houses the pituitary gland. Inferior to the sella turcica is the body of the sphenoid, which contains the second paranasal sinus—the **sphenoid sinus**. Laterally it consists of two sets of "wings"—the small **lesser wings** and the larger **greater wings**. Inferiorly, it has another set of "wings" called the **pterygoid processes**.

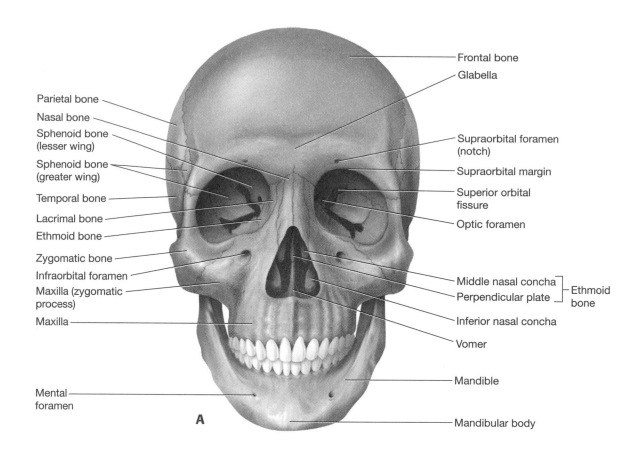

Parietal bone

Nasal bone

Sphenoid bone
(lesser wing)

Sphenoid bone
(greater wing)

Temporal bone

Lacrimal bone

Ethmoid bone

Zygomatic bone

Infraorbital foramen

Maxilla (zygomatic
process)

Maxilla

Mental
foramen

Frontal bone

Glabella

Supraorbital foramen
(notch)

Supraorbital margin

Superior orbital
fissure

Optic foramen

Middle nasal concha ⎤
                    ⎬ Ethmoid
Perpendicular plate ⎦ bone

Inferior nasal concha

Vomer

Mandible

Mandibular body

A

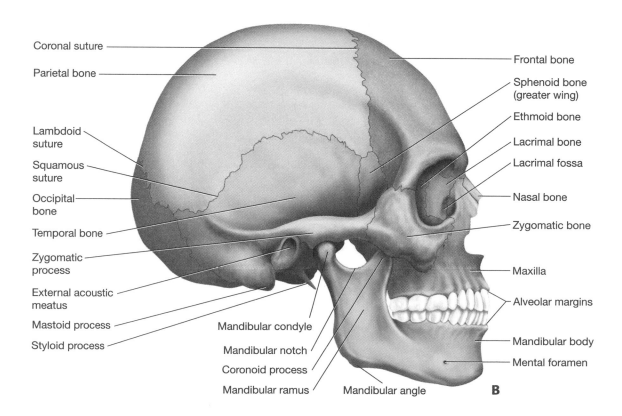

Coronal suture

Parietal bone

Lambdoid
suture

Squamous
suture

Occipital
bone

Temporal bone

Zygomatic
process

External acoustic
meatus

Mastoid process

Styloid process

Mandibular condyle

Mandibular notch

Coronoid process

Mandibular ramus

Mandibular angle

Frontal bone

Sphenoid bone
(greater wing)

Ethmoid bone

Lacrimal bone

Lacrimal fossa

Nasal bone

Zygomatic bone

Maxilla

Alveolar margins

Mandibular body

Mental foramen

B

FIGURE **8.4** The skull: (A) anterior view; (B) lateral view

EXPLORING ANATOMY & PHYSIOLOGY IN THE LABORATORY

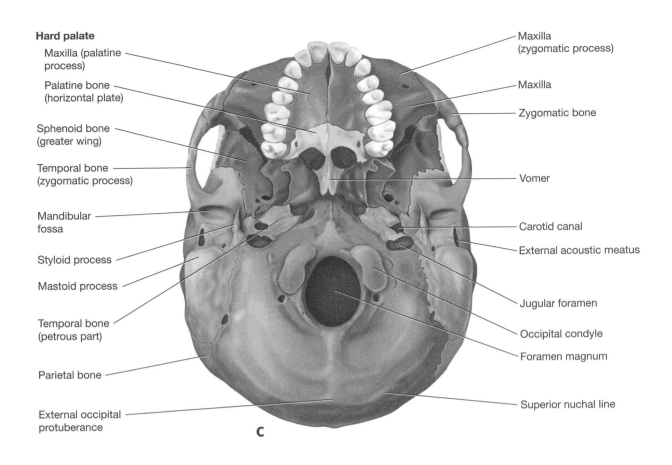

Hard palate

Maxilla (palatine process)

Palatine bone (horizontal plate)

Sphenoid bone (greater wing)

Temporal bone (zygomatic process)

Mandibular fossa

Styloid process

Mastoid process

Temporal bone (petrous part)

Parietal bone

External occipital protuberance

Maxilla (zygomatic process)

Maxilla

Zygomatic bone

Vomer

Carotid canal

External acoustic meatus

Jugular foramen

Occipital condyle

Foramen magnum

Superior nuchal line

C

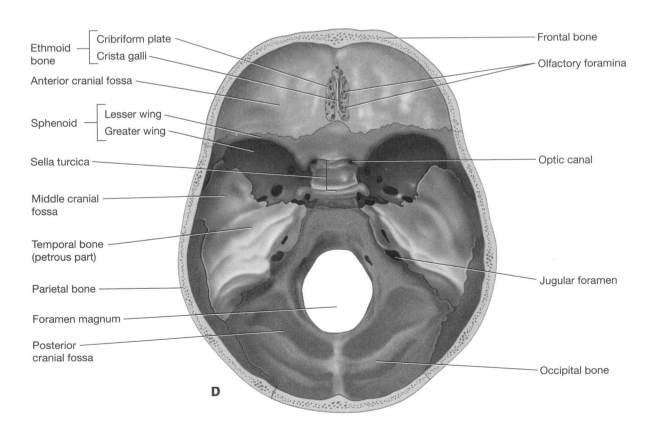

Ethmoid bone
- Cribriform plate
- Crista galli

Anterior cranial fossa

Sphenoid
- Lesser wing
- Greater wing

Sella turcica

Middle cranial fossa

Temporal bone (petrous part)

Parietal bone

Foramen magnum

Posterior cranial fossa

Frontal bone

Olfactory foramina

Optic canal

Jugular foramen

Occipital bone

D

FIGURE **8.4** The skull (cont.): (C) inferior view (mandible removed); (D) internal view (calvaria removed)

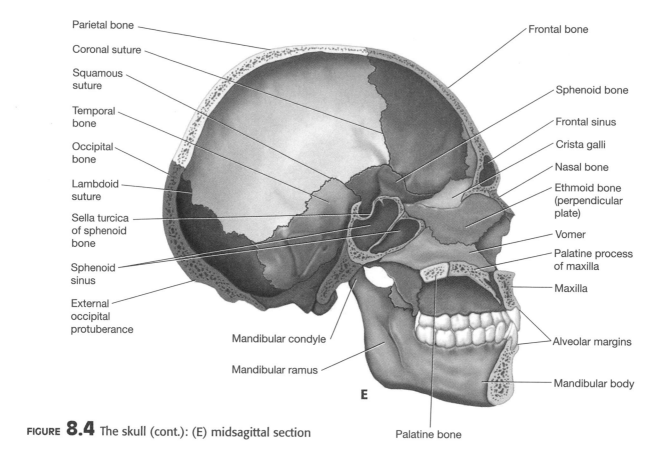

Parietal bone

Coronal suture

Squamous suture

Temporal bone

Occipital bone

Lambdoid suture

Sella turcica of sphenoid bone

Sphenoid sinus

External occipital protuberance

Mandibular condyle

Mandibular ramus

Frontal bone

Sphenoid bone

Frontal sinus

Crista galli

Nasal bone

Ethmoid bone (perpendicular plate)

Vomer

Palatine process of maxilla

Maxilla

Alveolar margins

Mandibular body

E

Palatine bone

FIGURE **8.4** The skull (cont.): (E) midsagittal section

6. **Ethmoid bone.** The complex ethmoid bone is the deepest cranial bone and the most difficult to see from standard views of the skull. Located anterior to the sphenoid bone and posterior to the nasal bones of the face, it contains a number of features, including the following:

a. Its superior surface, called the **cribriform plate**, forms the roof of the nasal cavity. It has small holes through which olfactory nerves pass, and a projection called the **crista galli**.

b. The lateral bodies of the ethmoid bone form part of the orbit and the walls of the nasal cavity. Internally, the bone contains numerous cavities called the **ethmoid sinuses**, which comprise the third set of paranasal sinuses. Extending medially from the lateral bodies are two projections into the nasal cavity—the **superior nasal conchae** and **middle nasal conchae**.

c. The middle portion of the ethmoid bone, called the **perpendicular plate**, forms the superior part of the bony nasal septum.

The 14 facial bones form the framework for the face, provide openings for ventilation and eating, and form cavities for the sense organs. These bones are the following:

1. **Mandible.** The mandible, or the lower jaw bone, consists of a central **body** and two "arms" called the **mandibular rami**. The mandibular rami turn superiorly at the **mandibular angle** and have two processes—an anterior process called the **coronoid process** and a posterior process called the **mandibular condyle**. The mandibular condyle fits into a depression in the temporal bone to form the **temporomandibular joint**.

2. **Maxillae.** The two fused maxillae are the upper jaw bones. They form part of the orbit and the anterior portion of the hard palate (via their **palatine processes**). Within their walls are cavities called the **maxillary sinuses**, the fourth and final set of paranasal sinuses. Laterally the maxillae have projections called **zygomatic processes** that form part of the cheekbone.

3. **Lacrimal bones.** The tiny lacrimal bones are located in the medial part of the orbit, where they form part of the structure that drains tears produced by the lacrimal gland of the eye.

4. **Nasal bones.** The two nasal bones form the anterior framework of the bridge of the nose.

5. **Vomer.** The single vomer forms the inferior portion of the bony nasal septum (best seen in **Figure 8.4E**).

6. **Inferior nasal conchae.** The small inferior nasal conchae form part of the lateral walls of the nasal cavity. As their name implies, they are located inferior to the middle nasal conchae.

7. **Palatine bones.** The two palatine bones form the posterior part of the hard palate (seen in **Figure 8.4C**) and the posterolateral walls of the nasal cavity.

8. **Zygomatic bones.** The two zygomatic bones form the bulk of the cheek and a significant portion of the "cheekbone" or zygomatic arch.

As you have just read, the **orbit** and the **nasal cavity** are complicated structures with contributions from several bones. The orbit is formed by parts of seven bones: the frontal bone, the maxilla, the sphenoid bone, the ethmoid bone, the lacrimal bone, the zygomatic bone, and a tiny piece of the palatine bone. The nasal cavity is formed by parts of six bones: the ethmoid bone, the maxilla, the palatine bone, the inferior nasal concha, the sphenoid bone, and the vomer.

 ## Procedure Cooperative Learning for the Skull

This exercise takes an approach called *cooperative learning*, in which you work with your lab partners to teach one other the bones and bone markings. The process may seem a bit confusing at first, but by the end of the first couple of rotations, it should move more quickly and smoothly. After you have completed the activity, answer Check Your Understanding questions 1–3 (p. 181).

1 Assemble into groups of a minimum of four students; five is optimum.

2 Distribute a skull to each member of the group, and assign each student one of the following five groups of bones and bone markings (the specific structures of each bone are listed on p. 156):

Group A: calvaria, base, and structures of the frontal bone and parietal bones

Group B: temporal bone and occipital bone structures

Group C: ethmoid bone and sphenoid bone structures

Group D: mandible and maxillary bone structures

Group E: remainder of the facial bones, orbit, sutural bones, and anterior and posterior fontanels.

3 Spend approximately 3 minutes learning the assigned structures on your own.

4 Then have each student spend approximately 1 to 2 minutes teaching the group his or her assigned structures.

5 Rotate the assigned structures clockwise so each student has a new set of structures to learn (the student that was assigned group A will take group B, and so on).

6 Spend approximately 2–3 minutes learning your newly assigned structures.

7 Have each student to spend approximately 1–2 minutes teaching the group his or her assigned structures.

8 Repeat this process (it begins to speed up significantly at this point) until each student has taught each group of structures once.

By the end of this activity, each group member will have learned and presented each group of structures.

The following is a list of bones and bones and bone markings of the skull that will be covered in this exercise.

Cranial Bones

Group A Structures:

1. Calvaria
2. Base of cranial cavity
   a. Anterior cranial fossa
   b. Middle cranial fossa
   c. Posterior cranial fossa
3. Frontal bone
   a. Frontal sinuses
   b. Glabella
   c. Supraorbital foramen
4. Parietal bones
   a. Coronal suture
   b. Sagittal suture
   c. Squamous suture
   d. Lambdoid suture

Group B Structures:

1. Temporal bones
   a. Zygomatic process
   b. Mandibular fossa
   c. External acoustic meatus
   d. Styloid process
   e. Mastoid process with mastoid air cells
   f. Carotid canal
   g. Jugular foramen
2. Occipital bone
   a. Occipital condyles
   b. Foramen magnum
   c. External occipital protuberance
   d. Superior nuchal line

Group C Structures:

1. Sphenoid bone
   a. Body
   b. Greater and lesser wings
   c. Optic foramen (canal)
   d. Superior orbital fissure
   e. Sella turcica
   f. Sphenoid sinus
2. Ethmoid bone
   a. Perpendicular plate
   b. Superior nasal conchae
   c. Middle nasal conchae
   d. Crista galli
   e. Cribriform plate
   f. Ethmoid sinuses

Facial Bones and Other Structures

Group D Structures:

1. Mandible
   a. Mandibular condyle
   b. Coronoid process
   c. Mandibular notch
   d. Angle of mandible
   e. Mandibular ramus
   f. Body of mandible
   g. Mental foramen
   h. Alveolar margin
2. Maxilla
   a. Palatine processes
   b. Infraorbital foramen
   c. Zygomatic processes
   d. Maxillary sinuses
   e. Alveolar margin

Group E Structures:

1. Palatine bones
2. Zygomatic bones and zygomatic arch
3. Lacrimal bones and lacrimal fossa
4. Nasal bones
5. Vomer
6. Inferior nasal conchae
7. Sutural (Wormian) bones
8. Auditory ossicles
9. Anterior fontanel
10. Posterior fontanel
11. Orbit

Your instructor may wish to omit certain structures included above or add structures not included in these lists. List any additional structures below:

 **Exercise 2**

Remainder of the Axial Skeleton

MATERIALS NEEDED

- Vertebral column, articulated
- Disarticulated vertebrae, sacrum, sternum, ribs, hyoid bone
- Skeleton, articulated

Another key component of the axial skeleton is the **vertebral column**, which consists of 22 vertebrae, the **sacrum**, and the **coccyx** (Figure 8.5). The vertebrae are divided into 7 **cervical vertebrae**, 12 **thoracic vertebrae**, and 5 **lumbar vertebrae**. Nearly all vertebrae share certain general features, including a posterior **spinous process**, an anterior **vertebral body**, two lateral **transverse processes**, and a central **vertebral foramen**.

The basic properties of each region of the vertebral column are as follows:

1. The seven cervical vertebrae are located in the neck (**Figure 8.5B**). All cervical vertebrae have holes in their transverse processes called **transverse foramina**. These foramina permit the passage of blood vessels called the *vertebral artery* and *vein*. In addition, the spinous processes of cervical vertebrae are often bifid or forked. Two cervical vertebrae are named differently than the others because of their unique features:

   a. **Atlas** (C1): The atlas is the first cervical vertebra that articulates with the occipital bone. It is easily identified because it has a large vertebral foramen, no body, and no spinous process (**Figure 8.5C**).

   b. **Axis** (C2): The axis is the second cervical vertebra. It is also easily identified by a superior projection called the **dens** (or the **odontoid process**; Figure 8.5D). The dens fits up inside the atlas to form the *atlanto-axial joint*, which allows rotation of the head.

2. The 12 thoracic vertebrae share the following common features (**Figure 8.5E**):

   a. The spinous processes are thin and point inferiorly.

   b. All have two *costal facets* that articulate with the ribs (there are 12 pairs of ribs).

   c. All have triangular vertebral foramina.

   d. If you look at a thoracic vertebra from the posterior side, it looks like a giraffe (seriously!).

3. The five lumbar vertebrae share the following common features (**Figure 8.5F**):

   a. All have a large, block-like body.

   b. The spinous processes are thick and point posteriorly.

   c. If you look at a lumbar vertebra from the posterior side, it looks like a moose (really!).

4. The sacrum consists of five fused vertebrae (**Figures 8.5G** and **8.5H**). You can see the fused spinous processes of the vertebrae on the sacrum's posterior side as the **median sacral crest**. Spinal nerves pass through holes called **sacral foramina,** which flank both sides of the sacral bodies. The lateral surfaces of the sacrum articulate with the hip bones to form the **sacroiliac joints**.

5. The coccyx consists of three to five small, fused vertebrae that articulate superiorly with the sacrum.

The remainder of the axial skeleton consists of the bones of the thoracic cavity (the **sternum** and the **ribs**, Figure 8.6A) and the **hyoid bone** in the neck (Figure 8.6B). The sternum is divided into three parts—the upper **manubrium**, the middle **body**, and the lower **xiphoid process**. The ribs are classified according to how they attach to the sternum: Ribs 1–7 are considered **true ribs** because they attach directly to the sternum by their own cartilage, ribs 8–10 are classified as **false ribs** because they attach to the cartilage of the true ribs rather than directly to the sternum, and ribs 11–12 are called **floating ribs** because they lack an attachment to the sternum.

The hyoid bone often is classified as a skull bone, although it does not articulate with any skull bone or any other bone. It is held in place in the superior neck by muscles and ligaments, and it helps to form part of the framework for the larynx (voice box). It also serves as an attachment site for the muscles of the tongue and aids in swallowing. When a person is choked manually, the hyoid bone is often broken (I would not recommend testing this on your lab partner!).

The bones and bone markings of the axial skeleton are listed with Exercise 3, where they will be included with another cooperative learning exercise.

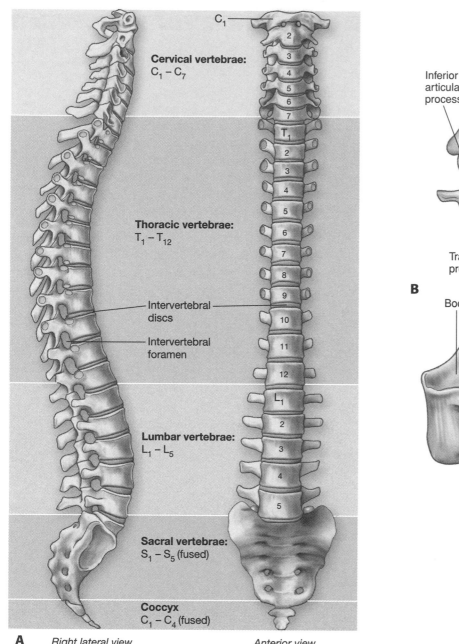

Cervical vertebrae:
$C_1 - C_7$

Thoracic vertebrae:
$T_1 - T_{12}$

Intervertebral discs

Intervertebral foramen

Lumbar vertebrae:
$L_1 - L_5$

Sacral vertebrae:
$S_1 - S_5$ (fused)

Coccyx
$C_1 - C_4$ (fused)

A   *Right lateral view*   *Anterior view*

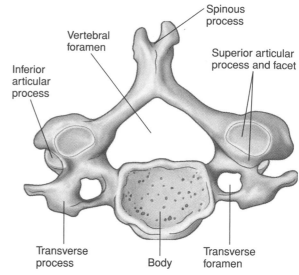

Spinous process

Vertebral foramen

Inferior articular process

Superior articular process and facet

Transverse process

Body

Transverse foramen

B

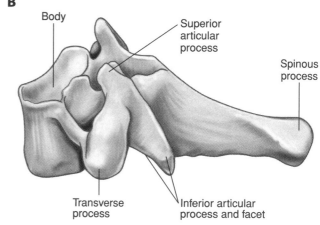

Body

Superior articular process

Spinous process

Transverse process

Inferior articular process and facet

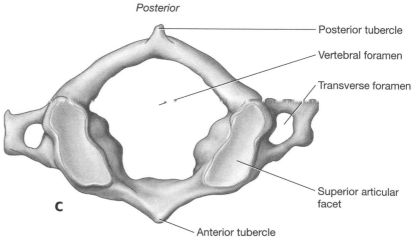

Posterior

Posterior tubercle

Vertebral foramen

Transverse foramen

Superior articular facet

C

Anterior tubercle

FIGURE 8.5 Vertebral column and vertebrae:
(A) vertebral column; (B) typical cervical vertebra,
superior and lateral views; (C) atlas (C1)

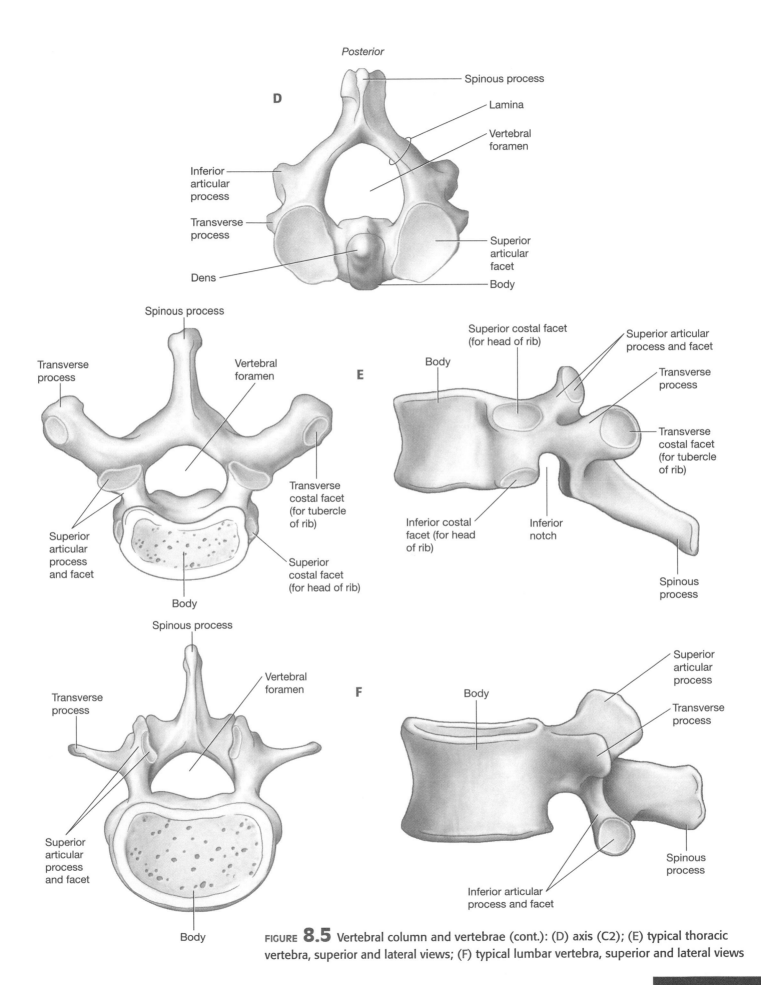

D

Posterior

Spinous process

Lamina

Vertebral foramen

Inferior articular process

Transverse process

Dens

Superior articular facet

Body

Spinous process

Transverse process

Vertebral foramen

Superior articular process and facet

Transverse costal facet (for tubercle of rib)

Body

Superior costal facet (for head of rib)

E

Superior costal facet (for head of rib)

Body

Superior articular process and facet

Transverse process

Transverse costal facet (for tubercle of rib)

Inferior costal facet (for head of rib)

Inferior notch

Spinous process

Spinous process

Transverse process

Vertebral foramen

Superior articular process and facet

Body

F

Body

Superior articular process

Transverse process

Inferior articular process and facet

Spinous process

FIGURE **8.5** Vertebral column and vertebrae (cont.): (D) axis (C2); (E) typical thoracic vertebra, superior and lateral views; (F) typical lumbar vertebra, superior and lateral views

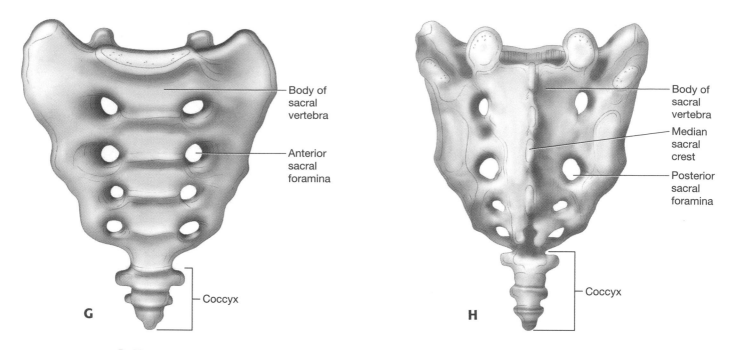

FIGURE **8.5** Vertebral column and vertebrae (cont.): (G) sacrum, anterior view; (H) sacrum, posterior view

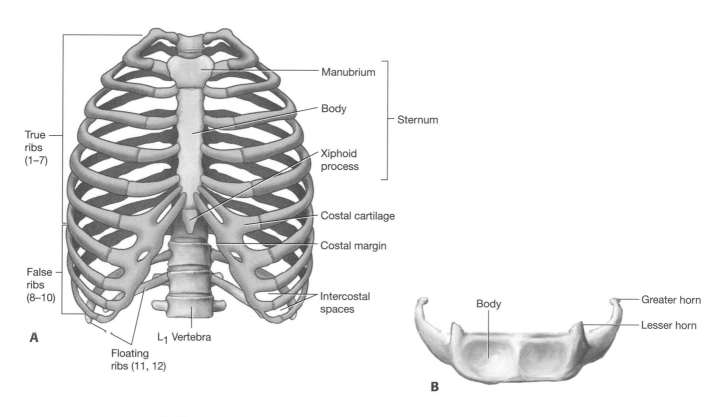

FIGURE **8.6** Remainder of the axial skeleton: (A) sternum and ribs; (B) hyoid bone

Exercise 3

The Appendicular Skeleton

MATERIALS NEEDED

- Disarticulated bones
- Skeleton, articulated
- Male and female pelvises

The bones of the appendicular skeleton can be divided into those of the **pectoral girdle**, the upper limb, the **pelvic girdle**, and the lower limb. The pectoral girdle consists of the two bones that frame the shoulder—the **scapula** and the **clavicle**. The scapula, shown in **Figure 8.7**, has a posterior ridge called the **spine** and a lateral depression called the **glenoid cavity**, which forms the shoulder joint with the humerus. Superiorly it has two projections— the anterior **coracoid process** and the posterior **acromion**, which forms a joint with the clavicle called the **acromioclavicular** (AC) joint.

The upper limb consists of the arm, the forearm, the wrist, and the hand. The only bone within the arm is the **humerus** (Figure 8.8). The humerus has a number of features, some of which include the following:

- At the proximal end of the humerus is a rounded **head** that fits into the glenoid cavity. Just lateral to the head is the **lesser tubercle**, separated from the larger **greater tubercle** by the **intertubercular sulcus**.
- The middle of the humerus features a projection called the **deltoid tuberosity**, where the deltoid muscle attaches.
- At the humerus' distal end we find two condyles—the medial **trochlea**, which is shaped like a spool of thread, and the lateral **capitulum**, which is ball-shaped. Just proximal to these condyles are indentations in the humerus where the bones of the forearm articulate—the anterior **coronoid fossa** and the posterior **olecranon fossa**.

The two forearm bones, shown in Figure 8.9, are the lateral **radius** and medial **ulna** (if you have a hard time remembering which is which, stand in anatomical position and take your radial pulse; it is on the lateral side of the forearm, just like the radius). The ulna is wide proximally where it articulates with the humerus and thin distally where it articulates with the bones of the wrist. Its proximal end has two processes—the large, posterior **olecranon process** and the smaller, anterior **coronoid process**—separated by a deep curve called the **trochlear notch**.

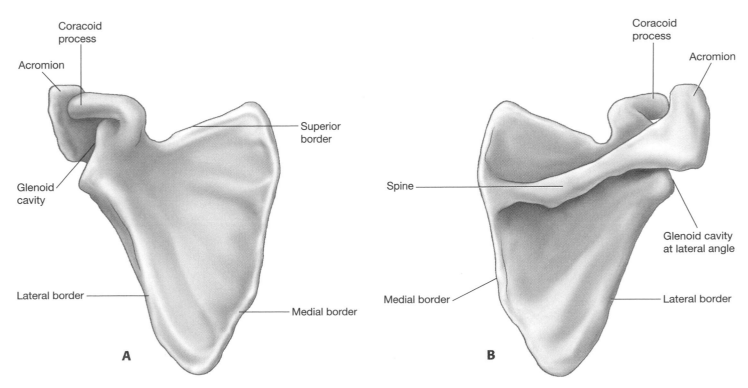

FIGURE 8.7 The right scapula: (A) anterior view; (B) posterior view

As its name implies, the trochlear notch fits around the trochlea of the humerus to form the elbow joint. The olecranon process is the actual "elbow bone" that you can feel on your posterior arm. The trochlear notch and the ulna's two processes form a "U" shape when the ulna is held on its side. This makes the ulna easy to differentiate from the radius.

The radius has a width distribution opposite from that of the ulna. It is skinny proximally and wide distally. Proximally it consists of a **radial head** that articulates with the ulna at the **proximal radioulnar joint**. Note in **Figure 8.9** that the two bones also articulate at

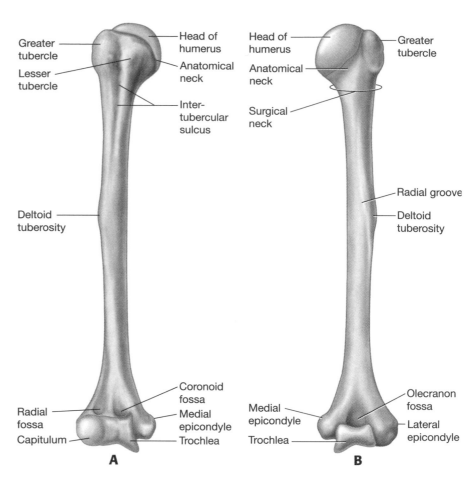

FIGURE **8.8** The right humerus: (A) anterior view; (B) posterior view

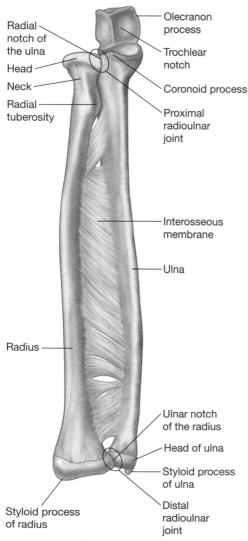

FIGURE **8.9** The right radius and ulna, anterior view

their distal ends at the **distal radioulnar joint**. Both the radius and the ulna have projections on their distal ends called **styloid processes**.

The wrist is composed of eight short bones called **carpals**, labeled individually in **Figure 8.10**. The carpals articulate with the radius and the ulna, as well as the five long bones in the hand called **metacarpals**. The metacarpals articulate distally with the fingers, which are formed from 14 long bones called **phalanges**. The second through fifth digits have three phalanges each (the *proximal*, *intermediate*, and *distal phalanges*); the thumb has only two (a *proximal* and a *distal phalanx*).

The pelvic girdle connects the lower limbs to the trunk, supports the pelvic organs, and transmits the weight of the trunk to the legs. It is formed by two **coxal bones**. Each coxal bone is also known as a **hemipelvis**. Three fused bones—the **ilium**, the **ischium**, and the **pubis**—make up each coxal bone (**Figure 8.11**). Notice that the lateral side of the hip (**Figure 8.11B**) has a place where all three bones come together to form a deep socket. This socket, called the **acetabulum**, forms the hip joint with the femur. Notice also that where the ischium and pubis meet there is a large hole called the **obturator foramen**. In a living person this hole is covered with a membrane and allows only small blood vessels and nerves to pass through.

The three bones of the hip have the following features:

1. **Ilium**. The ilium is the largest of the three bones. Its main portion is called the **body**, and its superior "wing" is called the **ala**. The ridge

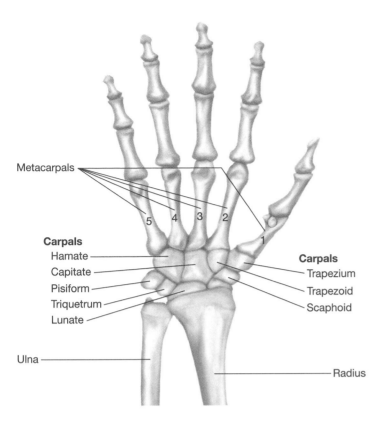

Metacarpals

5 4 3 2 1

Carpals
Hamate
Capitate
Pisiform
Triquetrum
Lunate

Carpals
Trapezium
Trapezoid
Scaphoid

Ulna

Radius

FIGURE **8.10** The right wrist and hand, anterior view

of the ala, called the **iliac crest**, is where you rest your hands when your hands are on your hips. At the anterior end of the crest we find a projection called the **anterior superior iliac spine**. A similar but smaller **posterior superior iliac spine** is located posteriorly. The posterior ilium also features a notch called the **greater sciatic notch**, which allows the large greater sciatic nerve to pass between the pelvis and the thigh.

2. **Ischium**. The ischium makes up the posteroinferior pelvis. It contains three features on its posterior side: the superior **ischial spine**, the middle **lesser sciatic notch**, and the thick, inferior **ischial tuberosity**. The ischial tuberosities are the "butt bones," the bones that bear your weight when you sit down.

3. **Pubis**. The pelvis' anterior portion is formed by the pubis or the **pubic bone**. The pubis consists of a **body** and two extensions called the **superior** and **inferior rami**. The bodies of the two pubic bones meet at a fibrocartilage pad called the **pubic symphysis**. The angle that these bones form when they meet, called the **pubic arch**, can help to determine the gender of a skeleton (see the "Hints and Tips" box).

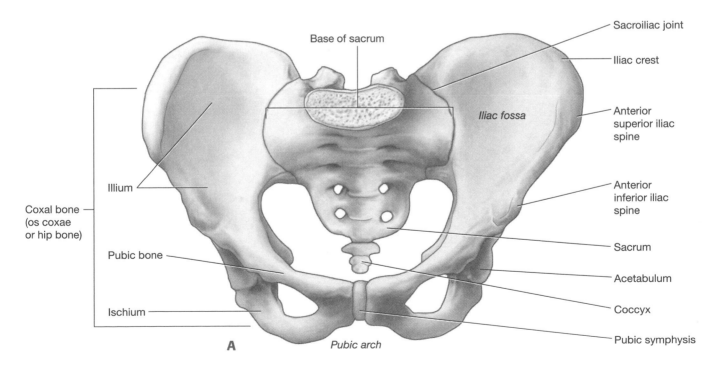

Base of sacrum

Sacroiliac joint

Iliac crest

Iliac fossa

Anterior superior iliac spine

Illium

Coxal bone (os coxae or hip bone)

Anterior inferior iliac spine

Pubic bone

Sacrum

Ischium

Acetabulum

Coccyx

A

Pubic arch

Pubic symphysis

FIGURE **8.11** The pelvis: (A) fully articulated pelvis

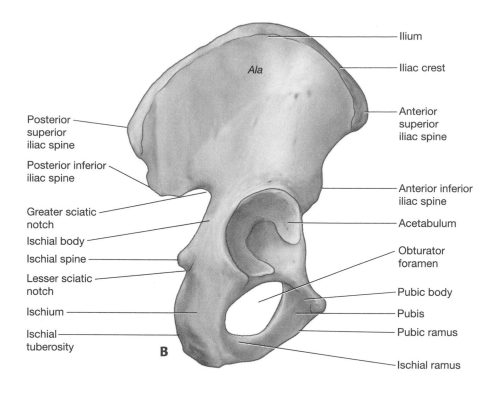

Ilium
Iliac crest
Anterior superior iliac spine

Ala

Posterior superior iliac spine
Posterior inferior iliac spine
Greater sciatic notch
Ischial body
Ischial spine
Lesser sciatic notch
Ischium
Ischial tuberosity

Anterior inferior iliac spine
Acetabulum
Obturator foramen
Pubic body
Pubis
Pubic ramus
Ischial ramus

B

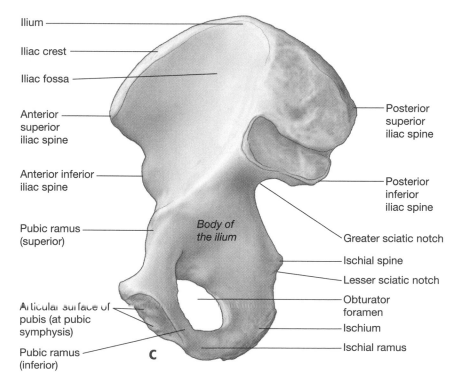

Ilium
Iliac crest
Iliac fossa
Anterior superior iliac spine
Anterior inferior iliac spine
Pubic ramus (superior)
Articular surface of pubis (at pubic symphysis)
Pubic ramus (inferior)

Body of the ilium

Posterior superior iliac spine
Posterior inferior iliac spine
Greater sciatic notch
Ischial spine
Lesser sciatic notch
Obturator foramen
Ischium
Ischial ramus

C

FIGURE 8.11 The pelvis (cont.):
(B) right hemipelvis, lateral view; (C) right hemipelvis, medial view

The lower limb consists of the thigh, the **patella** (kneecap), the leg, the ankle, and the foot. The thigh contains only a single bone, the large, heavy **femur** (**Figure 8.12**). Proximally, the femur articulates with the acetabulum at its rounded **head**. In the center of the femoral head we find a pit called the fovea capitis. This is the point of attachment for a ligament that holds the femur in the acetabulum. Just distal to the femoral head is the **neck** of the femur, the weakest part of the femur and the most common location for it to fracture (when the femoral neck fractures, it is usually called a "broken hip"). Where the femoral neck meets the femoral shaft, we find two large prominences—the anterolateral **greater trochanter** and the posteromedial **lesser trochanter**.

Distally, the femur expands into the **medial** and **lateral condyles**, which form the knee joint with the largest bone of the leg, the **tibia**. The tibia is flattened proximally at its articular surface, and its **medial** and **lateral condyles** fit together with those of the femur (**Figure 8.13**). Distally, the tibia articulates with a tarsal bone called the **talus**, with which it forms the ankle joint. At its terminal end is a projection called the **medial malleolus**, the medial ankle bone.

The other leg bone is the thin, lateral **fibula**. The proximal fibular end is called the **head**, and its distal end is the **lateral malleolus** (*lateral ankle bone*). The fibula doesn't articulate directly with either the femur or the talus. It does, however, articulate with the lateral side of the tibia at *proximal* and *distal tibiofibular joints*.

The ankle is composed of seven short bones called **tarsals**, labeled individually in **Figure 8.14**. The tarsals articulate with the five long bones in the foot called **metatarsals**. Like the bones of the fingers, the bones of the toes also consist of 14 **phalanges**. The second through fifth digits have three phalanges each (the *proximal*, *intermediate*, and *distal phalanges*); and the big toe or *hallux* has only two (a *proximal* and a *distal phalanx*).

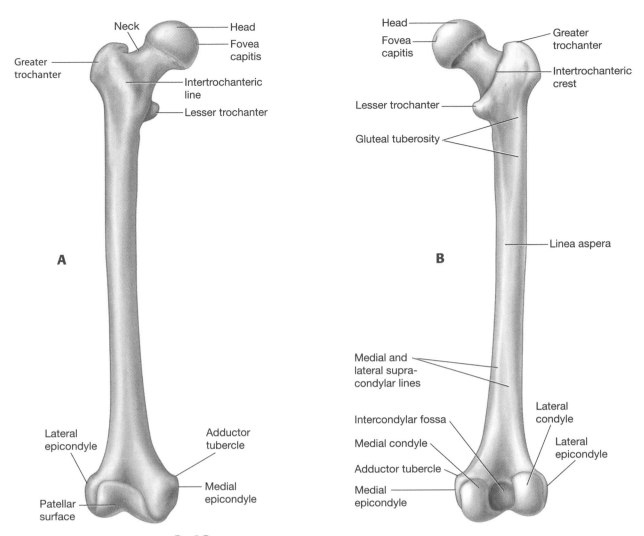

FIGURE 8.12 The right femur: (A) Anterior view; (B) Posterior view

HINTS & TIPS

How to Determine Gender Based Upon Pelvic Features

A skeleton's gender can be determined based upon the differences between the male and female pelvises. In general, the male and female pelvis can be distinguished by the features listed below:

Feature	Female Pelvis	Male Pelvis
Pelvic inlet shape	Wider and oval-shaped	Narrower and heart-shaped
Pubic arch	Wide angle	Narrow angle
Acetabulae	Farther apart	Closer together
Ischial tuberosities	Everted	Inverted
Coccyx	Straighter, more movable	Curved anteriorly, less movable

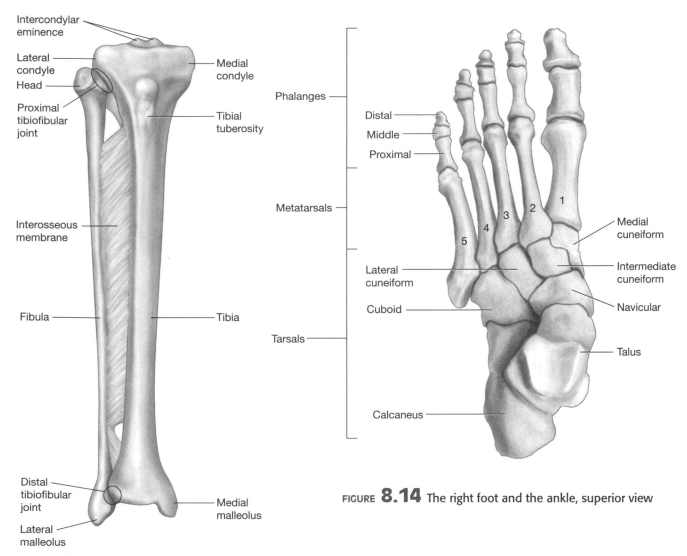

Intercondylar eminence

Lateral condyle

Head

Proximal tibiofibular joint

Interosseous membrane

Fibula

Distal tibiofibular joint

Lateral malleolus

Medial condyle

Tibial tuberosity

Tibia

Medial malleolus

FIGURE 8.13 The right tibia and the fibula, anterior view

Phalanges

Distal

Middle

Proximal

Metatarsals

Lateral cuneiform

Cuboid

Tarsals

Calcaneus

Medial cuneiform

Intermediate cuneiform

Navicular

Talus

FIGURE 8.14 The right foot and the ankle, superior view

Procedure Cooperative Learning for the Axial and Appendicular Skeletons

We will follow essentially the same procedure here as we did in Exercise 1, but with different groups of bones and bone markings. Note that this cooperative learning exercise includes the structures from Exercises 2 and 3. When you have completed this activity, answer Check Your Understanding questions 4 and 5 (p. 182).

1 Assemble into groups with a minimum of four students; five is optimum.

2 Distribute a bone or set of bones to each member of the group, and assign each student one of the following five groups of bones and bone markings (the specific structures of each bone are listed on pp. 167 and 168):

Group A: cervical, thoracic, and lumbar vertebrae and vertebral markings; ribs and rib markings; sternum structures and the hyoid bone

Group B: scapula structures, clavicle structures, humerus structures

Group C: radius and ulna markings, carpals, metacarpals, and phalanges

Group D: ilium, ischium, and pubis structures, and the difference between the male and female pelvises

Group E: femur, tibia, fibula structures, tarsals, metatarsals, phalanges.

3 Spend approximately 3 minutes learning the assigned structures on your own.

4 Ask each student to spend approximately 1 to 2 minutes teaching the group his or her assigned structures.

5 Rotate the assigned structures clockwise so each student has a new set of structures to learn (the student that was assigned group A will take group B, and so on).

6 Spend approximately 2–3 minutes learning your newly assigned structures.

7 Then ask each student to take approximately 1–2 minutes to teach the group his or her assigned structures.

8 Repeat this process (it begins to speed up significantly at this point) until each student has been taught each group of structures once. By the end of this "game," each group member will have learned and presented each group of structures.

The following is a list of bone and bone markings of the appendicular skeleton covered in this exercise.

Remaining Structures of the Axial Skeleton

Group A Structures:

1. Vertebrae
   a. Body
   b. Spinous process
   c. Vertebral foramen
   d. Vertebral arch
   e. Transverse processes
2. Cervical vertebrae
   a. Atlas
   b. Axis with dens
   c. Transverse foramina
3. Thoracic vertebrae
   a. Costal facets
4. Lumbar vertebrae
5. Sacrum
   a. Median sacral crest
   b. Posterior sacral foramina

6. Coccyx
7. Ribs
   a. True ribs
   b. False ribs
   c. Floating ribs
8. Sternum
   a. Manubrium
   b. Body
   c. Xiphoid process
9. Hyoid bone

Pectoral Girdle and Upper Limb

Group B Structures:

1. Clavicle
   a. Acromial end
   b. Sternal end
2. Scapula
   a. Acromion
   b. Spine
   c. Coracoid process
   d. Glenoid cavity (fossa)
   e. Lateral, medial, and superior border

3. Humerus
   a. Head
   b. Greater tubercle
   c. Lesser tubercle
   d. Intertubercular groove
   e. Deltoid tuberosity
   f. Medial and lateral epicondyles
   g. Capitulum
   h. Trochlea
   i. Olecranon fossa
   j. Coronoid fossa

Group C Structures:

1. Radius
   a. Head
   b. Neck
   c. Radial tuberosity
   d. Styloid process
2. Ulna
   a. Olecranon process
   b. Coronoid process
   c. Trochlear notch
   d. Styloid process
   e. Radial notch

3. Carpals
   a. Scaphoid
   b. Lunate
   c. Triquetrum
   d. Pisiform
   e. Trapezium
   f. Trapezoid
   g. Capitate
   h. Hamate
4. Metacarpals
5. Phalanges

Pelvic Girdle and Lower Limb

Group D Structures:

1. Ilium
   a. Iliac crest
   b. Anterior superior iliac spine
   c. Anterior inferior iliac spine
   d. Posterior superior iliac spine
   e. Posterior inferior iliac spine
   f. Greater sciatic notch
2. Ischium
   a. Ischial tuberosity
   b. Lesser sciatic notch
   c. Ischial spine
   d. Obturator foramen
3. Pubis
   a. Superior and inferior rami
   b. Pubic arch
   c. Pubic symphysis
4. Acetabulum
5. Male and female pelvises

Group E Structures:

1. Femur
   a. Head
   b. Neck
   c. Greater trochanter
   d. Lesser trochanter
   e. Linea aspera
   f. Lateral and medial epicondyles
   g. Medial and lateral condyles
2. Patella
3. Tibia
   a. Medial and lateral condyles
   b. Tibial tuberosity
   c. Medial malleolus
4. Fibula
   a. Head
   b. Lateral malleolus
5. Tarsals
   a. Talus
   b. Calcaneus
   c. Navicular
   d. Cuboid
   e. Cuneiforms
6. Metatarsals
7. Phalanges

Your instructor may wish to omit certain structures included above or add structures not included in these lists. List any additional structures below:

Exercise 4

More Practice

MATERIALS NEEDED

- Disarticulated bones in a box
- Skeleton, articulated
- Colored pencils

On average, the human body contains 206 bones and thousands of bone markings, processes, fossae, and foraminae. The previous cooperative learning exercises gave you a solid foundation on which to build your knowledge of the skeleton—but a good foundation is just a start. A lot more work is needed to actually build a house on that foundation. This exercise contains activities to help you build that "house" and practice what you learned in Exercises 1 through 3.

 ## Procedure Building a Skeleton

1 Obtain a set of disarticulated bones (real bones are best).

2 Assemble the bones into a full skeleton. (If you have an articulated vertebral column and rib cage, go ahead and use them.)

3 Be certain to keep your skeleton in anatomical position. **Figure 8.15** gives an overall "big picture" view of the skeleton that you may use for reference.

4 Assemble the bones into a full skeleton.

 ## Procedure Identifying Bones Blindly

1 Place your set of disarticulated bones in a box.

2 Working with your lab partner, close your eyes, reach into the box, grab a bone randomly, and attempt to identify it only by its feel.

3 If you are unable to identify the bone, have your lab partner give you hints to help you work out the bone's identity.

 ## Procedure Draw a Skeleton

Anatomy and physiology coloring books have always been a popular study tool because they engage several parts of the brain and facilitate more rapid learning and memorization. An even better memorization tool than coloring is drawing. Drawing even the crudest picture of an anatomical structure or physiological pathway engages multiple parts of the brain and greatly enhances memory.

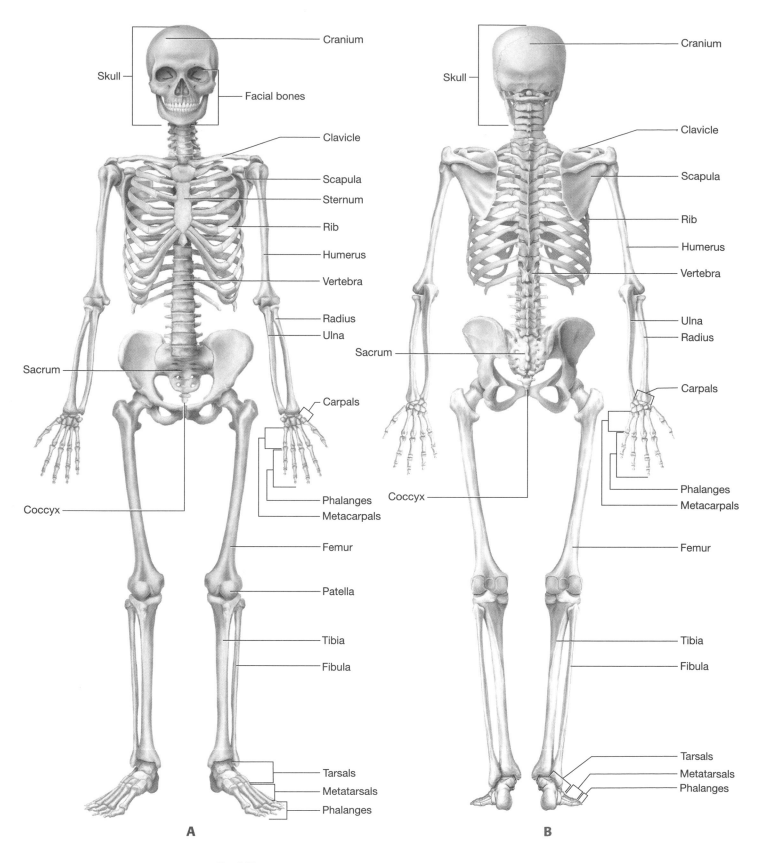

FIGURE **8.15** The articulated skeleton: (A) anterior view; (B) posterior view

EXPLORING ANATOMY & PHYSIOLOGY IN THE LABORATORY

1 Draw a skull and a skeleton in the spaces below.

2 Label and color your drawings.

These drawings are not meant to be works of art, so don't worry if you aren't even remotely skilled as an artist (however, if they turn out well, feel free to take them out of your lab book and put them on your refrigerator).

1. Skull: anterior view

2. Skull: lateral view

3. Skeleton: anterior view

EXPLORING ANATOMY & PHYSIOLOGY IN THE LABORATORY

NAME _____

SECTION _____ DATE _____

Check Your Recall

1 Label the following bones in Figure 8.16.

- Radius
- Clavicle
- Ilium
- Sternum
- Femur
- Tarsals
- Scapula
- Ulna
- Tibia
- Metacarpals
- Pubis
- Humerus
- Fibula

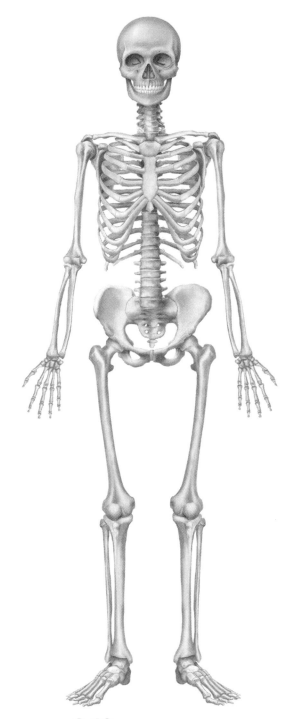

FIGURE **8.16** Anterior view of the skeleton

2 Label the following bones of the skull in **Figure 8.17.**

- Parietal bone
- Occipital bone
- Vomer
- Zygomatic bone
- Temporal bone
- Mandible
- Nasal bones
- Ethmoid bone
- Maxilla
- Palatine bone
- Sphenoid bone
- Frontal bone
- Lacrimal bone
- Inferior nasal concha

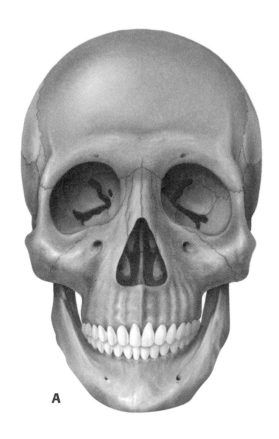

A

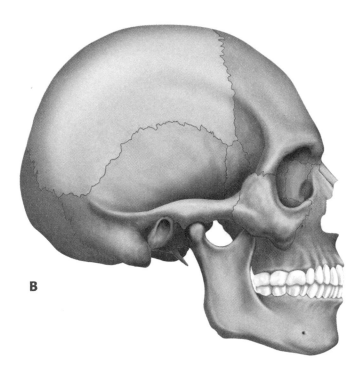

B

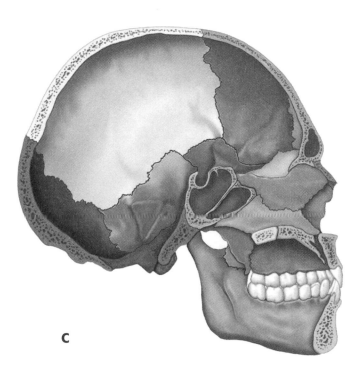

C

FIGURE 8.17 The skull: (A) anterior view; (B) lateral view; (C) midsagittal section

EXPLORING ANATOMY & PHYSIOLOGY IN THE LABORATORY

3 Label the following skull markings and processes in **Figure 8.18**.

- Sphenoid sinus
- Palatine process of maxilla
- Styloid process
- Zygomatic arch
- Foramen magnum
- Cribriform plate
- Frontal sinus
- External acoustic meatus
- Perpendicular plate of ethmoid bone
- Sella turcica
- Mastoid process

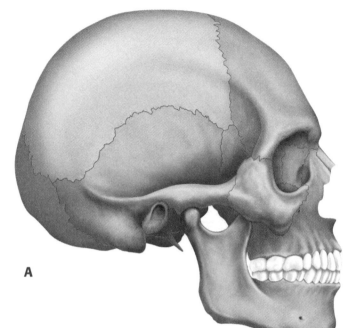

A

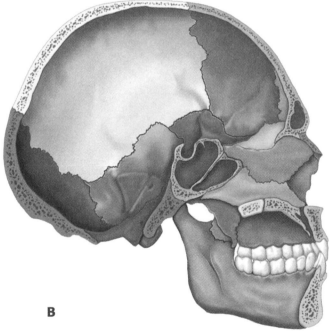

B

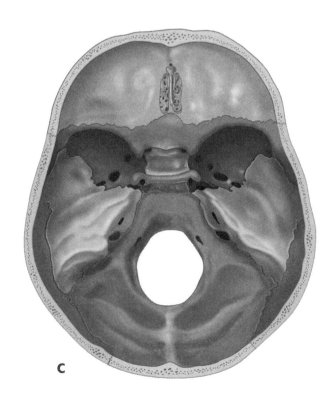

C

FIGURE 8.18 The skull: (A) lateral view; (B) midsagittal section; (C) internal view

4 Label the following parts of the
vertebral column in **Figure 8.19**.

- Cervical vertebrae
- Thoracic vertebrae
- Lumbar vertebrae
- Sacrum
- Coccyx

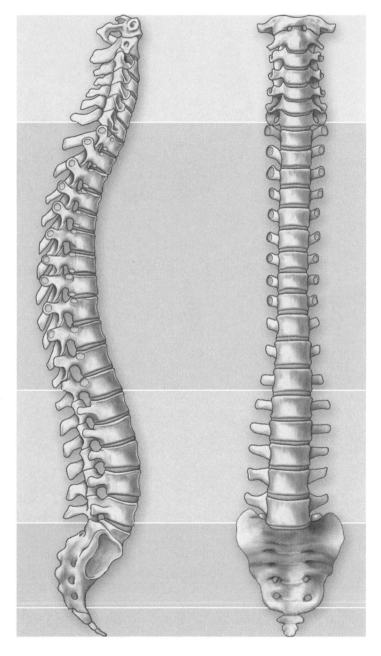

FIGURE **8.19** The vertebral column

5 Label the following parts of the axial skeleton in Figure 8.20.

- True ribs
- False ribs
- Floating ribs
- Manubrium
- Xiphoid process

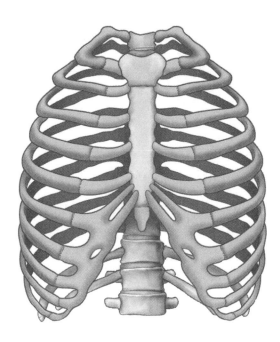

FIGURE **8.20** The remainder of the axial skeleton

6 Identify each of the following vertebrae in **Figure 8.21** as being cervical, thoracic, or lumbar. Additionally, explain how you arrived at this conclusion.

a. _____

b. _____

c. _____

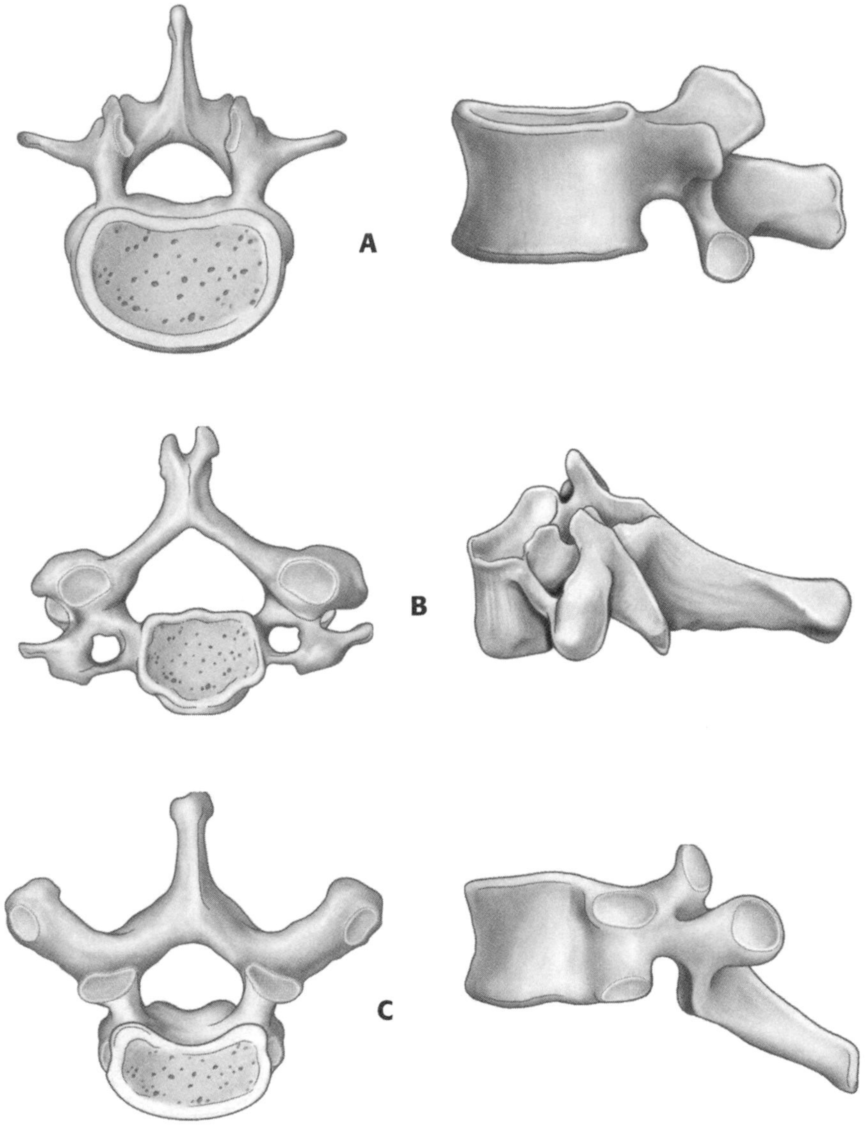

FIGURE **8.21** (A) unknown vertebra 1; (B) unknown vertebra 2; (C) unknown vertebra 3

7 Label the following parts of the scapula in **Figure 8.22**.

- Glenoid cavity
- Acromion
- Coracoid process
- Spine

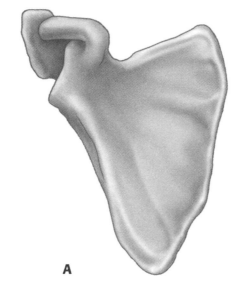
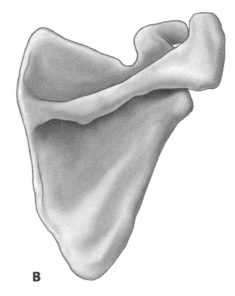

FIGURE 8.22 Scapula:
(A) anterior view;
(B) posterior view

A   B

8 Label the following parts of the upper limb in **Figure 8.23**.

- Head
- Greater tubercle
- Intertubercular sulcus
- Deltoid tuberosity
- Trochlea
- Capitulum
- Olecranon
- Trochlear notch
- Coronoid process
- Styloid processes

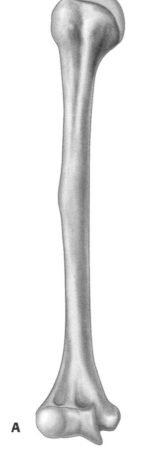
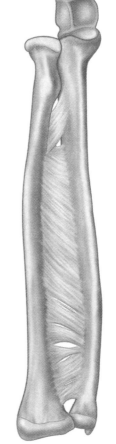

FIGURE 8.23 Upper limb:
(A) arm, anterior view;
(B) forearm, anterior view

A   B

9 Label the following parts of the hip bone (hemipelvis) in **Figure 8.24.**

- Ilium
- Pubis
- Ischium
- Ischial tuberosity
- Iliac crest
- Obturator foramen
- Acetabulum
- Greater sciatic notch
- Anterior superior iliac spine

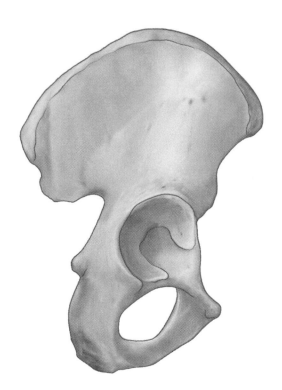

FIGURE **8.24** Right hemipelvis, lateral view

10 Label the following parts of the lower limb in **Figure 8.25.**

- Medial malleolus
- Greater trochanter
- Head
- Neck
- Lesser trochanter
- Medial and lateral condyles
- Lateral malleolus

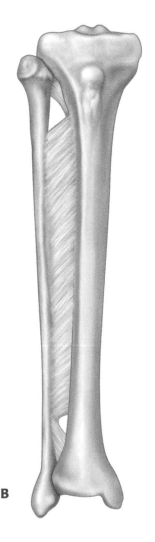

FIGURE **8.25** Lower limb:
(A) thigh, anterior view;
(B) leg, anterior view

A

B

EXPLORING ANATOMY & PHYSIOLOGY IN THE LABORATORY

NAME _____

SECTION _____ DATE _____

1 Which bones potentially could be involved in a fracture of the orbit? Why do you think orbital fractures are often difficult to fixate?

2 Your friend has suffered a "broken nose." Which bones potentially could be involved in a fracture involving the nose and the nasal cavity?

3 The bones of the fetal skull are not yet fused, giving the skull many "soft spots," or fontanels. Why do you think the bones of the fetal skull are not fused at birth?

4 Explain the reasons for the anatomical differences in the male and female pelvises.

5 Examine the following bones and figure out how you can determine if the bones are from the left or the right side of the body:

a. Femur

b. Tibia

c. Humerus

d. Scapula

Articulations

OBJECTIVES

Once you have completed this unit, you should be able to:

- Classify joints based upon structure and function.

- Identify examples of the different types of joints.

- Identify structures associated with synovial joints.

- Classify synovial joints according to range of motion.

- Identify structures of the knee joint.

- Demonstrate and describe motions allowed at synovial joints.

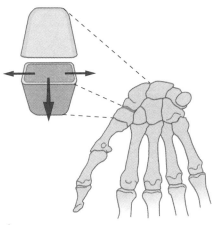

A Plane joint (intercarpal joint)

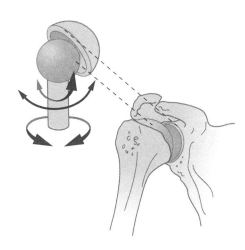

Pre-Lab Exercises | Complete the following exercises prior to coming to lab, using your textbook and lab manual for reference.

Pre-Lab Exercise 1

Key Terms

Table 9.1 lists the key terms with which you should be familiar before coming to lab.

TABLE 9.1	Key Terms

Term	Definition
Classes of Joints by Function	
Synarthrosis	_____
Diarthrosis	_____
Amphiarthrosis	_____
Classes of Joints by Structure	
Fibrous	_____
Cartilaginous	_____
Synovial	_____
Classes of Synovial Joints by Range of Motion	
Nonaxial	_____
Uniaxial	_____

Biaxial _____

_____

Multiaxial _____

_____

Classes of Synovial Joints by Structure

Plane _____

_____

Hinge _____

_____

Pivot _____

_____

Condyloid _____

_____

Saddle _____

_____

Ball and socket _____

_____

Types of Movement in Synovial Joints

Flexion _____

_____

Extension _____

_____

Abduction _____

_____

Adduction _____

_____

Circumduction _____

_____

Rotation _____

_____

Inversion _____

_____

Eversion _____

_____

Plantarflexion _____

_____

Dorsiflexion _____

_____

Pronation _____

_____

Supination _____

_____

Elevation _____

_____

Depression _____

_____

Opposition _____

_____

EXPLORING ANATOMY & PHYSIOLOGY IN THE LABORATORY

Pre-Lab Exercise 2
Structural Classes of Joints

Label and color the structural classes of joints depicted in **Figure 9.1** with the terms from Exercise 2. Use your text and Exercise 2 in this unit for reference.

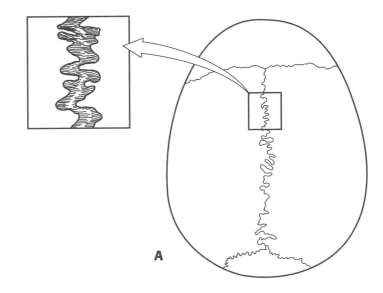

A

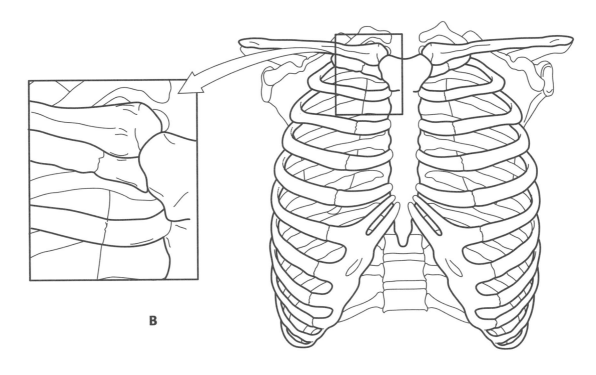

B

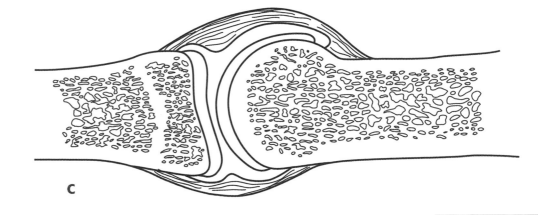

C

FIGURE **9.1** The three structural classes of joints:
(A) fibrous joint;
(B) cartilaginous joint;
(C) synovial joint

Pre-Lab Exercise 3

The Knee Joint

Label and color the knee joint depicted in **Figure 9.2** with the terms from Exercise 2. Use your text and Exercise 2 in this unit for reference.

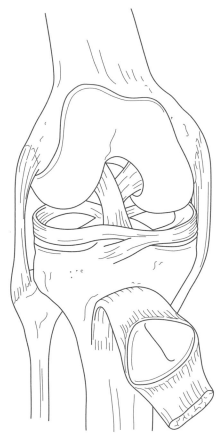

FIGURE **9.2** The knee joint

 Exercises

Most bones in the body form **articulations**, or **joints**, with another bone. These articulations allow you to perform the wide variety of motions involved in everyday movement. You probably don't realize just how many joints are articulating when you perform routine activities such as answering the phone, walking to the door, or holding a model in A & P lab. In the following exercises you will identify and classify joints according to structure, function, and amount of motion, examine the knee joint, and determine which joints are involved in simple everyday movements.

Exercise 1

Classification of Joints

MATERIALS NEEDED

- Skeleton, articulated
- Skull

Joints are classified according to both their structure and their function. Functionally, joints are classified as:

1. **synarthroses**—immovable joints,
2. **amphiarthroses**—joints that allow some motion, or
3. **diarthroses**—freely moveable joints.

EXPLORING ANATOMY & PHYSIOLOGY IN THE LABORATORY

Structurally, joints are classified as (Figure 9.3):

1. **Fibrous joints**. Fibrous joints consist of bones joined by short connective tissue fibers. Most fibrous joints allow no motion and are synarthroses.

2. **Cartilaginous joints**. Cartilaginous joints consist of bones united by cartilage rather than fibrous connective tissue. Most cartilaginous joints allow some motion and are amphiarthrotic.

3. **Synovial joints**. Synovial joints have a true joint cavity and consist of two bones whose articular ends are covered with hyaline cartilage. The joint cavity is lined by a **synovial membrane**, which secretes a watery fluid called **synovial fluid** similar in composition to blood plasma without the proteins. The fluid bathes the joint to permit frictionless motion. We discuss synovial joints further in Exercise 2.

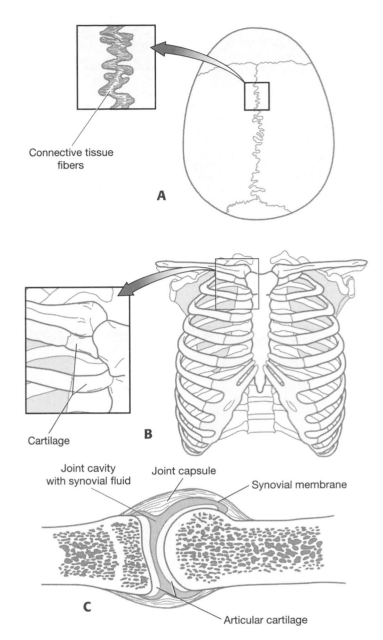

Connective tissue fibers

A

Cartilage

B

Joint cavity with synovial fluid   Joint capsule

Synovial membrane

C

Articular cartilage

FIGURE **9.3** The three classes of joints: (A) fibrous joint; (B) cartilaginous joint; (C) synovial joint

Procedure Classifying Joints by Structure and Function

Classify each joint listed in Table 9.2 by its structure. Then examine and manipulate the joint to determine the amount of motion allowed at the joint. When you have determined how much movement is allowed at the joint, classify it functionally. After you have completed the activity, answer Check Your Understanding question 1 (p. 201).

TABLE 9.2	Structural and Functional Classification of Joints		
Joint	Structural Classification	Amount of Motion	Functional Classification
Intervertebral joint			
Shoulder (glenohumeral) joint			
Intercarpal joint			
Coronal suture			
Pubic symphysis			
Interphalangeal joint			

Exercise 2

Synovial Joints

MATERIALS NEEDED

- Fresh pig or chicken joints
- Dissection trays and kits
- Skeleton, articulated
- Knee joint model

Synovial joints have a fluid-filled cavity lined by a synovial membrane. Features common to synovial joints are the following:

- **Joint capsule**. The joint capsule is made of dense irregular collagenous connective tissue and provides strength and structural reinforcement for the joint. It is lined by the synovial membrane.
- **Articular cartilage**. The articulating ends of the bones are covered with articular cartilage, which is usually hyaline cartilage. The cartilage provides a smooth, nearly frictionless surface for articulation.
- **Ligaments**. The bones in a synovial joint are held together by ligaments that further reinforce the joint. Some ligaments (called extrinsic ligaments) are within the joint cavity, whereas others (intrinsic ligaments) are embedded in the capsule (they are called "intrinsic" because they are an intrinsic part of the capsule).
- **Articular discs**. Also known as **menisci,** articular discs are fibrocartilage pads that improve the fit of two bones to prevent dislocation.

Synovial joints typically are surrounded by tendons that move the bones involved in the joint. The tendons generally are wrapped in a sheath of connective tissue in which they can slide with a minimum of friction. Fluid-filled sacs called **bursae** are often located between tendons and joints, and this also reduces friction.

Procedure Identifying Structures of Synovial Joints

Identify the following structures on fresh specimens such as pigs' feet. If fresh specimens are not available, use anatomical models instead. When you have completed the activity, answer Check Your Understanding questions 2 and 3 (pp. 201–202).

- Joint cavity
- Joint capsule
- Articular cartilage
- Articular discs (menisci)
- Synovial membrane
- Synovial fluid

- Bursae
- Tendon with tendon sheath
- Ligaments
  - Intrinsic
  - Extrinsic

Types of Synovial Joints

The range of motion of a synovial joint is described conventionally in terms of an invisible *axis* about which the bone moves. Synovial joints are classified according to the number of planes of motion in which the bones can move around this axis. The classes are as follows:

1. **Nonaxial joints.** As implied by its name, a nonaxial joint does not move around an axis. Instead, the bones in a nonaxial joint simply glide past one another. An example of a nonaxial joint is the vertebrocostal joint.

2. **Uniaxial joints.** Joints that allow motion in one plane or direction only are called uniaxial joints. A classic example is the elbow, which permits only flexion and extension.

3. **Biaxial joints.** Joints that allow motion in two planes are called biaxial joints. An example of a biaxial joint is the wrist, which allows both flexion/extension and abduction/adduction.

4. **Multiaxial joints.** The joints with the greatest range of motion are multiaxial joints, which allow motion in multiple planes. An example of a multiaxial joint is the shoulder joint.

In addition to this classification scheme, synovial joints are classified according to their structure. The structural classes are illustrated in **Figure 9.4** and include the following:

▶ **Plane**. The bones of plane joints have flat articular surfaces that allow the bones to glide past one another.

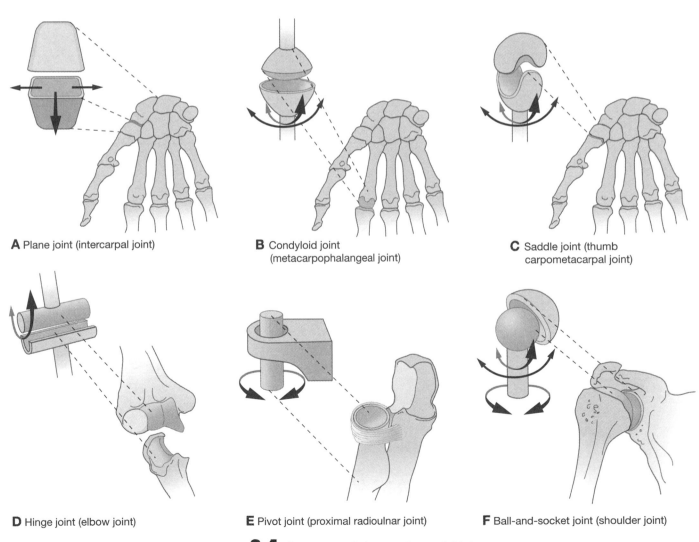

A Plane joint (intercarpal joint)

B Condyloid joint (metacarpophalangeal joint)

C Saddle joint (thumb carpometacarpal joint)

D Hinge joint (elbow joint)

E Pivot joint (proximal radioulnar joint)

F Ball-and-socket joint (shoulder joint)

FIGURE **9.4** The structural classes of synovial joints

- ▸ **Condyloid.** Condyloid joints consist of one bone that fits into the concave surface of another bone.
- ▸ **Saddle.** Note in **Figure 9.4** that saddle joints somewhat resemble condyloid joints but permit a greater range of motion.
- ▸ **Hinge.** In a hinge joint, the bones fit together much like the hinge of a door. Generally, the convex articular surface of one bone fits into a concave articular surface of another bone.
- ▸ **Pivot.** In a pivot joint, one bone rotates or "pivots" around another bone. Generally, pivot joints consist of one bone with a rounded projection that fits into a groove of another bone.
- ▸ **Ball and socket.** Ball and socket joints are named for the rounded, ball-like end of one bone that fits into the concave "socket" of another bone.

 Procedure Classifying Synovial Joints

For each of the following joints, first list the joint's structural classification in Table 9.3. Then obtain an articulated skeleton so you can manipulate each joint and determine if the joint is nonaxial, uniaxial, biaxial, or multiaxial.

TABLE **9.3**	Classification of Synovial Joints	
Joint	**Structural Classification**	**Range of Motion (nonaxial, uniaxial, biaxial, or multiaxial)**
Intercarpal joint		
Proximal radioulnar joint		
Radiocarpal joint		
Thumb carpometacarpal joint		
Interphalangeal joint		
Knee joint		
Atlantoaxial joint		
Hip joint		

Knee Joint

The knee joint, illustrated in **Figure 9.5**, is a modified hinge joint. It is stabilized by intrinsic ligaments (which are part of the joint capsule), extrinsic ligaments, and by the **medial** and **lateral menisci**. Four important extrinsic ligaments include the following:

1. **Anterior cruciate ligament (ACL).** The anterior cruciate ligament, or ACL, extends from the anterior tibial plateau to the lateral femoral condyle. Its function is to prevent hyperextension of the knee.

2. **Posterior cruciate ligament (PCL).** The posterior cruciate ligament, or PCL, extends from the posterior tibial plateau to the medial femoral condyle. It crosses under the ACL, and the two together form an "X." The PCL prevents posterior displacement of the tibia on the femur.

3. **Medial collateral and lateral collateral ligaments.**
   The medial and lateral collateral ligaments (MCL and LCL) extend from the medial tibia and the lateral fibula to the femur, respectively. They resist varus and valgus (medial and lateral) stresses.

 ## Procedure Identifying Structures of the Knee Joint

Identify the following structures of the knee joint on anatomical models or fresh specimens. Check off each structure as you identify it.

▶ Joint capsule
▶ Ligaments
   • Lateral collateral ligament
   • Medial collateral ligament
   • Anterior cruciate ligament (ACL)
   • Posterior cruciate ligament (PCL)
   • Patellar ligament (tendon)
▶ Menisci
   • Medial meniscus
   • Lateral meniscus
▶ Medial and lateral femoral condyles
▶ Tibial plateau

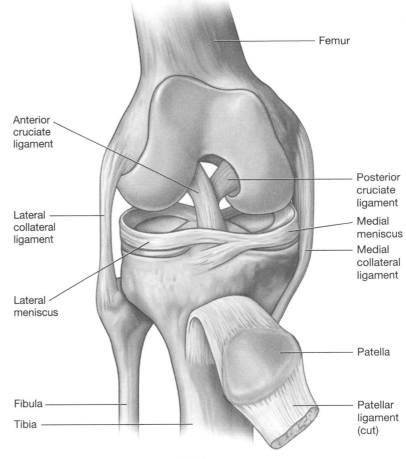

FIGURE **9.5** The knee joint

 ## Procedure Testing the Integrity of the ACL and PCL

One of the more common knee injuries is a torn ACL and/or PCL. To assess the integrity of the ACL and PCL, tests known as the **anterior drawer** and **posterior drawer** are performed, respectively. A normal result of both tests is a minimum amount of motion when the tibia is moved anteriorly and posteriorly. When you have completed the tests, answer Check Your Understanding question 4 (p. 202).

1 Have your partner sit with the knees bent at 90 degrees and relaxed.

2 Grasp your partner's leg with both hands around the proximal tibia and fibula.

3 Gently pull the leg anteriorly, being careful not to extend the knee joint.

   ▶ Amount of motion:

   ▶ Which ligament was being assessed (ACL or PCL)?

4 Now gently push the leg posteriorly. Be careful not the flex the knee joint.

   ▶ Amount of motion:

   ▶ Which ligament was being assessed (ACL or PCL)?

Motions of Synovial and Cartilaginous Joints

Each time you move your body in a seemingly routine fashion (such as walking or climbing stairs), you are producing motion at a tremendous number of joints. Many possible motions can occur at synovial and cartilaginous joints. These motions are illustrated in **Figure 9.6.**

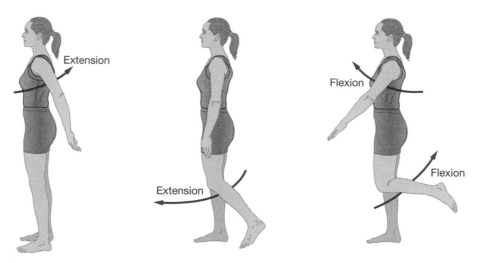

A Angular movements: Extension and flexion at the shoulder and knee

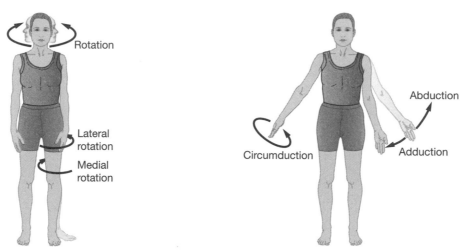

B Rotation of the head, neck, and lower limb

C Angular movements: adduction, abduction, and circumduction of the upper limb at the shoulder

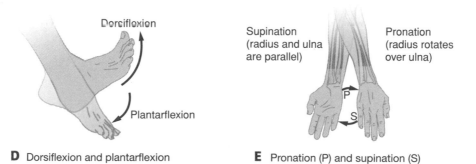

D Dorsiflexion and plantarflexion

E Pronation (P) and supination (S)

Figure **9.6** Motions of synovial and cartilaginous joints

In the following procedure you will determine which joints you are moving with some commonly performed actions and which motions are occurring at each joint.

Procedure Identifying Joint Motions of Common Movements

Team up with a partner, and have your partner perform each of the following actions. Watch carefully as the actions are performed, and list the joints that are in motion.

Ask your instructor if he or she wants you to use the technical name or the common name for each joint (e.g., glenohumeral versus shoulder joint). Some of the joints, such as the hip joint and knee joint, will be obvious. Others, such as the radioulnar joint, the fingers and toes, and the sacroiliac joint, are less obvious and easily overlooked.

Once you have listed the joints being used, determine which motions are occurring at each joint. Keep in mind the type and the range of motion of each joint as you answer each question.

1 Walking up stairs:

Joints moving: Motions occurring:

2 Doing jumping jacks:

Joints moving: Motions occurring:

3 Answering the telephone:

Joints moving: Motions occurring:

4 Jumping rope:

Joints moving: Motions occurring:

NAME _____

SECTION _____ DATE _____

 Check Your Recall

1 Joints that permit no motion are called

- a. synarthroses.
- b. amphiarthroses.
- c. cartilaginous joints.
- d. diarthroses.

2 Freely moveable joints are called

- a. cartilaginous joints.
- b. synarthroses.
- c. amphiarthroses.
- d. fibrous joints.

3 Most cartilaginous joints allow _____ and are _____ joints.

- a. no motion; synarthrotic
- b. some motion; diarthrotic
- c. no motion; amphiarthrotic
- d. some motion; amphiarthrotic

4 Synovial joints are filled with _____, which _____ the joint.

- a. serous fluid; lubricates
- b. synovial fluid; lubricates
- c. serous fluid; increases range of motion of
- d. synovial fluid; increases range of motion of

5 Mark the following statements as true (T) or false (F). If the statement is false, correct it so it is a true statement.

a. The articulating ends of the bones in a synovial joint are covered in a synovial membrane. _____

b. Articular discs improve the fit between two bones in a synovial joint. _____

c. The joint capsule of a synovial joint is lined with a synovial membrane. _____

d. Ligaments provide a smooth, nearly frictionless surface for articulation. _____

e. Fluid-filled sacs called bursae often lie between tendons and the joint capsule of a synovial joint. _____

6 Define the following terms:

Nonaxial joint _____

_____

Uniaxial joint _____

_____

Biaxial joint _____

_____

Multiaxial joint _____

_____

7 Which of the following describes a plane joint correctly?
   a. The convex articular surface of one bone fits into a concave articular surface of another bone.
   b. One bone rotates around another bone.
   c. The flat articular surfaces of two bones glide past one another.
   d. The rounded, ball-like end of one bone fits into a concave depression of another bone.

8 Which of the following correctly describes a pivot joint?
   a. The convex articular surface of one bone fits into a concave articular surface of another bone.
   b. One bone rotates around another bone.
   c. The flat articular surfaces of two bones glide past one another.
   d. The rounded, ball-like end of one bone fits into a concave depression of another bone.

9 Label the following parts of the knee joint in **Figure 9.7.**
- Lateral collateral ligament
- Medial collateral ligament
- Anterior cruciate ligament
- Posterior cruciate ligament
- Patellar ligament
- Medial meniscus
- Lateral meniscus

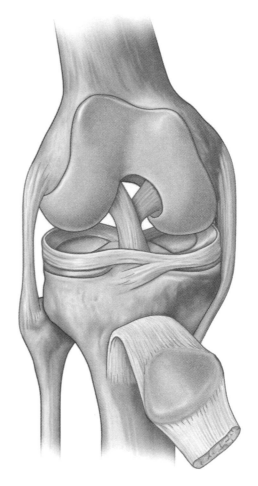

FIGURE **9.7** The knee joint

10 Match the following terms with the correct description.

a. Flexion _____   A. Movement of a body part toward the midline

b. Circumduction _____   B. Movement around a central axis

c. Adduction _____   C. Turning the palm over to face down

d. Extension _____   D. Decreasing the angle between two bones

e. Dorsiflexion _____   E. Movement of a body part away from the midline

f. Abduction _____   F. Increasing the angle between two bones

g. Rotation _____   G. Movement of the ankle that decreases the angle between the foot and leg

h. Pronation _____   H. Movement in a circle

NAME _____

SECTION _____ DATE _____

 Check Your Understanding | **Critical Thinking and Application Questions**

1 Why do fibrous and cartilaginous joints have no joint cavity ?

2 Compare the structure of the hip joint and the shoulder joint. Which joint do you think would be more likely to dislocate? Why?

3 The term "double-jointed" describes individuals who have an abnormally large range of motion in a given joint. This excess range of motion is not due to the presence of a second joint but, instead, weakness of the ligaments surrounding the joint. Why would weakness of ligaments lead to a greater range of motion at a joint?

4 Your friend is on the basketball court when she suddenly hears a loud popping sound and hyperextends her knee. What likely has happened? What results would you see from the anterior drawer test?

Muscle Tissue

OBJECTIVES

Once you have completed this unit, you should be able to:

- Describe the microscopic anatomy of skeletal muscle fibers.

- Identify structures of skeletal muscle fibers.

- Identify structures associated with the neuromuscular junction.

- Describe the sequence of events of the skeletal muscle fiber action potential and the sliding filament mechanism of contraction.

- Describe the length–tension relationship of skeletal muscle fibers.

- Distinguish between smooth and cardiac muscle tissue.

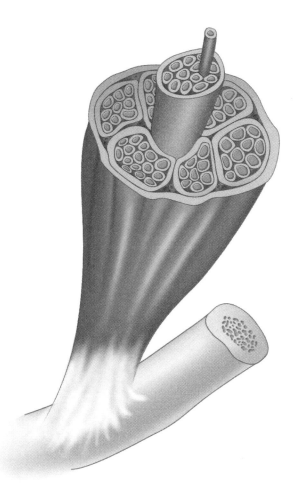

Pre-Lab Exercise 1
Key Terms

Table 10.1 lists the key terms with which you should be familiar before coming to lab.

TABLE 10.1	Key Terms

Term	Definition
General Skeletal Muscle Structures	
Epimysium	_____ _____
Fascicle	_____ _____
Perimysium	_____ _____
Muscle fiber	_____ _____
Endomysium	_____ _____
Structures of the Skeletal Muscle Fiber	
Sarcolemma	_____ _____
T-tubule	_____ _____
Sarcoplasmic reticulum	_____ _____
Terminal cisternae	_____ _____
Myofibril	_____ _____

Myofilament _____

_____

Thick filament _____

_____

Thin filament _____

_____

Components of the Sarcomere

A band _____

_____

I band _____

_____

Z disc _____

_____

H zone _____

_____

M line _____

_____

Terms of Muscle Physiology

Action potential _____

_____

Crossbridge cycle _____

_____

Excitation-contraction coupling _____

_____

Sliding filament mechanism _____

_____

Length-tension relationship _____

_____

Pre-Lab Exercise 2

Basic Skeletal Muscle Anatomy

Label and color the basic structures of skeletal muscles depicted in **Figure 10.1** with the terms from Exercise 1. Use your text and Exercise 1 in this unit for reference.

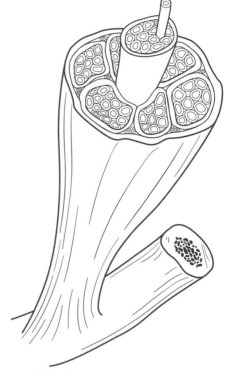

FIGURE **10.1** Basic skeletal muscle structure

Pre-Lab Exercise 3

Muscle Fiber Microanatomy

Label and color the microscopic anatomy of the skeletal muscle fiber depicted in **Figure 10.2** with the terms from Exercise 1. Use your text and Exercise 1 in this unit for reference.

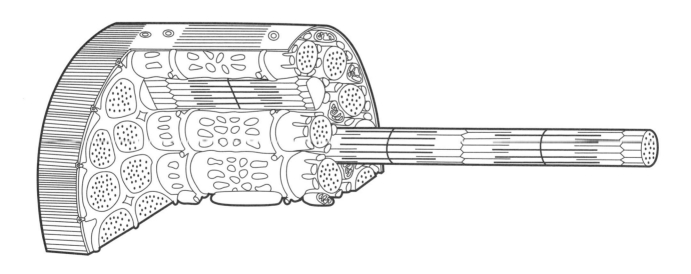

FIGURE **10.2** A skeletal muscle fiber

EXPLORING ANATOMY & PHYSIOLOGY IN THE LABORATORY

Pre-Lab Exercise 4

Neuromuscular Junction

Label and color the microscopic anatomy of the neuromuscular junction depicted in **Figure 10.3** with the terms from Exercise 2. Use your text and Exercise 2 in this unit for reference.

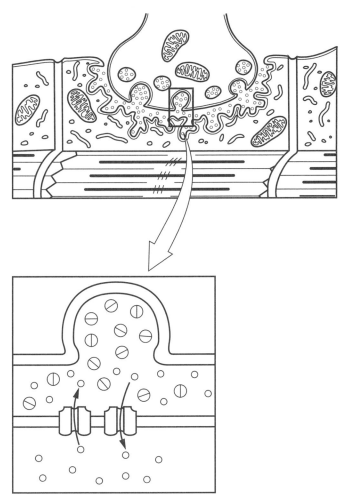

FIGURE **10.3** The neuromuscular junction

 Pre-Lab Exercise 5

Types of Muscle Tissue

You learned about the three types of muscle tissue in Unit 5 (p. 98). Review the properties of each type of muscle tissue by filling in Table 10.2

TABLE **10.2**	The Three Types of Muscle Tissue				
Muscle Type	Striated or Nonstriated	One or Multiple Nuclei	Size and Shape of Cells	Voluntary or Involuntary	Special Features
Cardiac muscle					
Skeletal muscle					
Smooth muscle					

Exercises

You already began to explore muscle tissue in Unit 5 (p. 98), where you were introduced to the three types of muscle tissue: skeletal muscle, smooth muscle, and cardiac muscle. In this unit we delve deeper into muscle tissue's structure and function. Our primary focus will be on skeletal muscle tissue to prepare us for our examination of skeletal muscles in Unit 11. The structure of skeletal muscle cells is one of the most complicated topics you will study in this book, but it can become easier if you remember that their structure follows their function, all the way down to the tiny proteins that make up the majority of a muscle fiber.

The first two exercises in this unit explore the gross structure and the microscopic structure of skeletal muscles and the nerve cells that supply them. In the third exercise we will apply what we learned about skeletal muscle structure to skeletal muscle physiology. In the final exercise we will examine the structure of smooth and cardiac muscle tissue to help you differentiate these tissue types on microscope slides.

Exercise 1

Skeletal Muscle Anatomy

MATERIALS NEEDED

- Skeletal muscle models
- Skeletal muscle fiber model
- Skeletal muscle tissue slide
- Light microscope with oil immersion objective
- Oil
- Colored pencils

Skeletal muscle fibers are long, cylindrical cells that are wrapped by their plasma membrane, known as the **sarcolemma** (**Figures 10.5** and **10.6**). About 80% of the skeletal muscle fiber's cytoplasm, also called its **sarcoplasm**, is filled with small cylindrical organelles called **myofibrils**. The remainder of the sarcoplasm contains abundant mitochondria, multiple nuclei, and a modified endoplasmic reticulum called the **sarcoplasmic reticulum** (SR), which wraps around the myofibrils.

As we will discuss shortly, the SR stores calcium ions that are required for muscle to contract. Note that at certain points along the myofibril, the SR swells to form **terminal cisternae**. Running down the middle of each terminal cistern is an inward extension of the sarcolemma known as a **T-tubule**. A group of two terminal cisternae and a T-tubule is called a **triad**.

Myofibrils are composed of protein subunits called **myofilaments**. The two types of myofilaments involved in contraction are: (a) **thick filaments**, composed of the contractile protein **myosin**, and (b) **thin filaments**, composed of the contractile protein **actin** and the regulatory proteins **troponin** and **tropomyosin**. The arrangement of myofilaments within the myofibrils is what gives skeletal muscle its

A skeletal muscle is composed of groups of skeletal muscle cells, also called **muscle fibers**, arranged into groups called **fascicles** (**Figure 10.4**). The muscle as a whole is covered by a connective tissue sheath called the **epimysium**, which blends with the thick, superficial fascia that binds muscles into groups. The epimysium also blends with the fibers of tendons and aponeuroses, which connect the muscle to bones or soft tissue. Each fascicle is surrounded by another connective tissue sheath called the **perimysium**. Individual skeletal muscle fibers are surrounded by their extracellular matrix, known as the **endomysium**.

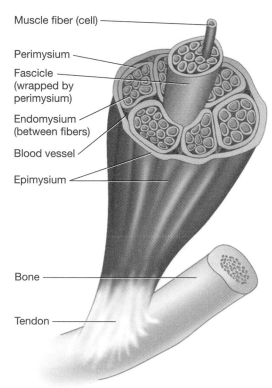

FIGURE 10.4 Basic skeletal muscle structure

characteristic **striated** appearance. The dark regions of the striations, called **A bands**, are dark because this is where the thick and thin filaments overlap. The light regions, called **I bands**, appear light because they contain only thin filaments.

On closer inspection, there are more lines and bands than simply the A and I bands. For example, bisecting the A band is a lighter region called the **H zone** (which itself is bisected by a line called the **M line**). Also, bisecting the I band is a dark line called the **Z disc**. The terms of this alphabet soup are used to describe the fundamental unit of contraction: the **sarcomere**. A sarcomere, defined as the space from one Z-disc to the next Z-disc, consists of a full A band and two half-I bands.

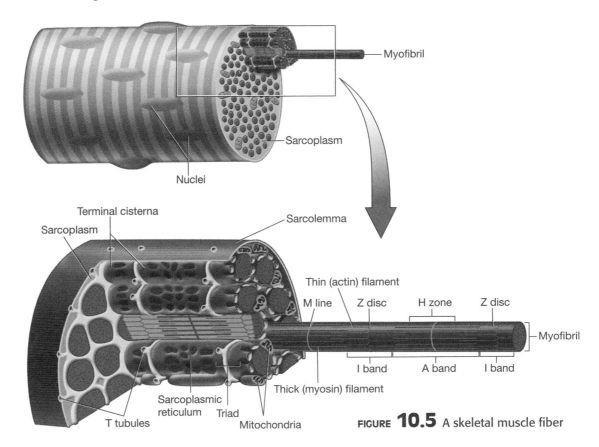

FIGURE 10.5 A skeletal muscle fiber

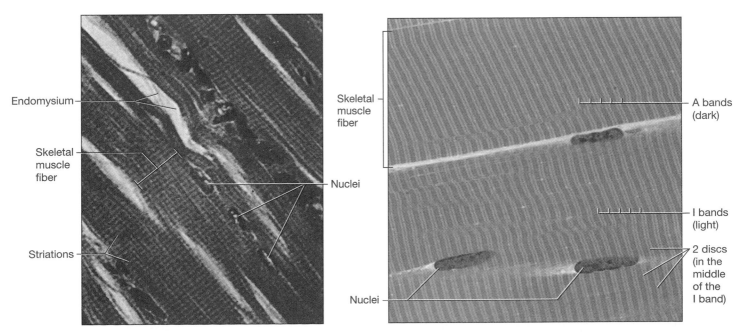

FIGURE 10.6 Skeletal muscle tissue, photomicrograph (A) 40× objective; (B) 100× objective (oil immersion)

Procedure Model Inventory

Identify the following structures of skeletal muscle and skeletal muscle fibers on models and diagrams, using your textbook and this unit for reference. As you examine the anatomical models and diagrams, record on the model inventory in Table 10.3 the name of the model and the structures you were able to identify.

Skeletal Muscle Anatomy

1. Connective tissue coverings
   a. Epimysium
   b. Perimysium
2. Fascicle
3. Muscle fiber (cell)
4. Tendon
5. Aponeurosis

Muscle Fiber Microanatomy

1. Sarcolemma
2. Sarcomere
   a. A band
   b. I band
   c. H zone
   d. Z disc
   e. M line
3. Transverse tubules
4. Sarcoplasmic reticulum
5. Terminal cisternae
6. Triad
7. Endomysium

TABLE **10.3**	Model Inventory for Skeletal Muscle Anatomy and Microanatomy
Model/Diagram	**Structures Identified**

Procedure Microscopy of Skeletal Muscle Tissue

View a prepared slide of skeletal muscle tissue.

1 First examine the slide with the regular high-power objective of your light microscope.

2 Draw and color what you see, and label your drawing with the terms indicated.

3 Then switch to an oil-immersion lens (your instructor may have one set up as a demonstration) and identify the structures of the sarcomere.

High-Power (40×) Objective:

1. Striations
   a. A band
   b. I band
2. Sarcolemma
3. Nuclei
4. Endomysium

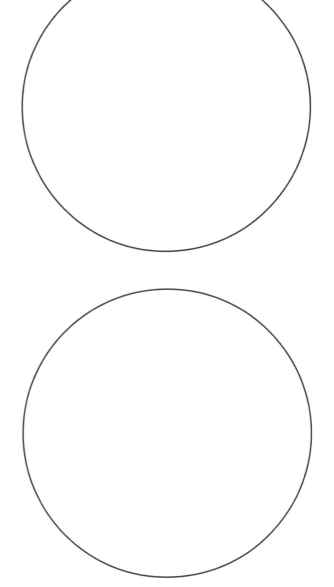

Oil-Immersion Lens:

1. Sarcomere
2. A band
3. I band
4. Z disc
5. H zone
6. M line

Exercise 2

The Neuromuscular Junction

MATERIALS NEEDED

- Neuromuscular junction models
- Neuromuscular junction slide
- Light microscope
- Colored pencils

Skeletal muscle fibers are voluntary, which means that they contract under the conscious control of the nervous system. Each skeletal muscle fiber has a **motor neuron** that triggers it to contract. The place where the nerve meets the muscle fiber is called the **neuromuscular junction** (NMJ) (**Figure 10.7**). The neuromuscular junction consists of three parts:

1. **Axon terminal.** Each motor neuron splits into many swollen axon terminals, each of which contacts one muscle fiber. Each axon terminal has **synaptic vesicles** in its cytosol that contain the neurotransmitter **acetylcholine.**

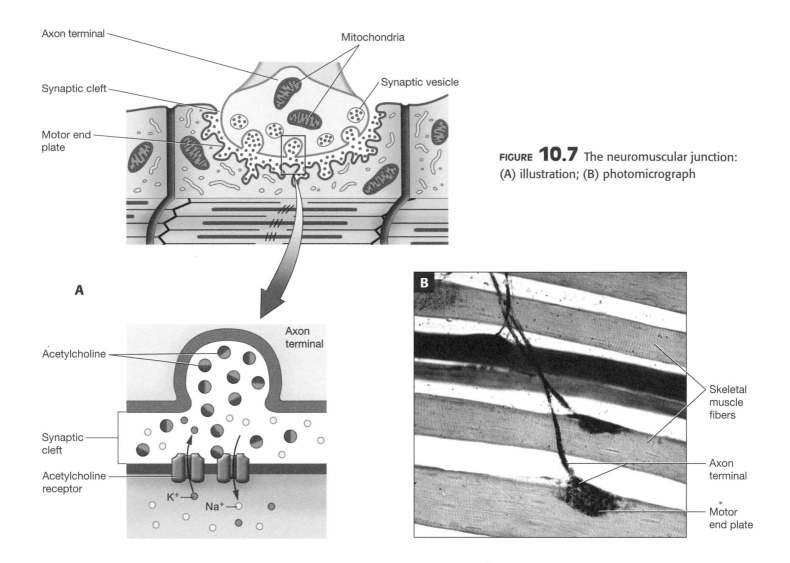

FIGURE **10.7** The neuromuscular junction:
(A) illustration; (B) photomicrograph

2. **Synaptic cleft.** The axon terminal doesn't come into direct contact with the muscle fiber; instead, there is a space between the axon terminal and the muscle fiber called the synaptic cleft.

3. **Motor end plate.** The motor end plate is a specialized region of the sarcolemma that contains **acetylcholine receptors**.

In the next exercise we will examine the physiology of the NMJ and its role in muscle contraction.

 Procedure Model Inventory

Identify the following structures of the neuromuscular junction on models and diagrams, using your textbook and this unit for reference. As you examine the anatomical models and diagrams, record on the model inventory in Table 10.3 the name of the model and the structures that you were able to identify (note that this is the model inventory in Exercise 1).

◗ Motor neuron
◗ Axon terminal
◗ Synaptic vesicles
◗ Synaptic cleft
◗ Motor end plate
◗ Acetylcholine receptors

Procedure Microscopy of the Neuromuscular Junction

Obtain a prepared slide of a neuromuscular junction. Use your colored pencils to draw what you see, and label your drawing with the following terms.

1. Motor neuron

2. Axonal terminal

3. Skeletal muscle fiber

4. Striations

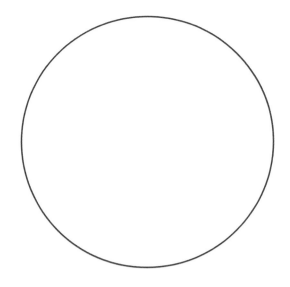

Exercise 3

Muscle Physiology

MATERIALS NEEDED

▶ Hand dynamometer
▶ Model/diagram of the contraction cycle
▶ Colored pencils

Skeletal muscle contraction begins when a skeletal muscle fiber is stimulated by a motor neuron, in a relationship known as **excitation–contraction coupling**. Notice in **Figure 10.8** what happens when a nerve impulse (**neuronal action potential**) reaches the axon terminal: Acetylcholine is released into the synaptic cleft by exocytosis of the synaptic vesicles. Acetylcholine then diffuses across the synaptic cleft and binds to acetyl-choline receptors in the motor end plate, which triggers a series of electrical events across the sarcolemma of the muscle fiber collectively called a **muscle action potential**. A muscle action potential begins when acetylcholine opens sodium channels in the motor end plate, which causes sodium ions to rush into the sarco-plasm. As the cell accumulates positive charges, the electrical difference across the sarcolemma becomes more positive, a change called **depolarization**.

The depolarization that begins at the motor end plate propagates through the sarcolemma and dives into the interior of the muscle fiber along the T-tubules. As the action potential spreads along the T-tubules, calcium channels open in the adjacent terminal cisternae and calcium ions flood the sarcoplasm. Before calcium ions enter the sarcoplasm, actin and myosin are prevented from interacting by the regulatory protein tropomyosin. When calcium ions flood the sarcoplasm, they bind to the regulatory protein troponin, which causes the tropomyosin to shift its position away from the actin-myosin binding site. This allows a contraction to begin.

A skeletal muscle fiber contracts by a mechanism known as the **sliding filament mechanism**, in which the thick and thin filaments slide past one another during a contraction. The steps of the sliding filament mechanism are as follows (**Figure 10.9**):

1. ATP is hydrolyzed, and this energy "cocks" a portion of the myosin molecule known as the myosin **crossbridge** into an upright position. The ADP and P_i remain associated with the myosin crossbridge.

2. Actin binds to the myosin crossbridge.

3. ADP and P_i dissociate from the myosin head, and the myosin crossbridge moves forward into its relaxed position. As it moves, it pulls the actin strand with it in a motion known as a **power stroke**.

4. A new ATP molecule binds to myosin, which causes actin and myosin to dissociate. ATP is hydrolyzed again, the myosin crossbridge is re-cocked, and the cycle, known as a **crossbridge cycle**, repeats.

1. A neuronal action potential travels through the motor neuron to the axon terminals.

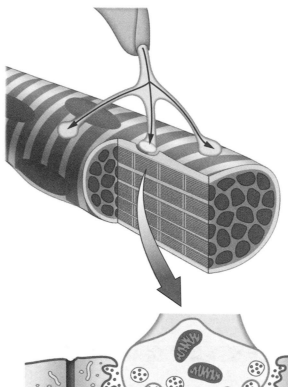

2. The action potential triggers release of acetylcholine from the synaptic vesicles by exocytosis.

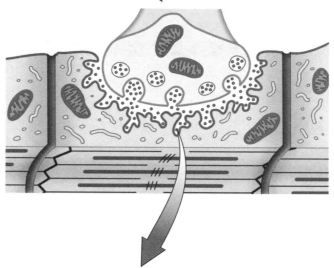

3. Acetylcholine binds acetylcholine receptors in the motor end plate and triggers a muscle action potential.

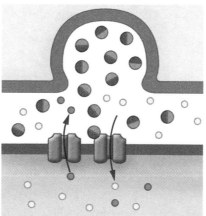

FIGURE **10.8** Events at the neuromuscular junction

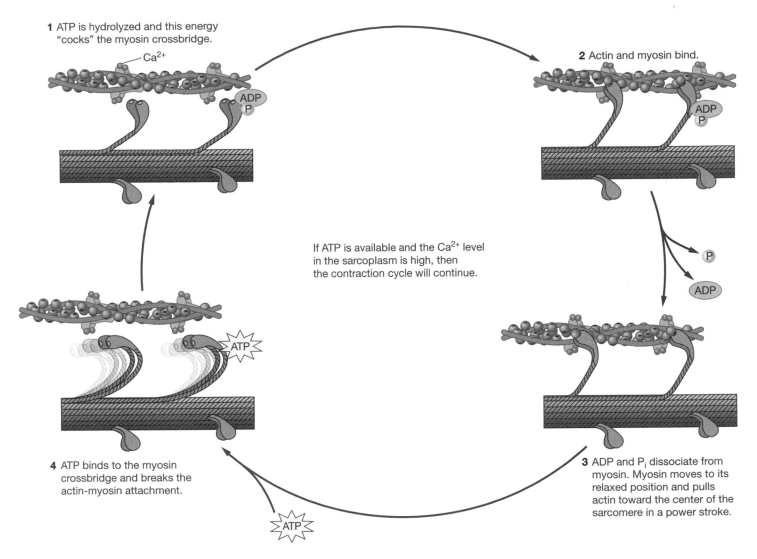

1 ATP is hydrolyzed and this energy "cocks" the myosin crossbridge.

Ca²⁺

ADP
P

2 Actin and myosin bind.

ADP
P

If ATP is available and the Ca²⁺ level in the sarcoplasm is high, then the contraction cycle will continue.

P

ADP

ATP

4 ATP binds to the myosin crossbridge and breaks the actin-myosin attachment.

ATP

3 ADP and P$_i$ dissociate from myosin. Myosin moves to its relaxed position and pulls actin toward the center of the sarcomere in a power stroke.

FIGURE **10.9** A single crossbridge cycle showing the sliding filament mechanism of contraction

This cycle will continue as long as acetylcholine remains in the synaptic cleft (provided that calcium ions and ATP remain available). Therefore, to terminate a contraction, the body must degrade the acetylcholine with an enzyme called **acetylcholinesterase.** This causes the muscle action potential to enter a phase known as **repolarization,** during which potassium channels open and potassium ions exit the muscle fiber. As the fiber repolarizes, calcium ions are pumped back into the sarcoplasmic reticulum and tropomyosin returns to its original position.

 Procedure Time to Trace

Now let's trace the entire series of events from the time a neuronal action potential reaches the axon terminal to the end of a muscle contraction when tropomyosin returns to the position where it blocks actin's binding site. As you trace, you will include the events that occur at the neuromuscular junction, the electrical changes that occur across the sarcolemma, the events of the crossbridge cycle, and the events that terminate the muscle contraction. Hints are included to ensure that you stay on the right track as you trace. When you have completed the activity, answer Check Your Understanding questions 1 through 3 (pp. 227–228).

Start: neuronal action potential reaches the axon terminal →

_____ →

_____ →

acetylcholine binds to receptors on the motor end plate →

_____ →

_____ →

_____ →

_____ →

calcium binds to troponin →

_____ →

_____ →

_____ →

_____ →

crossbridge cycle repeats →

_____ →

_____ →

_____ →

tropomyosin moves back into its position blocking actin's binding site **End**

Procedure Diagramming Your Tracing

Take the pathway that you just traced and turn it into a diagram that contains the neuromuscular junction, a skeletal muscle fiber, and the thick and thin filaments in a myofibril. Your diagram does not have to be a work of art. A simple schematic representation of each structure is all that is necessary. You may use the figures in this unit for reference, but it is best to compose and organize your diagram in a way that makes sense to you so you may return to it for study purposes.

Procedure Measuring the Length-Tension Relationship

The amount of tension that can be generated during a contraction is influenced by how much the thick and thin filaments overlap when the muscle fiber is at rest—a relationship known as the **length–tension relationship** (Figure 10.10). The overlap of the thick and thin filaments is determined by the length of the sarcomeres. When the muscle is stretched, the sarcomeres are longer and have less overlap. When the muscle is shortened, the sarcomeres are shorter and have more overlap.

Notice in **Figure 10.10** that the muscle fiber can generate the maximum amount of tension when its sarcomeres are at about 100%–120% of their natural length. Conversely, the amount of tension generated during a contraction diminishes greatly both when the muscle is overly stretched and when it is excessively shortened.

In this experiment you will be measuring the ability of your digital flexor muscles to generate tension at different resting muscle lengths. You will take your measurements with an instrument called a **hand dynamometer**, which measures the force that you apply when it is squeezed.

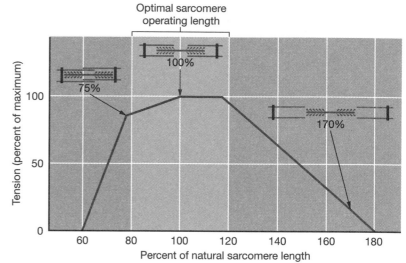

FIGURE 10.10 The length-tension relationship

1 Position your arm and wrist as shown in **Figure 10.11A** with your wrist flexed at a 90° angle.

2 Grip the dynamometer and squeeze it with your maximum effort. Record in Table 10.4 the amount of force you generated.

3 Position your arm and wrist as shown in **Figure 10.11B**, with your wrist extended at a 0° angle.

4 Grip the dynamometer and squeeze it with your maximum effort. Record in Table 10.4 the amount of force that you generated.

5 Position your arm and wrist as shown in **Figure 10.11C** with your wrist hyperextended at a −90° angle.

6 Grip the dynamometer and squeeze it with your maximum effort. Record in Table 10.4 the amount of force that you generated.

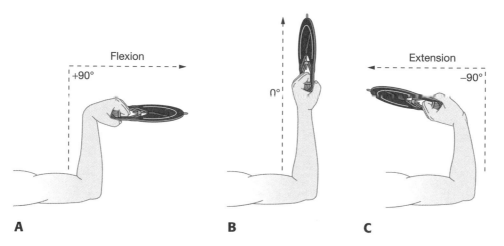

FIGURE 10.11 Wrist positions for the dynamometer experiment: (A) flexed at 90°; (B) extended at 0°; (C) hyperextended at −90°

TABLE 10.4	Results of Dynamometer Experiment
Wrist Position	**Amount of Force Generated**
Flexed at 90°	
Extended at 0°	
Hyperextended at −90°	

7 Interpret your results:

   a. In which wrist position were the muscles' sarcomeres overly stretched?

   b. In which wrist position were the muscles' sarcomeres excessively shortened?

   c. Which wrist position yielded the most forceful contraction? Why?

 Exercise 4

Smooth and Cardiac Muscle Tissues

MATERIALS NEEDED

▶ Smooth muscle slide
▶ Cardiac muscle slide
▶ Light microscope
▶ Colored pencils

Although smooth and cardiac muscle tissues differ from skeletal muscle tissue in many ways, all three share the same basic function: to generate tension. This functional similarity brings with it structural similarities, as well. Cardiac muscle tissue is structurally the most similar to skeletal muscle tissue. As you can see in **Figure 10.12A**, cardiac muscle tissue is also striated and consists of alternating A bands and I bands arranged into sarcomeres.

The two tissue types have notable differences, though. For one, the cells of cardiac muscle tissue, called **cardiac myocytes**, are shorter than skeletal muscle fibers, branching, and typically have only a single nucleus. In addition, adjacent cardiac myocytes are joined by specialized structures called **intercalated discs**, which contain gap junctions and desmosomes. The intercalated discs connect all cardiac myocytes physically and electrically and allow the heart to contract as a unit. Finally, cardiac myocytes are involuntary and **autorhythmic**: They depolarize and contract without stimulation from the nervous system because of specialized cardiac myocytes called **pacemaker cells**.

Smooth muscle tissue, shown in **Figure 10.12B**, is not striated and, therefore, has no A bands, I bands, or sarcomeres. Smooth muscle cells tend to be long, thin, and spindle-shaped with a single, centrally located nucleus. All smooth muscle cells are involuntary, and most require extrinsic innervation to contract. Smooth muscle is found in many places in the body, including the lining of all hollow organs, the arrector pili muscles in the dermis, and the iris and ciliary muscle of the eye.

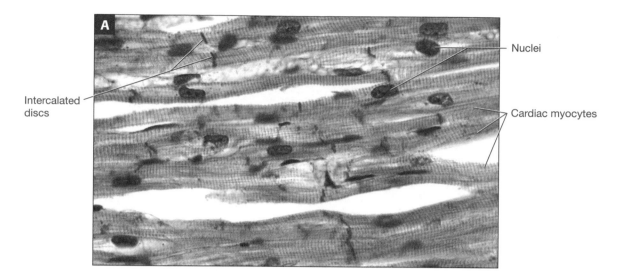

Intercalated
discs

Nuclei

Cardiac myocytes

Nuclei of
smooth
muscle
fibers

Smooth
muscle
tissue

FIGURE **10.12** Other types of muscle tissue: (A) cardiac muscle tissue; (B) smooth muscle tissue

How to Distinguish Between Types of Muscle Tissue

With a bit of practice, distinguishing between the three types of muscle tissue on microscope slides becomes fairly easy. Following are some hints to help you differentiate these three tissues:

Striations If striations are present, the muscle is either skeletal or cardiac. If there are no striations, it is smooth muscle.

Shape of the cells Skeletal muscle fibers are long and thin, whereas cardiac myocytes are short, wide, and branching. Smooth muscle cells are thin and flat. My past students have remarked that smooth muscle cells somewhat resemble stratified squamous epithelium.

Location of the nucleus Each skeletal muscle fiber has multiple nuclei located next to the sarcolemma. Cardiac myocytes usually have a single, large nucleus located among the myofibrils. The single, flattened nucleus of smooth muscle cells is usually in the middle of the cell.

Other features

▶ The intercalated discs of cardiac muscle tissue appear as black lines oriented parallel to the striations. No other tissue type has a similar feature, so intercalated discs are an immediate giveaway that you are looking at cardiac muscle tissue.

▶ Smooth muscle tissue is typically found on a slide with other types of tissue because it lines hollow organs. Scroll through the slide to see if you can find any additional tissue types, such as epithelial and/or connective tissue. If these other tissue types are on top of or below the muscle cells, there is a good chance you are examining smooth muscle tissue.

Procedure Microscopy of Cardiac and Smooth Muscle Tissue

1 Obtain prepared slides of cardiac and smooth muscle tissue.

2 Use your colored pencils to draw what you see, and label your drawing with the following terms.

After you have completed the activity, answer Check Your Understanding question 4 (p. 228).

1. Cardiac muscle
   a. Striations
      (1) A band
      (2) I band
   b. Sarcolemma
   c. Nucleus
   d. Intercalated disc
   e. Cardiac myocyte

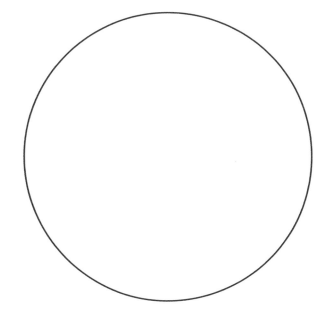

2. Smooth Muscle
   a. Nucleus
   b. Smooth muscle cell
   c. Blood vessels (or other tissues)

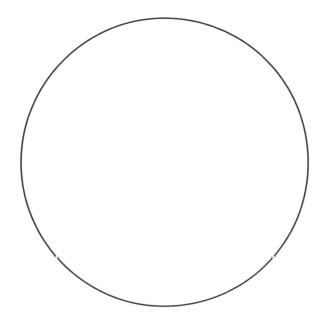

NAME _____

SECTION _____ DATE _____

Check Your Recall

1 Label the following terms on **Figure 10.13.**

- Epimysium
- Muscle fiber
- Fascicle
- Endomysium
- Perimysium
- Tendon

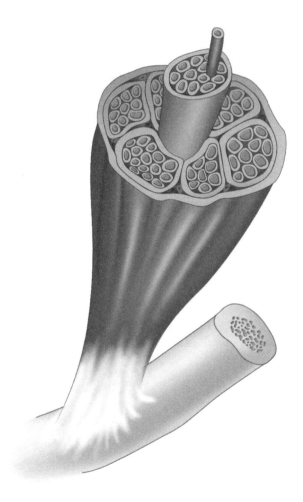

FIGURE **10.13** Basic skeletal muscle structure

2 Label the following terms on **Figure 10.14.**

- Myofibril
- Sarcolemma
- T-tubule
- Sarcoplasmic reticulum
- A band
- I band
- Z disc
- Sarcomere
- Thick filament
- Thin filament

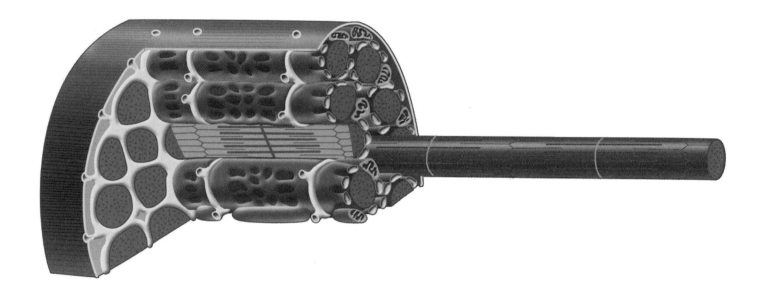

FIGURE **10.14** A skeletal muscle fiber

3 The striations in skeletal muscle fibers are attributable to

a. light and dark pigments found in the sarcoplasm.
b. the arrangement of thick and thin filaments in the myofibril.
c. overlapping Z-discs.
d. overlapping adjacent skeletal muscle fibers.

4 The sarcoplasmic reticulum

a. stores sodium ions.
b. forms from inward extensions of the sarcolemma.
c. stores calcium ions.
d. is made of contractile proteins.

EXPLORING ANATOMY & PHYSIOLOGY IN THE LABORATORY

5 Fill in the blanks: Thick filaments are composed of the protein _____, and thin filaments are composed of the proteins _____, _____, and _____.

6 Which of the following is *not* a part of the neuromuscular junction?
   a. Axon terminal
   b. Dendritic receptor
   c. Synaptic cleft
   d. Motor end plate

7 Excitation-contraction coupling refers to
   a. the muscle fiber's depolarization.
   b. the relationship between length of the sarcomere and the tension generated.
   c. the sliding of thick and thin filaments past one another.
   d. the fact that a muscle must be excited by a neuron to contract.

8 Number the following events of skeletal muscle contraction in the proper order, with number 1 for the first event and number 9 for the last event.

   _____ Actin binds to the myosin crossbridge and a power stroke occurs as ADP and P_i dissociate from myosin.

   _____ The terminal cisternae release calcium ions into the sarcoplasm.

   _____ The sarcoplasmic reticulum removes calcium from the sarcoplasm and tropomyosin returns to its original position.

   _____ The energy from ATP hydrolysis "cocks" the myosin crossbridge into an upright position.

   _____ Acetylcholinesterase degrades acetylcholine in the synaptic cleft, potassium channels open in the sarcolemma, and the muscle fiber repolarizes.

   _____ Calcium ions bind to troponin which pulls tropomyosin away from actin.

   _____ Acetylcholine binds to receptors on the motor end plate and opens sodium channels.

   _____ ATP binds and causes the actin and myosin crossbridge to dissociate from one another, after which it is hydrolyzed and myosin is re-cocked. The cycle repeats.

   _____ Sodium ions flood the sarcoplasm and initiate an action potential that spreads along the sarcolemma and into the muscle fiber along the T-tubules.

9 Maximal tension in a skeletal muscle contraction is generated

 a. when the sarcomeres are at about 100–120% of their natural length.
 b. when the muscle fiber is stretched before the contraction.
 c. when the muscle fiber is shortened before the contraction and there is maximal overlap of the filaments.
 d. when the sarcoplasmic reticulum releases fewer calcium ions.

10 Mark the following statements as applying to *skeletal* muscle tissue, *cardiac* muscle tissue, *smooth* muscle tissue, or any combination of the three.

_____ The primary function is to generate tension.

_____ The tissue lines hollow organs.

_____ The cells are uninucleate.

_____ The cells are involuntary.

_____ The cells are autorhythmic.

_____ The cells must be stimulated by a neuron to contract.

_____ The cells are striated in appearance.

_____ The cells lack striations and sarcomeres.

_____ The cells are multinucleate.

Check Your Understanding | **Critical Thinking and Application Questions**

1 The drug Botox® is made from the toxin of the bacterium *Clostridium botulinum*. It is injected subdermally to minimize fine lines and creases in the face. The toxin prevents motor neurons from releasing acetylcholine. How would this produce the desired cosmetic results? What could happen if this toxin were absorbed systemically?

2 **Myasthenia gravis** is an autoimmune disease in which the immune system produces antibodies that attack the acetylcholine receptor on the motor end plate. What symptoms would you expect with this disease? Explain.

3 The condition **rigor mortis** develops several hours after death because of a lack of ATP. It is characterized by muscular rigidity. Explain why a lack of ATP would cause the sustained muscle contraction of rigor mortis.

4 How does the structure of cardiac myocytes and intercalated discs follow the function of cardiac muscle tissue and the heart?

The Muscular System

OBJECTIVES

Once you have completed this unit, you should be able to:

- Identify muscles of the upper and lower limbs, trunk, head, and neck.

- Describe the origin, insertion, and action of selected muscles.

- List the muscles required to perform common movements.

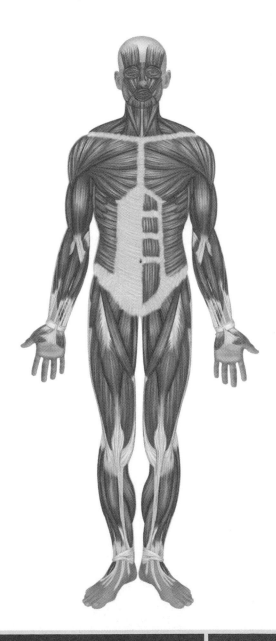

 ## Pre-Lab Exercise 1

Key Terms

Table 11.1 lists the key terms with which you should be familiar before coming to lab. For the skeletal muscles in this table, describe the muscles' location and appearance. Please note that this list is not all-inclusive, as the terminology for the skeletal muscles is extensive.

TABLE **11.1**	Key Terms

Term   **Definition/Description**

General Terms

Origin _____

_____

Insertion _____

_____

Muscle action _____

_____

Skeletal Muscles of the Head, Neck, and Thorax

Muscles of facial expression _____

_____

Muscles of mastication _____

_____

Sternocleidomastoid _____

_____

Trapezius _____

_____

Erector spinae _____

_____

Intercostal muscles _____

_____

Diaphragm _____

_____

Rectus abdominis _____

_____

Skeletal Muscles That Move the Upper Limb

Deltoid _____

_____

Latissimus dorsi _____

_____

Pectoralis major _____

_____

Rotator cuff muscles _____

_____

Biceps brachii _____

_____

Triceps brachii _____

_____

Brachioradialis _____

_____

Wrist/digit flexors _____

_____

Wrist/digit extensors _____

_____

Skeletal Muscles That Move the Lower Limb

Iliopsoas _____

_____

Gluteus muscles _____

_____

Sartorius _____

_____

Quadriceps femoris group _____

_____

Adductor group _____

_____

Hamstrings group _____

_____

Gastrocnemius _____

_____

Soleus _____

_____

Ankle/foot flexors _____

_____

Ankle/foot extensors _____

_____

✎ Pre-Lab Exercise 2

Skeletal Muscle Anatomy

Label and color the muscles depicted in **Figure 11.1** with the terms from Exercise 1. Use your text and Exercise 1 in this unit for reference.

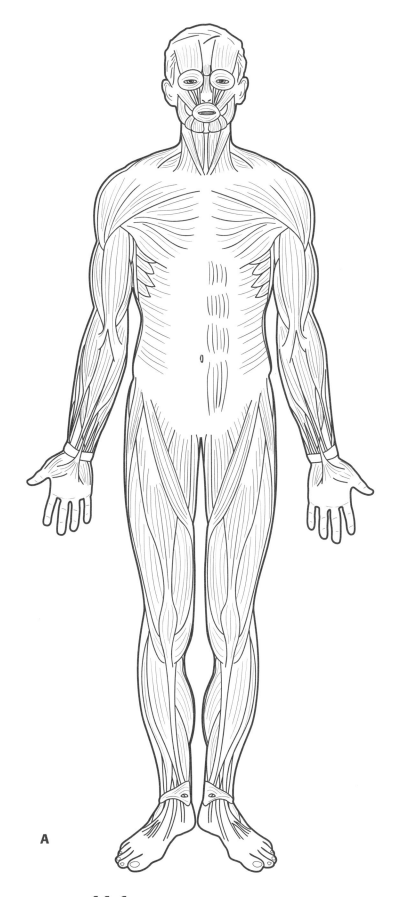

FIGURE **11.1** Human musculature: (A) anterior view

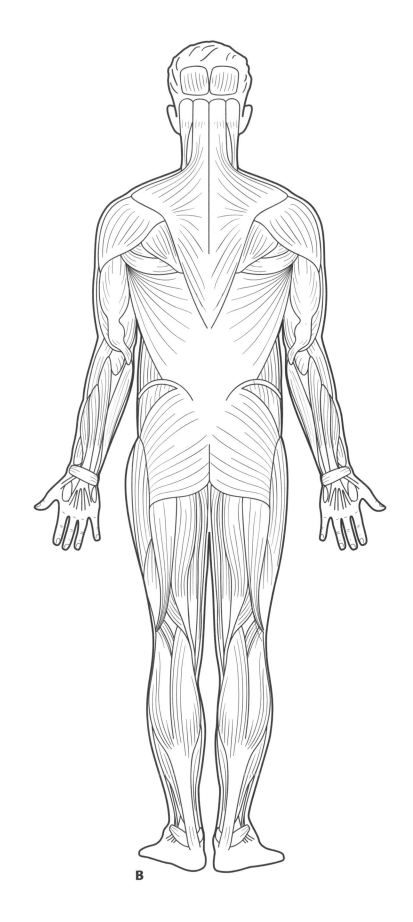

B

FIGURE **11.1** Human musculature (cont.): (B) posterior view

EXPLORING ANATOMY & PHYSIOLOGY IN THE LABORATORY

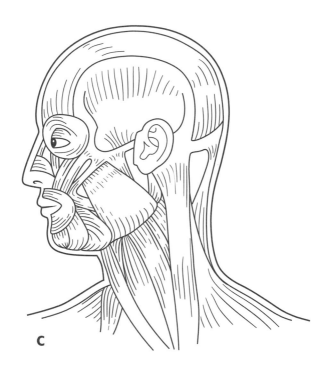

C

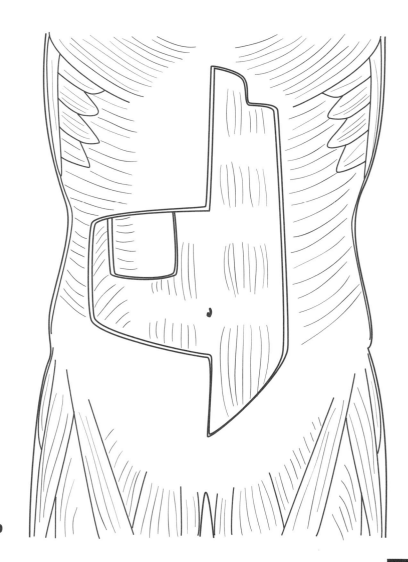

FIGURE 11.1
Human musculature (cont.):
(C) lateral view of the face;
(D) abdominal muscles

D

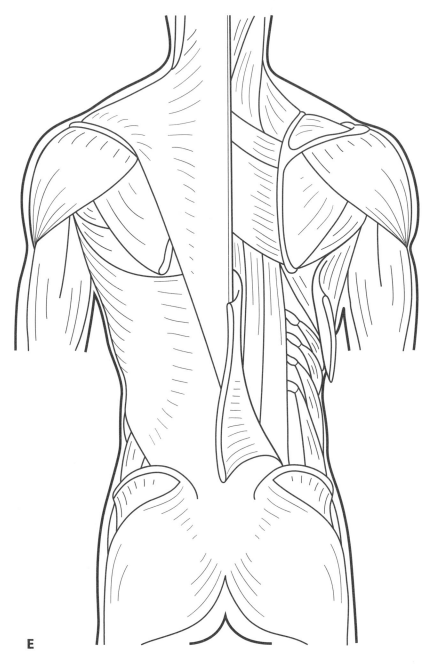

E

FIGURE **11.1** Human musculature (cont.): (E) posterior torso

✏️ Pre-Lab Exercise 3

Muscle Origins, Insertions, and Actions

Complete Table 11.2 with the origins, insertions, and *main* actions for the listed muscles. You may notice in your textbook that the origin and insertion of a muscle often are extensive. For example, the latissimus dorsi muscle originates from T7–L5, the lower three or four ribs, the inferior angle of the scapula, and the thoracolumbar fascia. This obviously is quite a bit to write and to remember, so I suggest shortening it to something that makes sense to you (e.g., "lower back and scapula"). You also will notice that most muscles have more than one action, which you may wish to simplify as well. If your instructor prefers that you to learn the more technical version, learning a simplified version first may be helpful.

EXPLORING ANATOMY & PHYSIOLOGY IN THE LABORATORY

TABLE **11.2** | Muscle Origins, Insertions, and Actions

Muscle	Origin	Insertion	Action(s)
Muscles of the Head and Neck			
Sternocleidomastoid m.			
Trapezius m.			
Muscles of the Thorax, Abdomen, and Back			
Pectoralis minor m.			
Rectus abdominis m.			
External oblique m.			
Internal oblique m.			
Serratus anterior m.			
Tranversus abdominis m.			
Erector spinae m.			
Deltoid m.			
Pectoralis major m.			
Latissimus dorsi m.			
Muscles of the Upper Limb			
Biceps brachii m.			
Triceps brachii m.			
Brachialis m.			
Brachioradialis m.			

(Continued)

TABLE **11.2**	Muscle Origins, Insertions, and Actions (cont.)		
Muscle	**Origin**	**Insertion**	**Action(s)**
Muscles of the Lower Limb			
Iliopsoas m.			
Gluteus maximus m.			
Gluteus medius m.			
Sartorius m.			
Tensor fascia lata m.			
Rectus femoris m.			
Vastus medialis m.			
Vastus intermedius m.			
Vastus lateralis m.			
Adductor group			
Gracilis m.			
Biceps femoris m.			
Semitendinosus m.			
Semimembranosus m.			
Gastrocnemius m.			
Soleus m.			
Tibialis anterior m.			

EXPLORING ANATOMY & PHYSIOLOGY IN THE LABORATORY

Exercises

The human body has nearly 700 **skeletal muscles** that range dramatically in size and shape from the large, triangular trapezius muscle to the tiny corrugator supercilii muscle. Luckily for you, we will be learning only about 70 muscles rather than the full 700.

A muscle begins at its **origin,** which is generally the more stationary part, and attaches to its **insertion,** which is generally the part that the muscle moves. We are most familiar with muscles that insert into and move bones. Many muscles, however, insert into structures other than bones. Examples include the diaphragm, which inserts into its own central tendon, and the muscles of facial expression, which insert into skin or other muscles.

The exercises in this unit help you to become familiar with the main muscle groups and their component muscles. The first exercise introduces you to muscle anatomy, the second exercise focuses on muscle origins and insertions, and the third and final exercise explores muscle actions.

 Exercise 1

Skeletal Muscles

MATERIALS NEEDED

> Muscle models: upper limb, lower limb, trunk, head, neck

In this exercise, we divide skeletal muscles into muscle groups that have similar functions. For example, we classify the latissimus dorsi muscle with the muscles of the upper limb rather than the muscles of the back or trunk because it moves the upper limb. We will use the following groupings of skeletal muscles in this unit:

1. **Muscles that move the head, neck, and face.** We can subdivide the muscles that move the head, neck, and face into the muscles of **facial expression,** the muscles of **mastication,** and the muscles that move the head. The muscles of facial expression control the various facial expressions that humans are capable of making, and they all insert into skin or other muscles. Examples include the **orbicularis oculi,** which closes and squints the eye, the **orbicularis oris,** which purses the lips, and the **zygomaticus major** and **minor,** which pull the corners of the mouth laterally during smiling. The other muscles of facial expression are illustrated in **Figure 11.2.** The muscles of mastication are involved in chewing. They include the **masseter,** which is a thick muscle over the lateral jaw, and the fan-shaped **temporalis muscle** that rests over the lateral skull.

The two muscles that we discuss that move the head are the **trapezius,** which holds the head upright and hyperextends the neck (note that it also elevates the shoulders), and the **sternocleidomastoid,** a strap-like muscle in the neck that flexes the head and neck, rotates the neck laterally, and flexes the neck laterally.

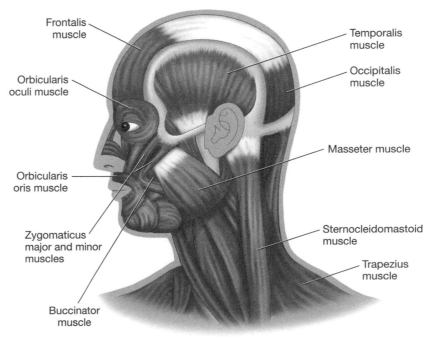

FIGURE 11.2 Facial musculature, lateral view

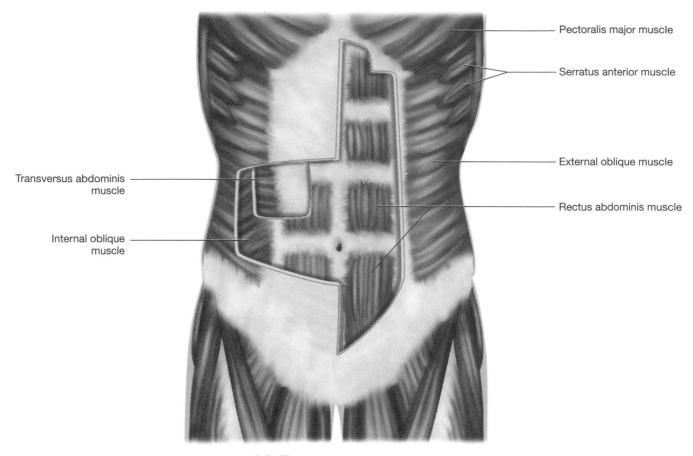

Transversus abdominis muscle

Internal oblique muscle

Pectoralis major muscle

Serratus anterior muscle

External oblique muscle

Rectus abdominis muscle

FIGURE 11.3 Muscles of the trunk, anterior view

2. **Muscles that move the trunk.** The muscles that move the trunk include the muscles of the thorax, muscles of the abdominal wall, and postural muscles of the back (**Figures 11.3** and **11.4**).

   a. The muscles of the thorax are generally involved in the muscle movements that produce ventilation. The small **pectoralis minor** draws the scapula anteriorly and the rib cage superiorly during forced inspiration and expiration; the **internal** and **external intercostal** muscles are located between the ribs and are involved in both quiet and forced inspiration and expiration; and the circular **diaphragm** is the main muscle for inspiration. Note that these muscles are not visible in Figure 11.3.

   b. The abdominal muscles move the vertebral column and increase intra-abdominal pressure. The **rectus abdominis** is the central and superficial muscle that flexes the vertebral column; the **internal** and **external obliques** are located laterally and rotate the vertebral column; and the deep **transversus abdominis** compresses the abdominal contents to increase intra-abdominal pressure.

   c. There are three postural muscles of the back, all of which are part of the long muscle known as the **erector spinae**. The three components of the erector spinae include the lateral **iliocostalis**, the middle **longissimus**, and the medial **spinalis**.

3. **Muscles that move the shoulder.** Only three muscles are "prime movers" of the shoulder: the **deltoid**, or "shoulder muscle," which is the prime abductor of the shoulder; the **pectoralis minor**, or "chest muscle," which is the prime muscle of shoulder flexion; and the posterior **latissimus dorsi**, which is the prime muscle of shoulder extension. The **serratus anterior**, located deep to the pectoralis major, assists in shoulder abduction and raising the shoulder. The other shoulder muscles, the **rotator cuff muscles**, include the **infraspinatus**, the **subscapularis**, the **supraspinatus**, and the **teres minor**. The tendons of these muscles unite with the fibrous capsule of the shoulder joint, and their main function is to reinforce this capsule to prevent dislocation of the humerus from the scapula.

4. **Muscles that move the forearm and wrist.** The four main muscles that move the forearm are the anterior **biceps brachii**, the anterior and deep **brachialis**, the anterior and lateral **brachioradialis**, and the posterior **triceps brachii**

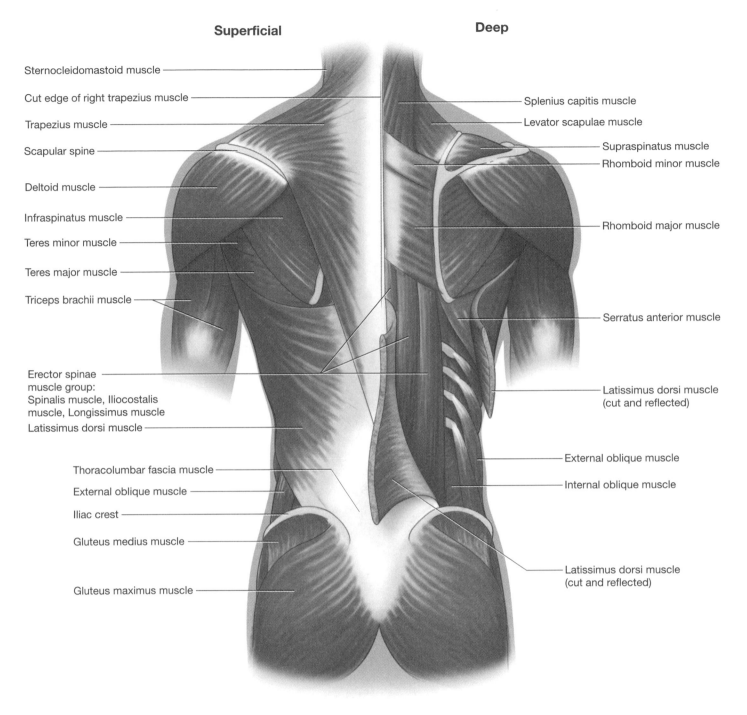

Superficial

Sternocleidomastoid muscle

Cut edge of right trapezius muscle

Trapezius muscle

Scapular spine

Deltoid muscle

Infraspinatus muscle

Teres minor muscle

Teres major muscle

Triceps brachii muscle

Erector spinae
muscle group:
Spinalis muscle, Iliocostalis
muscle, Longissimus muscle

Latissimus dorsi muscle

Thoracolumbar fascia muscle

External oblique muscle

Iliac crest

Gluteus medius muscle

Gluteus maximus muscle

Deep

Splenius capitis muscle

Levator scapulae muscle

Supraspinatus muscle

Rhomboid minor muscle

Rhomboid major muscle

Serratus anterior muscle

Latissimus dorsi muscle
(cut and reflected)

External oblique muscle

Internal oblique muscle

Latissimus dorsi muscle
(cut and reflected)

FIGURE **11.4** Muscles of the trunk, posterior view

(**Figures 11.5** and **11.6**). The first three muscles play a role in forearm flexion, and the triceps brachii extends the forearm. The remainder of the muscles of the upper limb act on the wrist, the hand, and the digits. Muscles that flex the wrist and the digits, located on the anterior forearm, include the **flexor carpi radialis, flexor carpi ulnaris, flexor digitorum superficialis,** and **flexor digitorum profundus** muscles. Muscles that extend the wrist and the digits, located on the posterior forearm, include the **extensor carpi radialis longus, extensor digitorum,** and **extensor carpi ulnaris.**

5. **Muscles that move the hip and knee.** Muscles that move the hip joint exclusively include the deep **iliopsoas** (not visible in Figure 11.5), the anterior **pectineus,** the anterior **adductor group,** and the posterior **gluteus maximus** and **gluteus medius.** The iliopsoas is the prime flexor of the hip joint, and is assisted by the pectineus. As implied by their names, the muscles in the adductor group are the prime adductors of the hip joint. The gluteus maximus is the prime extensor

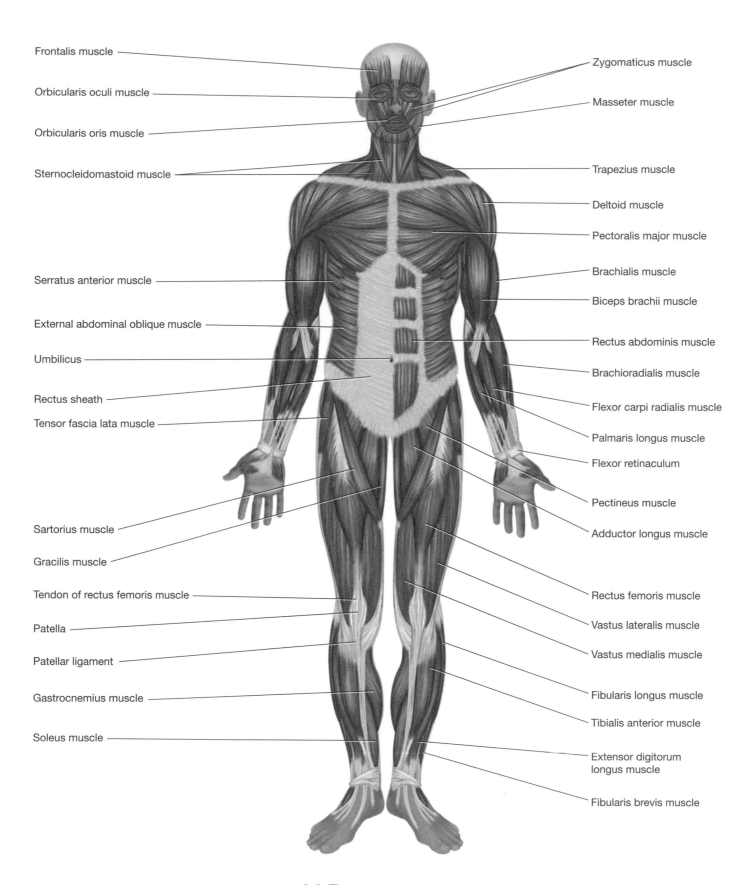

Frontalis muscle

Orbicularis oculi muscle

Orbicularis oris muscle

Sternocleidomastoid muscle

Serratus anterior muscle

External abdominal oblique muscle

Umbilicus

Rectus sheath

Tensor fascia lata muscle

Sartorius muscle

Gracilis muscle

Tendon of rectus femoris muscle

Patella

Patellar ligament

Gastrocnemius muscle

Soleus muscle

Zygomaticus muscle

Masseter muscle

Trapezius muscle

Deltoid muscle

Pectoralis major muscle

Brachialis muscle

Biceps brachii muscle

Rectus abdominis muscle

Brachioradialis muscle

Flexor carpi radialis muscle

Palmaris longus muscle

Flexor retinaculum

Pectineus muscle

Adductor longus muscle

Rectus femoris muscle

Vastus lateralis muscle

Vastus medialis muscle

Fibularis longus muscle

Tibialis anterior muscle

Extensor digitorum longus muscle

Fibularis brevis muscle

FIGURE 11.5 Muscles of the body, anterior view

EXPLORING ANATOMY & PHYSIOLOGY IN THE LABORATORY

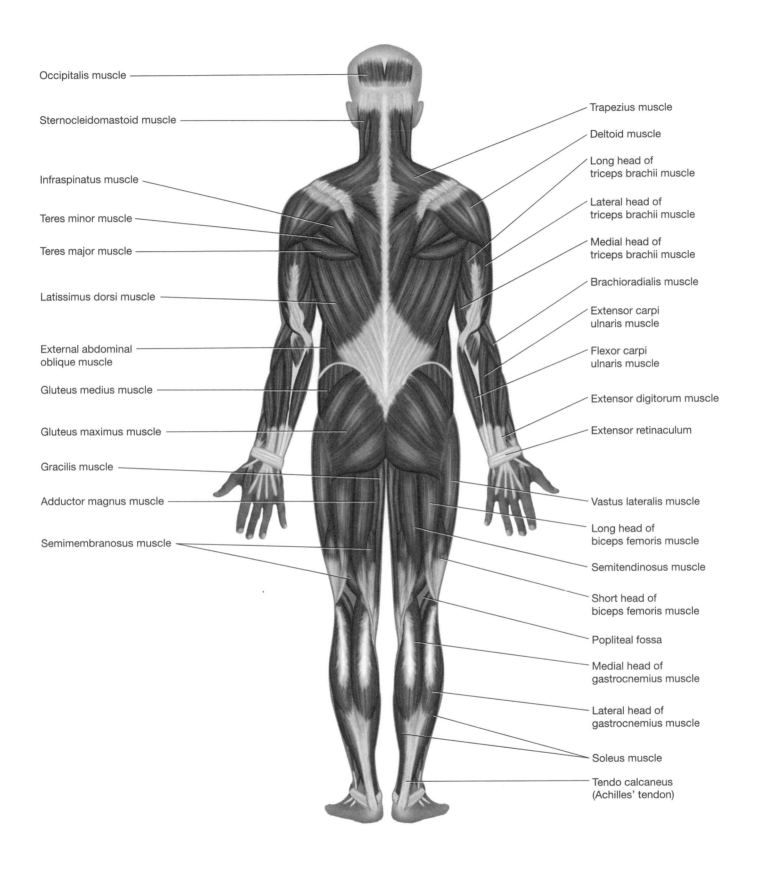

Occipitalis muscle

Sternocleidomastoid muscle

Infraspinatus muscle

Teres minor muscle

Teres major muscle

Latissimus dorsi muscle

External abdominal oblique muscle

Gluteus medius muscle

Gluteus maximus muscle

Gracilis muscle

Adductor magnus muscle

Semimembranosus muscle

Trapezius muscle

Deltoid muscle

Long head of triceps brachii muscle

Lateral head of triceps brachii muscle

Medial head of triceps brachii muscle

Brachioradialis muscle

Extensor carpi ulnaris muscle

Flexor carpi ulnaris muscle

Extensor digitorum muscle

Extensor retinaculum

Vastus lateralis muscle

Long head of biceps femoris muscle

Semitendinosus muscle

Short head of biceps femoris muscle

Popliteal fossa

Medial head of gastrocnemius muscle

Lateral head of gastrocnemius muscle

Soleus muscle

Tendo calcaneus (Achilles' tendon)

FIGURE **11.6** Muscles of the body, posterior view

of the hip joint, and the gluteus medius is a major abductor of the hip joint. The remainder of the thigh muscles generally move both the knee and hip joints. The anterior thigh muscles include the four muscles of the **quadriceps femoris group** (**rectus femoris**, **vastus lateralis**, **vastus intermedius**, and **vastus medialis** muscles), the **sartorius**, and the medial **gracilis**. Most of these muscles have the dual function of flexion at the hip joint and extension at the knee joint. The posterior thigh muscles include the three muscles of the **hamstrings group**: the lateral **biceps femoris** and the medial **semitendinosus** and **semimembranosus** muscles. These three muscles produce both extension at the hip joint and flexion at the knee joint.

6. **Muscles that move the ankle and foot.** The most obvious muscle of the posterior leg is the large **gastrocnemius** muscle, also known as the "calf muscle." This two-headed muscle originates on the distal femur and inserts into the posterior calcaneus via the **calcaneal tendon**, more commonly called the **Achilles' tendon**. Deep to the gastrocnemius is the **soleus** muscle, which unites with the gastrocnemius and contributes to the calcaneal tendon. Together these two muscles are the prime muscles that produce plantarflexion at the ankle joint. The gastrocnemius also produces some flexion at the knee joint. Deep to these two muscles are muscles that plantarflex the foot and the digits, as well as the **tibialis posterior**, which is the main inverter of the foot. On the lateral leg we find two muscles, the **fibularis longus** and **fibularis brevis** that are the main everters of the foot. Anteriorly are the extensors such as the **extensor digitorum longus**, which extends the toes, and the **tibialis anterior**, which dorsiflexes the foot and the ankle joint.

Procedure Model Inventory for the Skeletal Muscles

Identify the following muscles on models and diagrams, using your textbook and this unit for reference. As you examine the anatomical models and diagrams, record on the model inventory in Table 11.3 the name of the model and the structures you were able to identify. When you have completed the activity, answer Check Your Understanding question 1 (p. 257).

Muscles That Move the Head, Neck, and Face

1. Muscles of facial expression
   a. Epicranius m. (frontalis m. and occipitalis m.)
   b. Orbicularis oculi m.
   c. Zygomaticus m.
   d. Buccinator m.
   e. Orbicularis oris m.
2. Muscles of mastication
   a. Temporalis m.
   b. Masseter m.
3. Muscles of head and neck
   a. Platysma m.
   b. Sternocleidomastoid m.
   c. Trapezius m.

Muscles That Move the Trunk

1. Muscles of the thorax
   a. Pectoralis minor m.
   b. External intercostals m.
   c. Internal intercostals m.
   d. Diaphragm m.
2. Muscles of the abdominal wall
   a. Rectus abdominis m.
   b. External obliques m.
   c. Internal obliques m.
   d. Transversus abdominis m.
3. Postural muscles of the back
   a. Erector spinae m.
      (1) Iliocostalis m.
      (2) Longissimus m.
      (3) Spinalis m.

Muscles That Move the Shoulder

1. Muscles of the shoulder
   a. Deltoid m.
   b. Pectoralis major m.
   c. Latissimus dorsi m.
   d. Serratus anterior m.
2. Rotator cuff muscles
   a. Infraspinatus m.
   b. Subscapularis m.
   c. Supraspinatus m.
   d. Teres minor m.

Muscles That Move the Forearm and Wrist

1. Muscles of the arm
   a. Biceps brachii m.
   b. Triceps brachii m.
   c. Brachialis m.
   d. Brachioradialis m.
2. Pronator teres m.
3. Flexor carpi radialis m.
4. Flexor carpi ulnaris m.
5. Flexor digitorum superficialis m.
6. Flexor digitorum profundus m.
7. Extensor carpi radialis longus m.
8. Extensor digitorum m.
9. Extensor carpi ulnaris m.

Muscles That Move the Hip and Knee

1. Muscles of the pelvic girdle
   a. Iliopsoas m. (Iliacus m. and Psoas major m.)
   b. Gluteus maximus m.
   c. Gluteus medius m.
2. Muscles of the thigh
   a. Sartorius m.
   b. Tensor fascia lata m.

   c. Pectineus m.
   d. Quadriceps femoris group
      (1) Rectus femoris m.
      (2) Vastus medialis m.
      (3) Vastus intermedius m.
      (4) Vastus lateralis m.

   e. Adductor group
      (1) Adductor magnus m.
      (2) Adductor longus m.
      (3) Adductor brevis m.
   f. Hamstring muscles
      (1) Biceps femoris m.
      (2) Semimembranosus m.
      (3) Semitendinosus m.

Muscles That Move the Ankle and Foot

1. Gastrocnemius m.
2. Soleus m.
3. Flexor digitorum longus m.

4. Fibularis (peroneus) longus m.
5. Fibularis (peroneus) brevis m.
6. Tibialis posterior m.

7. Tibialis anterior m.
8. Extensor digitorum longus m.

Your instructor may wish to omit certain muscles included above or add muscles not included in these lists. List any additional structures below:

TABLE **11.3**	Model Inventory for Skeletal Muscles
Model/Diagram	**Structures Identified**

Exercise 2

Muscle Origins and Insertions

MATERIALS NEEDED

- Small skeleton
- Modeling clay

Before you can fully understand a muscle's actions, you must understand a muscle's origin and insertion. As we discussed in the introduction to this unit, the **origin** of a muscle is the part from which it originates, which generally is the more stationary part, and the muscle typically **inserts** into the part that it moves. For example, you can see in **Figure 11.7** that the biceps brachii muscle originates on the coracoid process of the scapula and crosses the elbow joint, where it inserts into the radial tuberosity. Notice also that the triceps brachii originates from the humerus and the inferior scapula and crosses the elbow joint to insert into the olecranon process.

Once you determine a muscle's origin and insertion, figuring out its actions becomes easy. Let's examine the biceps brachii and triceps brachii muscles again. The biceps brachii inserts into the radial tuberosity, so we know that it will move the forearm. Given how it crosses the anterior elbow joint, we can conclude that it will cause forearm flexion at the elbow joint. The triceps brachii inserts into the olecranon process, so we know that it also will move the forearm. Given how it crosses the posterior elbow joint, we can conclude that it will cause forearm extension at the elbow joint. Now wasn't that easy?

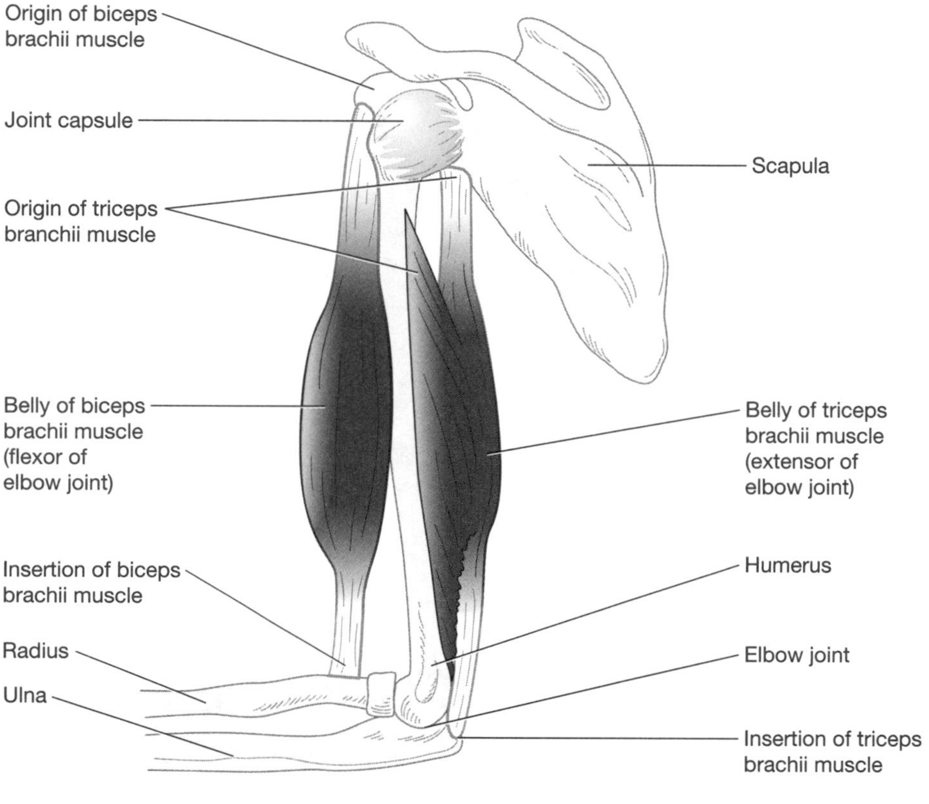

Origin of biceps brachii muscle

Joint capsule

Origin of triceps branchii muscle

Belly of biceps brachii muscle (flexor of elbow joint)

Insertion of biceps brachii muscle

Radius

Ulna

Scapula

Belly of triceps brachii muscle (extensor of elbow joint)

Humerus

Elbow joint

Insertion of triceps brachii muscle

FIGURE 11.7 Origin and insertion of the biceps brachii and triceps brachii, posterior view

 ## Procedure Build Muscles

In this exercise you will use small skeletons and modeling clay to build specific muscle groups. This may sound easy, but there is a catch: You must determine the actions of the muscle by looking only at the origin and insertion of each muscle that you build. Use Pre-Lab Exercise 3 as a guide to each muscle's origin and insertion.

1 Obtain a small skeleton and five colors of modeling clay.

2 Build the indicated muscles, using a different color of clay for each muscle. As you build, pay careful attention to the origin and insertion of each muscle.

3 Determine the primary actions for each muscle that you have built by looking *only* at the origin and insertion. Record this information in Tables 11.4–11.6.

a. Group 1:

TABLE **11.4**	Muscles and Actions for Group 1
Muscle	**Actions**
Biceps femoris m.	
Semitendinosus m.	
Semimembranosus m.	
Gracilis m.	
Adductor group mm.	

b. Group 2:

TABLE **11.5**	Muscles and Actions for Group 2
Muscle	**Actions**
Biceps brachii m.	
Triceps brachii m.	
Brachioradialis m.	
Deltoid m.	
Latissimus dorsi m.	

c. Group 3:

TABLE **11.6**	Muscles and Actions for Group 3
Muscle	**Actions**
Pectoralis major m.	
Pectoralis minor m.	
Trapezius m.	
Masseter m.	
Temporalis m.	

Exercise 3

Muscle Actions

This exercise should look familiar, as you did a similar exercise in Unit 9 (Articulations, p. 195). As before, you will be performing various activities to demonstrate which joints are moving with each activity. In this exercise, however, rather than focusing on the motions that occur at each joint, you will be determining *which muscles* are causing the motions at each joint. Following is an example:

Walking in Place

Part 1

First list all motions occurring at each joint (as before, the joints are listed with their common names, and the lists are certainly *not* all inclusive):

- Vertebral column and neck: extension
- Shoulder: flexion/extension
- Elbow: flexion
- Hip: flexion/extension, abduction/adduction
- Knee: flexion/extension
- Ankle: dorsiflexion/plantarflexion, inversion/eversion

Part 2

Now refer to Table 11.2 in Pre-Lab Exercise 3 to determine which muscles are causing each joint to move. Please note that these lists are far from complete. With nearly 700 muscles in the human body, a complete list could take all day!

- Vertebral column and neck
  - Extension: erector spinae and trapezius mm.
- Shoulder
  - Flexion: pectoralis major m.
  - Extension: latissimus dorsi m.
- Elbow
  - Flexion: biceps brachii, brachialis, and brachioradialis mm.
- Hip
  - Flexion: iliopsoas, sartorius mm.
  - Extension: biceps femoris, semitendinosus, semimembranosus, and gluteus maximus mm.
  - Abduction: tensor fasciae latae, sartorius, gluteus medius mm.
  - Adduction: gracilis, adductor mm.
- Knee
  - Flexion: biceps femoris, semitendinosus, semimembranosus, gastrocnemius mm.
  - Extension: rectus femoris, vastus medialis, vastus intermedius, vastus lateralis mm.
- Ankle
  - Dorsiflexion: tibialis anterior m.
  - Plantarflexion: gastrocnemius, soleus mm.
  - Inversion: tibialis anterior, tibialis posterior mm.
  - Eversion: fibularis brevis and fibularis longus mm.

Okay, so now you're thinking, "Wow, that's a lot of work!" Following are a few hints that will make this task less daunting:

▶ You will notice that the following activities are the same as the activities from Exercise 3 in Unit 9. To save yourself some work, reference this exercise on page 195 to complete part one of this exercise.

▶ As before, perform the indicated activity. If you don't, you'll miss some things!

▶ Unless your instructor asks you to do otherwise, try to keep it simple. List only the main muscles for each action. For example, when listing the muscles to flex the hip, we easily could include 10 different muscles. To make it simpler, list only those muscles we discussed that flex the hip as their *primary action* (as in the example on page 248).

Ready to give it a try?

 Procedure Determine the Muscles Involved in Common Movements

Determine the joints that are being moved for part 1, and then determine the muscles that are moving the joints for Part 2. Refer to Unit 9 (p. 195) for help. After you have completed the activity, answer Check Your Understanding questions 2 through 4 (pp. 257–258).

Climbing the Stairs

Part 1

Part 2

Doing Jumping Jacks

Part 1

Part 2

Answering the Telephone

Part 1

Part 2

Jumping Rope

Part 1

Part 2

EXPLORING ANATOMY & PHYSIOLOGY IN THE LABORATORY

NAME _____

SECTION _____ DATE _____

 Check Your Recall

1 Label the following muscles on **Figure 11.8.**

- Temporalis
- Sternocleidomastoid
- Frontalis
- Masseter
- Orbicularis oris
- Zygomaticus major and minor
- Trapezius
- Orbicularis oculi
- Occipitalis

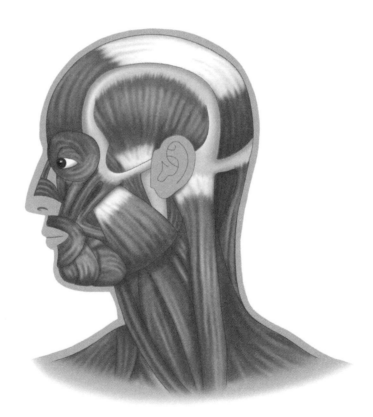

FIGURE **11.8** Facial musculature, lateral view

2 Label the following muscles on **Figure 11.9.**

- Rectus abdominis
- Serratus anterior
- External oblique
- Transversus abdominis
- Internal oblique

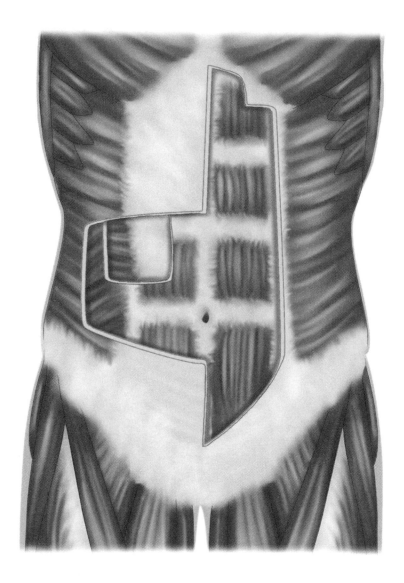

FIGURE **11.9** Muscles of the trunk, anterior view

EXPLORING ANATOMY & PHYSIOLOGY IN THE LABORATORY

3 Label the following muscles on **Figure 11.10.**

- Trapezius
- Latissimus dorsi
- Deltoid
- Erector spinae
- Triceps brachii
- Gluteus medius
- Gluteus maximus
- Supraspinatus
- Infraspinatus

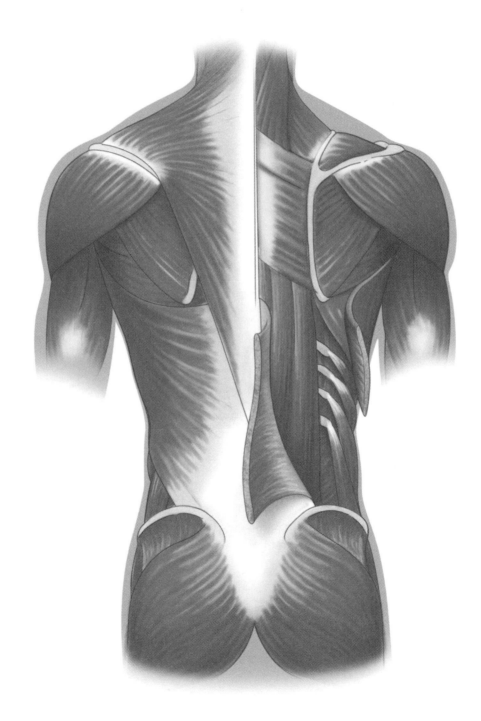

FIGURE **11.10** Muscles of the trunk, posterior view

4 Label the following muscles on **Figure 11.11**.

- Pectoralis major
- Biceps brachii
- Vastus lateralis
- Sartorius
- Brachioradialis
- Deltoid
- Rectus femoris
- Tibialis anterior
- Vastus medialis
- Gracilis

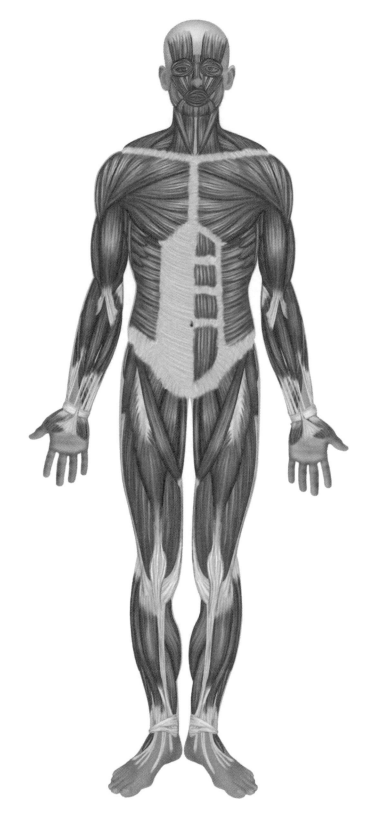

FIGURE 11.11 Muscles of the body, anterior view

EXPLORING ANATOMY & PHYSIOLOGY IN THE LABORATORY

5 Label the following muscles on **Figure 11.12.**
  - Gluteus maximus
  - Semimembranosus
  - Biceps femoris
  - Triceps brachii
  - Gastrocnemius
  - Semitendinosus

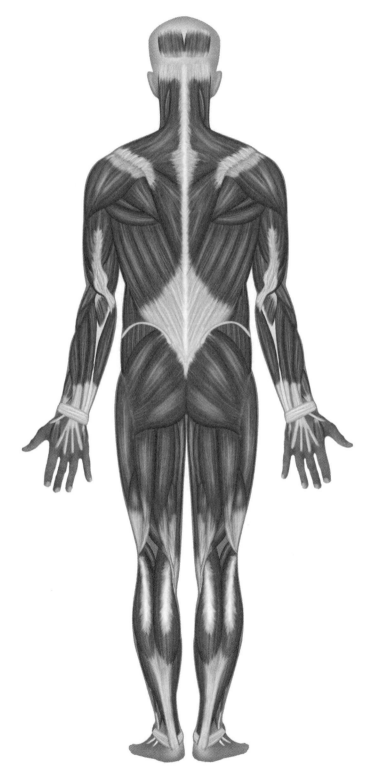

FIGURE **11.12** Muscles of the body, posterior view

6 Fill in the blanks: A muscle's _____ is generally the more stationary part, and a muscle _____ into the part that it moves.

7 Which of the following muscles is *not* a forearm flexor at the elbow?
   a. Brachioradialis
   b. Triceps brachii
   c. Biceps brachii
   d. Brachialis

8 Which of the following muscles does *not* extend the hip and flex the knee?
   a. Sartorius
   b. Biceps femoris
   c. Semimembranosus
   d. Semitendinosus

9 Which of the following muscles is the prime abductor of the arm at the shoulder joint?
   a. Pectoralis major
   b. Latissimus dorsi
   c. Pectoralis minor
   d. Deltoid

10 Which of the following muscles is primarily a postural muscle?
   a. Rectus abdominis
   b. Transversus abdominis
   c. Erector spinae
   d. Latissimus dorsi

Check Your Understanding | **Critical Thinking and Application Questions**

1 Muscle strain, or "pulling a muscle," may result from overuse injuries or from trauma. Typically muscle strain results in pain around the muscle with movement and with pressure. Predict which muscle or muscles may be strained from pain in each of the following locations:

a. Lateral thigh

b. Posterior arm

c. Lateral neck

d. Abdomen just medial to the midline

e. Anterolateral leg

f. Thoracic region just medial to the vertebral column

2 Neuromuscular diseases may lead to a condition called *drop foot*, in which a patient is unable to dorsiflex his or her foot. Which muscle(s) do you think is/are involved in this condition? Which muscles may have to compensate for lack of dorsiflexion during walking? Explain.

3 One of the most commonly performed orthopedic surgeries is rotator cuff repair. Often this involves removing and reattaching the deltoid muscle at its insertion. What functional impairment would be present after this procedure? What exercises could a physical therapist use to recondition the deltoid muscle?

4 Radical mastectomy is a type of surgery for breast cancer that involves removing all of the breast tissue, the lymph nodes, and the pectoralis major muscle. What functional impairment would be present following this surgery? Which other muscles might a physical therapist train to enable the patient to regain some of this lost function?

Nervous Tissue

OBJECTIVES

Once you have completed this unit, you should be able to:

- Describe the microanatomy of nervous tissue and the parts of a synapse.

- Identify structures of the neuron and neuroglial cells.

- Describe the sequence of events of a synapse and a neuronal action potential.

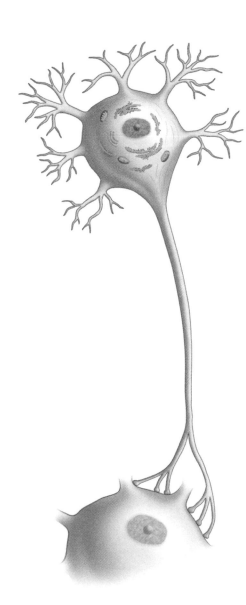

Pre-Lab Exercises
Complete the following exercises prior to coming to lab, using your textbook and lab manual for reference.

Pre-Lab Exercise 1
Key Terms

Table 12.1 lists the key terms with which you should be familiar before coming to lab.

TABLE **12.1**	Key Terms

Term　　　　　　**Definition**

Parts of the Neuron

Neuron　　_____

_____

Cell body　_____

_____

Axon　　　_____

_____

Dendrite　_____

_____

Nissl body　_____

_____

Myelin sheath　_____

_____

Node of Ranvier _____

_____

Neuroglial Cells

Oligodendrocyte _____

_____

Schwann cell　_____

_____

Satellite cell　_____

_____

Astrocyte _____

_____

Microglial cell _____

_____

Ependymal cell _____

_____

Structures of the Synapse

Synapse _____

_____

Axon terminal _____

_____

Synaptic vesicle _____

_____

Synaptic cleft _____

_____

Presynaptic neuron _____

_____

Postsynaptic neuron _____

_____

Neurophysiology Terms

Action potential _____

_____

Depolarization _____

_____

Repolarization _____

_____

Local potential _____

_____

Hyperpolarization _____

_____

Pre-Lab Exercise 2
Nervous Tissue Microanatomy

Label and color the microscopic anatomy of nervous tissue depicted in **Figure 12.1** with the terms from Exercise 1. Use your text and Exercise 1 in this unit for reference.

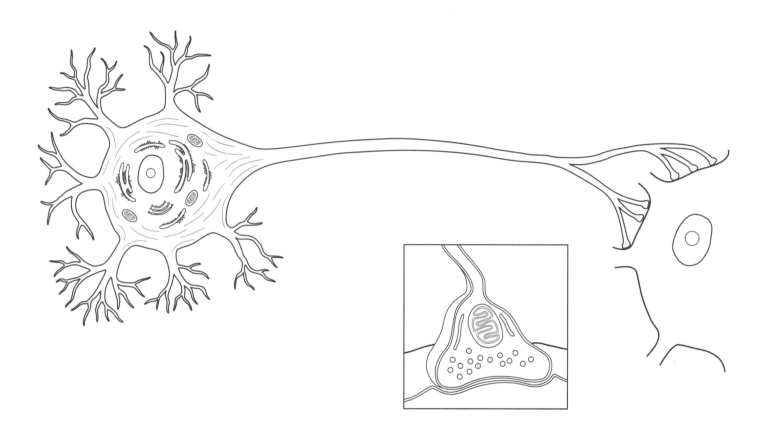

FIGURE **12.1** A multipolar neuron

Exercises

We begin our study of the nervous system in this unit, and we will explore it further over the next three units. The nervous system is one of the major homeostatic systems in the body. It regulates cellular activities by sending nerve impulses, also called **action potentials**, via cells called **neurons**. Neurons are supported by smaller cells called **neuroglial cells**, and together these two cell types make up **nervous tissue**.

In the following exercises you will examine the cells and structure of nervous tissue. Then you will apply what you learned about the structure of these cells to their physiology with a tracing exercise.

 ## Exercise 1

Neurons and Neuroglia

MATERIALS NEEDED

- Neuron models
- Synapse models
- Modeling clay in four colors
- Nervous tissue slide
- Myelin sheath slide
- Light microscope
- Colored pencils

There are two types of cells within nervous tissue: **neurons** and **neuroglial cells** (**Figure 12.2**). Neurons are large cells that transmit and generate messages in the form of nerve impulses or **neuronal action potentials**. Although they vary widely in size and structure, most have the following features in common (**Figure 12.3**):

1. **Cell body.** The cell body is the biosynthetic center of the neuron, containing the nucleus and many of the organelles. Its cytoskeleton consists of densely packed microtubules and **neurofilaments** that compartmentalize the rough endoplasmic reticulum into dark-staining structures called **Nissl bodies**. Cell bodies are often found in clusters. In the central nervous system, clusters of cell bodies are called **nuclei**, and in the peripheral nervous system, such clusters are called **ganglia**.

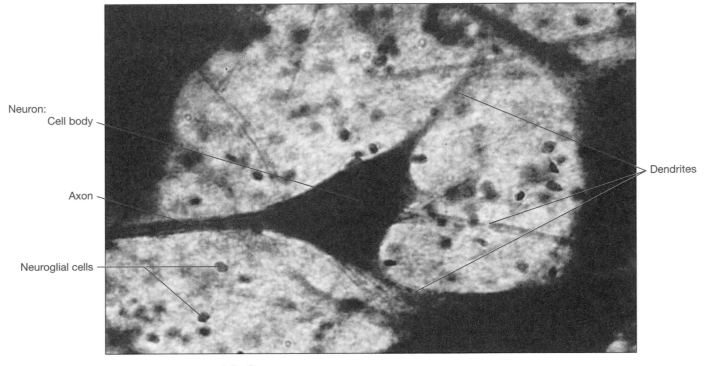

FIGURE 12.2 Nervous tissue, photomicrograph (motor neuron smear)

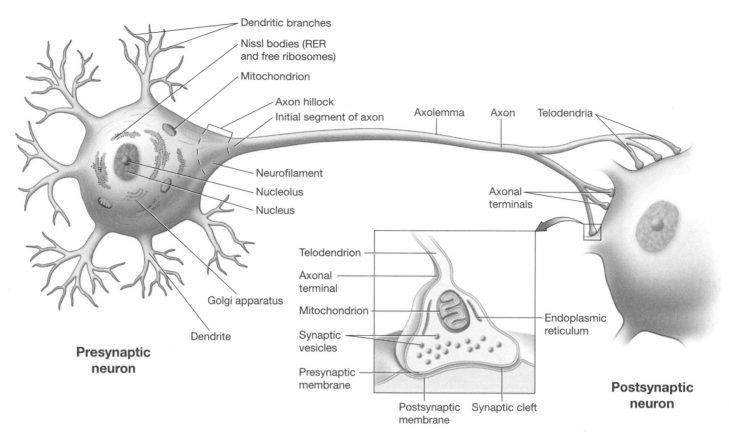

Dendritic branches
Nissl bodies (RER and free ribosomes)
Mitochondrion
Axon hillock
Initial segment of axon
Axolemma
Axon
Telodendria
Neurofilament
Nucleolus
Nucleus
Axonal terminals
Golgi apparatus
Dendrite

Presynaptic neuron

Telodendrion
Axonal terminal
Mitochondrion
Synaptic vesicles
Presynaptic membrane
Endoplasmic reticulum
Postsynaptic membrane
Synaptic cleft

Postsynaptic neuron

FIGURE **12.3** A multipolar neuron

2. **Axon.** A single axon exits the cell body to transmit messages in the form of action potentials to other neurons, muscles, and glands. An axon generates action potentials across its plasma membrane, known as the **axolemma,** at the **axon hillock**. The action potential propagates the length of the axon and down its multiple terminal branches, called **telodendria**. At the end of each telodendrion is an **axon terminal**. Recall from Unit 10 that the axon terminal contains synaptic vesicles with neurotransmitters that send messages to the axon's target.

3. **Dendrites.** Most neurons have one or more branching processes called dendrites, which receive messages from other neurons. They transmit these messages to the neuron's cell body, but they are not capable of generating action potentials.

Neurons are classified structurally on the basis of the number of processes extending from the cell body. The most common type of neuron is the large **multipolar neuron,** which has three or more processes—specifically, one axon and two or more dendrites. Most multipolar neurons resemble highly branched trees and have hundreds to thousands of dendrites.

Both **bipolar neurons** and **pseudounipolar neurons** are rarer than multipolar neurons and are associated with the senses. Bipolar neurons have one axon and one dendrite and are found in special sense organs such as the olfactory epithelium and the retina of the eye. Pseudounipolar neurons have only one process and are found in the skin, where they detect and transmit sensations such as pain and temperature. A short distance from the cell body, the process branches like a "T" into a **central process,** which travels from the cell body to the CNS, and a **peripheral process,** which travels from a sensory receptor to the cell body. Both the central process and the peripheral process can generate action potentials and are therefore considered axons.

Neurons communicate with one another and with their target cells at a junction called a **synapse**. In Unit 10 you learned about a type of synapse called the *neuromuscular junction* (NMJ). The synapse between two neurons is similar to the NMJ and consists of three parts:

1. **Presynaptic neuron.** The neuron that sends the message is called the presynaptic neuron. It contains neurotransmitters packaged in synaptic vesicles in its axon terminals.

EXPLORING ANATOMY & PHYSIOLOGY IN THE LABORATORY

2. **Synaptic cleft.** As we saw in the NMJ, the two neurons in a neuronal synapse are separated by a small space. This space is filled with extracellular fluid through which the neurotransmitters diffuse when they are released from the synaptic vesicles.

3. **Postsynaptic neuron.** The neuron that receives the message is the postsynaptic neuron. Its plasma membrane contains receptors for neurotransmitters. When the neurotransmitters bind to these receptors, the membrane potential of the postsynaptic neuron changes.

Neuroglial cells are much smaller than neurons, and they outnumber neurons about 50 to 1—no small feat considering that the nervous system contains about a trillion neurons! Neuroglial cells play a variety of roles, including the following:

1. **Schwann cells.** Schwann cells form a structure called the **myelin sheath** around the axons of certain neurons in the PNS (**Figures 12.4** and **12.5**). As you can see in **Figure 12.4A**, Schwann cells can myelinate only one axon, and they do so by wrapping clockwise around the axon. The outer edge of the Schwann cell, called the **neurilemma**, contains most of its cytoplasm and the nucleus. The myelin sheath protects and insulates the axons and speeds up conduction of action potentials. Because the sheath is made up of individual neuroglial cells, there are small gaps between the cells where the plasma membrane of the axon is exposed. These gaps are called **nodes of Ranvier**, and the myelin-covered segments between the nodes are called **internodes**.

2. **Satellite cells.** Satellite cells surround the cell bodies of neurons in the PNS. These cells are believed to enclose and support the cell bodies, although their precise function is unknown.

3. **Astrocytes.** Astrocytes are the most numerous neuroglial cell type in the CNS. These star-shaped cells have many functions, including anchoring neurons and blood vessels in place and regulating the extracellular environment of the brain. In addition, they facilitate the formation of the **blood-brain-barrier**, which is created by tight junctions in the brain capillaries that prevent many substances in the blood from entering the brain tissue.

4. **Oligodendrocytes.** Oligodendrocytes, shown in **Figure 12.4B**, have long extensions that wrap around the axons of certain neurons in the CNS to form the myelin sheath. Note that one oligodendrocyte can myelinate several axons.

5. **Microglial cells.** The small microglial cells are very active phagocytes that clean up debris surrounding the neurons.

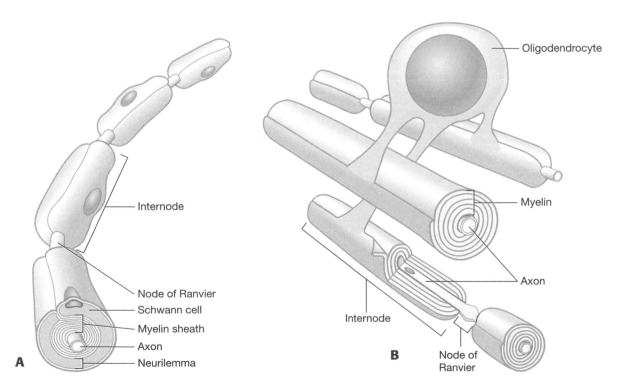

FIGURE 12.4 The myelin sheath: (A) Schwann cells around a PNS axon; (B) an oligodendrocyte around multiple CNS axons

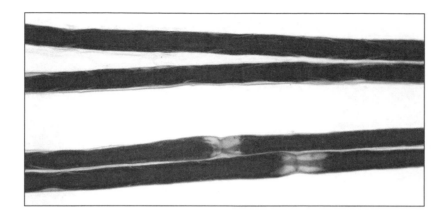

FIGURE **12.5** Myelin sheath, photomicrograph

6. **Ependymal cells.** The ciliated ependymal cells line the hollow spaces of the brain and spinal cord. They assist in forming the fluid that bathes the brain and spinal cord, **cerebrospinal fluid,** and circulate it with their cilia.

 ## Procedure Model Inventory

Identify the following structures of nervous tissue on models and diagrams using your textbook and this unit for reference. As you examine the anatomical models and diagrams, record on the model inventory in Table 12.2 the name of the model and the structures you were able to identify. When you have completed the activity, answer Check Your Understanding question 1 (p. 275).

1. Neuron
   a. Axon
      (1) Axolemma
      (2) Axon hillock
      (3) Telodendria
   b. Dendrite(s)
   c. Cell body (soma)
   d. Nissl bodies
   e. Neurofilaments

2. Neuroglial cells
   a. Oligodendrocytes
   b. Schwann cells
   c. Satellite cells
   d. Microglial cells
   e. Astrocytes
   f. Ependymal cells

3. Other structures
   a. Myelin sheath
   b. Node of Ranvier

   c. Nucleus
   d. Ganglion

4. Parts of the synapse
   a. Presynaptic neuron
   b. Axon terminal
   c. Synaptic vesicle
   d. Synaptic cleft
   e. Postsynaptic neuron
   f. Neurotransmitter receptors

TABLE **12.2** Model Inventory for Nervous Tissue	
Model/Diagram	**Structures Identified**

Procedure Building a Myelin Sheath

In this exercise you will demonstrate the difference between the methods by which Schwann cells and oligodendrocytes myelinate the axons of neurons in the peripheral nervous system and the central nervous system, respectively. Both types of cells create the myelin sheath by wrapping themselves around the axon repeatedly (as many as 100 times). Recall, however, that Schwann cells can myelinate only one axon, while oligodendrocytes send out "arms" to myelinate several axons in its vicinity. When you have completed the activity, answer Check Your Understanding question 2 (p. 275).

1 Obtain four colors of modeling clay (blue, green, yellow, and red, if available).

2 Use the following color code to build three central nervous system (CNS) axons, one peripheral nervous system (PNS) axon, one oligodendrocyte, and two Schwann cells:

   a. CNS axons: blue
   b. PNS axon: green
   c. Oligodendrocyte: yellow
   d. Schwann cells: red

3 Build your oligodendrocyte so it reaches out to myelinate the three CNS axons.

4 Build your Schwann cells so they myelinate the single PNS axon, making sure to leave a gap for the node of Ranvier.

 Procedure Microscopy of Nervous Tissue

Let's now identify some of the structures from the previous procedures on microscope slides. You will examine two tissue sections:

1. **Motor neuron smear.** On this slide you should see many multipolar neurons surrounded by many much smaller neuroglial cells (they may look like purple dots; refer to **Figure 12.2**). As you scan the slide with the low power objective, find a well-stained neuron and move the objective lens to high power. Look for Nissl bodies, and try to identify the single axon and the axon hillock. (This may be difficult, so don't get discouraged if all of the processes look similar.)

2. **Myelin sheath.** A longitudinal section of axons stained with a stain specific for the components of myelin (often osmium) is useful for seeing the sheath itself and also the nodes of Ranvier (refer back to **Figure 12.5**). You will likely need to use the high power objective to see the nodes.

Obtain prepared slides of a motor neuron smear and the myelin sheath. Use your colored pencils to draw what you see, and label your drawing with the following terms.

1. Nervous Tissue
   a. Cell body
   b. Axon
   c. Dendrites
   d. Neuroglial cells

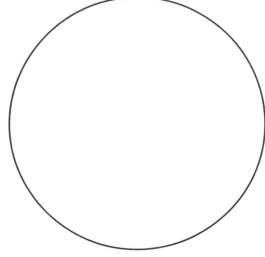

2. Myelin Sheath

    a. Axon with myelin sheath

    b. Node of Ranvier

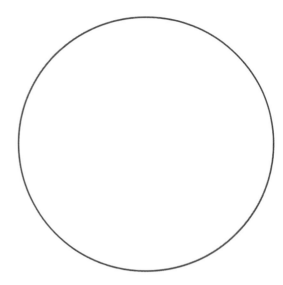

 Exercise 2

Nervous Tissue Physiology

We introduced some of the basic principles of nervous system physiology and electrophysiology in Unit 10, with muscle tissue physiology, and also earlier in this unit. In Unit 10 you learned that muscle fibers have excitable plasma membranes that can be stimulated by a neuron to elicit a muscle fiber action potential. Neurons also have excitable plasma membranes, and the sequence of events of a neuronal action potential closely resembles the sequence of events of a muscle fiber action potential. In addition, like a muscle fiber, a neuron does not generate action potentials spontaneously—postsynaptic neurons must be stimulated first by presynaptic neurons.

Despite these similarities, the physiology of the two cell types has key differences. One key difference between the two is that stimulation of a neuron doesn't always lead to an action potential. Recall that two neurons communicate at a synapse, and that the presynaptic neuron releases neurotransmitters that bind to receptors on the postsynaptic membrane. These neurotransmitters cause one of two effects: (1) it may be excitatory and cause a small depolarization in the postsynaptic membrane that makes an action potential more likely, or (2) it may be inhibitory and cause a small **hyperpolarization** in the postsynaptic membrane that makes an action potential less likely. Both effects are termed **local potentials**.

If a postsynaptic neuron is stimulated to depolarize enough times by one or more presynaptic neurons, these small depolarizations will **summate** and spread through the neuron's plasma membrane. When the depolarizations spread to the axon hillock and reach sufficient magnitude, a value called **threshold**, an action potential (or *nerve impulse*) is generated. The steps of a neuronal action potential are as follows:

1. Voltage-gated sodium channels in the axolemma open and sodium ions flood the cell, causing depolarization.

2. Voltage-gated sodium channels close and voltage-gated potassium channels open.

3. Potassium ions exit the cell, causing repolarization.

4. Potassium ions continue to exit the cell even after the cell returns to its resting membrane potential, causing hyperpolarization (the cell becomes more negative than at rest).

Once the action potential is generated at the axon hillock, it is **propagated** down the length of the axon to its axon terminals in much the same way that electricity is conducted along a wire. The action potential triggers exocytosis of synaptic vesicles at the axon terminals, and neurotransmitter molecules are released into the synaptic cleft. The effects of these neurotransmitters will depend upon the type of neurotransmitter that is released and the type of target cell (another neuron, a muscle cell, or a gland cell). If the target cell is another postsynaptic neuron, the neurotransmitters will cause a local potential in the postsynaptic neuron's membrane.

Procedure Time to Trace

Let's now trace the entire series of events through a three-neuron loop, including the steps of synaptic transmission between neurons 1 and 2, the steps of the action potential that occur in neuron 2, the propagation of the action potential down the axon of neuron 2, and the steps of synaptic transmission that occur between neuron 2 and 3. Hints are included to ensure that you stay on the right track as you trace. When you have completed the activity, answer Check Your Understanding questions 3 and 4 (p. 276).

Start: neuronal action potential reaches the axon terminal of neuron 1 →

_____ →

_____ →

neurotransmitters bind to the postsynaptic membrane of neuron 2 and generate an excitatory local potential →

_____ →

_____ →

axon hillock is depolarized to threshold →

_____ →

_____ →

_____ →

_____ →

the action potential is propagated along the axon →

_____ →

_____ →

_____ →

neurotransmitters cause an excitatory or inhibitory local potential in the postsynaptic membrane of neuron 3 **End**

Procedure Diagramming Your Tracing

Take the pathway that you just traced and turn it into a diagram that contains the three neurons. Your diagram does not have to be a work of art—a simple schematic representation of each neuron and each process is all that is necessary. You may use the figures in this unit and your text for reference, but it is best to compose and organize your diagram in a way that makes sense to you so you may return to it for study purposes.

NAME _____

SECTION _____ DATE _____

 Check Your Recall

1 Label the following structures on **Figure 12.6**.

- Axon
- Dendrites
- Cell body
- Axon hillock

- Axon terminals
- Telodendria
- Nissl bodies
- Presynaptic neuron

- Synaptic cleft
- Postsynaptic neuron
- Synaptic vesicles

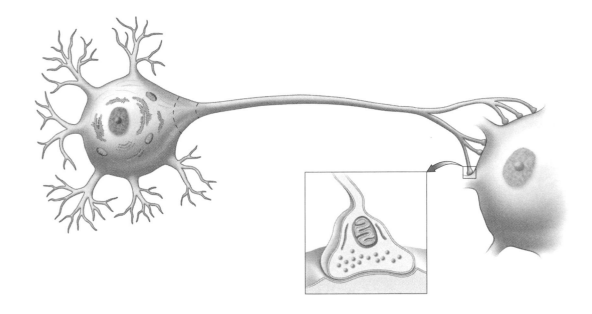

FIGURE **12.6** A neuron

2 The neuron pictured in number one is a

a. pseudounipolar neuron.
b. bipolar neuron.
c. multipolar neuron.
d. unipolar neuron.

3 Matching: Match the neuroglial cell with its correct function.

Oligodendrocytes   _____   A. Create the myelin sheath in the PNS

Astrocytes   _____   B. Ciliated cells in the CNS that form and circulate cerebrospinal fluid

Microglial cells   _____   C. Surround the cell bodies of neurons in the PNS

Schwann cells   _____   D. Anchor neurons and blood vessels, maintain extracellular environment around neurons, assist in the formation of the blood-brain-barrier

Satellite cells   _____

Ependymal cells   _____   E. Phagocytic cells of the CNS

F. Form the myelin sheath in the CNS

4 What is the function of the myelin sheath?

5 Label the following structures on **Figure 12.7**.

- Myelin
- Oligodendrocyte
- Schwann cell
- Neurilemma
- Node of Ranvier
- Internode
- Axon

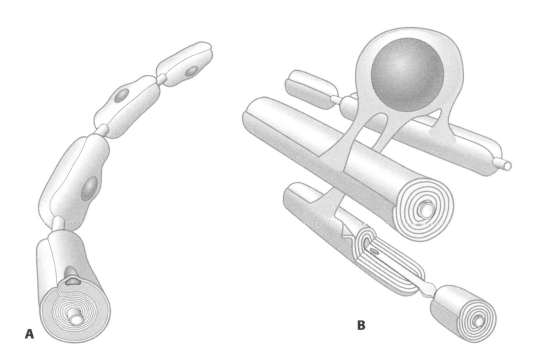

A

B

FIGURE **12.7** The myelin sheath

6 Which part of the neuron is capable of generating an action potential?

a. Dendrite
b. Axon
c. Myelin sheath
d. Cell body

7 Put the following events in the correct order, placing number 1 next to the first event, number 2 next to the second event, and so on.

_____ Voltage-gated sodium channels close and voltage-gated potassium channels open.

_____ Local depolarizations summate at the axon hillock and threshold is reached.

_____ Potassium ions continue to exit the cell, causing hyperpolarization.

_____ Voltage-gated sodium channels open and the axon depolarizes.

_____ Potassium ions exit the cell, causing repolarization.

8 Neurotransmitters that bind the postsynaptic membrane generally generate a/an

a. local potential.
b. action potential.
c. resting membrane potential.
d. pacemaker potential.

9 Where are synaptic vesicles located?

a. Axon terminals
b. Dendrites
c. Cell body
d. Both a and b are correct.
e. All of the above.

10 What triggers exocytosis of synaptic vesicles?

a. Arrival of a local potential at the cell body
b. Hyperpolarization of the postsynaptic membrane
c. Neurotransmitters binding to the postsynaptic membrane
d. Arrival of an action potential at the axon terminal

Check Your Understanding | Critical Thinking and Application Questions

1 Neurons are amitotic, which means that, after a certain stage, they do not divide further. Cancerous cells are characterized by a rapid rate of mitosis. Considering this, of which cell types (neurons or neuroglia) must brain tumors be composed? Why?

2 Multiple sclerosis is a *demyelinating* disease, in which the patient's immune system attacks and destroys the myelin sheath in the central nervous system. What types of symptoms would you expect from such a disease? Why? Would Schwann cells or oligodendrocytes be affected?

3 Local anesthetic drugs such as lidocaine block sodium channels in the neuron plasma membrane. What effect would this have on action potential generation and conduction?

4 Certain inhaled anesthetic agents are thought to open Cl^- channels in the membranes of postsynaptic neurons in the brain, an effect that causes inhibitory local potentials. Why might this action put a person "to sleep" during anesthesia?

Central Nervous System

OBJECTIVES

Once you have completed this unit, you should be able to:

- Describe the gross anatomy of the brain.

- Identify structures of the brain.

- Identify structures of the spinal cord.

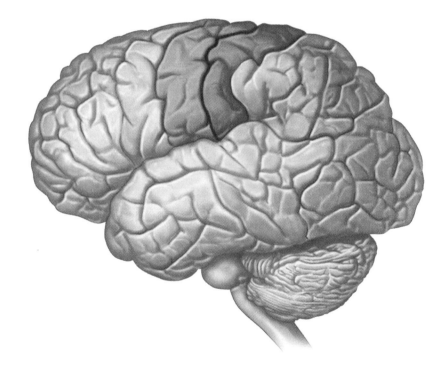

Pre-Lab Exercises
Complete the following exercises prior to coming to lab, using your textbook and lab manual for reference.

Pre-Lab Exercise 1
Key Terms

Table 13.1 lists the key terms with which you should be familiar before coming to lab.

TABLE **13.1**	Key Terms

Term **Definition**

Structures of the Brain

Cerebral hemispheres _____

_____

Cerebral lobes: frontal lobe, parietal lobe, temporal lobe, occipital lobe, insula _____

_____

_____

Cerebral cortex _____

_____

Basal nuclei _____

_____

Corpus callosum _____

_____

Diencephalon _____

_____

Thalamus _____

_____

Hypothalamus _____

_____

Epithalamus _____

_____

Brainstem _____

_____

EXPLORING ANATOMY & PHYSIOLOGY IN THE LABORATORY

Midbrain _____

_____

Pons _____

_____

Medulla oblongata _____

_____

Cerebellum _____

_____

Vermis _____

_____

Arbor vitae _____

_____

Dura mater _____

_____

Arachnoid mater _____

_____

Pia mater _____

_____

Ventricles _____

_____

Dural sinuses _____

_____

Structures of the Spinal Cord

Gray matter horns: anterior, lateral, and posterior horns _____

_____

_____

Funiculi: anterior, lateral, and posterior funiculi _____

_____

Nerve roots: anterior (ventral) root and posterior (dorsal) root _____

_____

Conus medullaris _____

_____

Filum terminale _____

_____

Cauda equina _____

_____

Central canal _____

_____

Pre-Lab Exercise 2

Brain Anatomy

Label and color the diagrams of the brain depicted in Figures 13.1, 13.2, 13.3, and 13.4 with the terms from Exercise 1.
Use your text and Exercise 1 in this unit for reference.

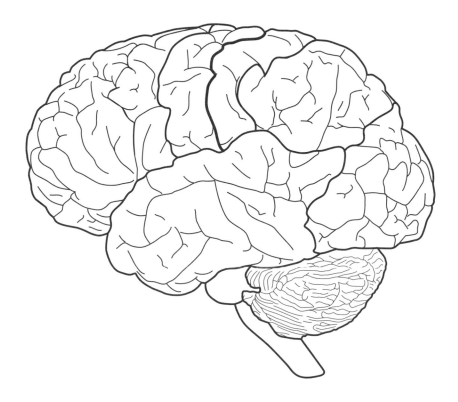

FIGURE **13.1** Brain, lateral view

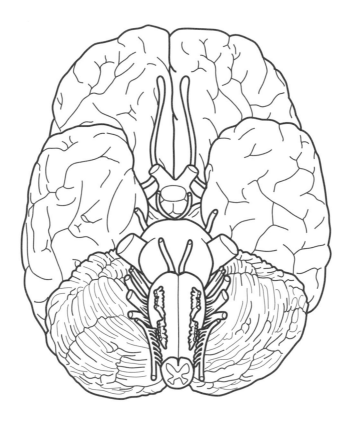

FIGURE **13.2** Brain, inferior view

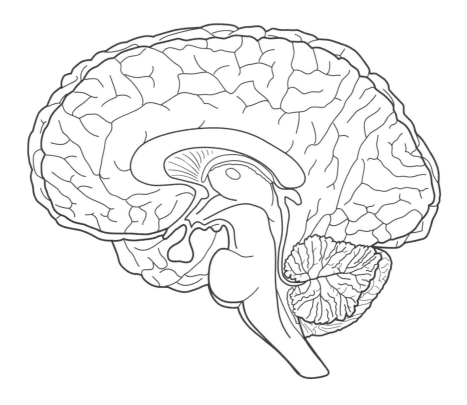

FIGURE **13.3** Brain, midsagittal section

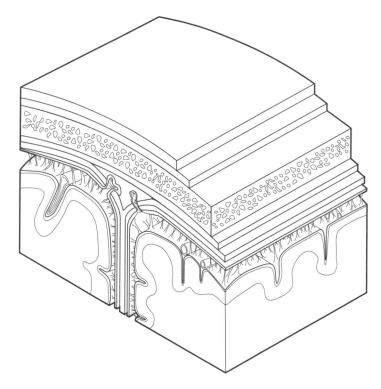

FIGURE **13.4** Brain and meninges, frontal and parasagittal section

☑ Pre-Lab Exercise 3
Spinal Cord Anatomy

Label and color the diagrams of the spinal cord depicted in **Figures 13.5** and **13.6** with the terms from Exercise 2. Use your text and Exercise 2 in this unit for reference.

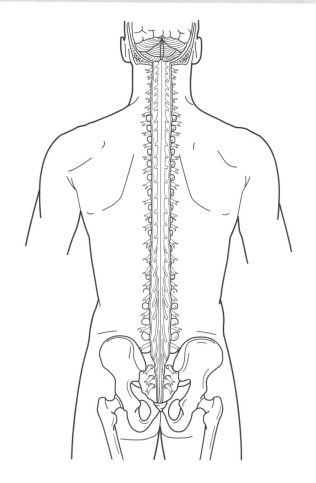

FIGURE **13.5** Brain and spinal cord, posterior view

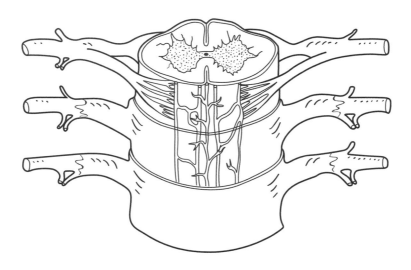

FIGURE **13.6** Spinal cord, transverse section

Exercises

The **central nervous system** (**CNS**) consists of two major parts: the **brain** and the **spinal cord**. The anatomy of the brain and the spinal cord may seem complex, but really they simply consist of a hollow, folded tube with an enlarged end (the brain). Together these two organs control many of our homeostatic functions, our sensation, our movement, and our higher brain functions. The anatomy of both organs will be explored in the three exercises in this unit.

Exercise 1

Anatomy of the Brain

MATERIALS NEEDED

- Brain models: whole and sectioned
- Ventricle models
- Brainstem models
- Dural sinus model

The brain (**Figure 13.7**) is divided into four regions: the **cerebral hemispheres** (collectively called the **cerebrum**), the **diencephalon**, the **brainstem**, and the **cerebellum**. As just discussed, the brain is a hollow organ with spaces called **ventricles**, filled with a fluid similar to plasma called **cerebrospinal fluid** (**CSF**). As you can see in **Figure 13.8**, the largest of the ventricles, called the **lateral ventricles**, are located in the right and left cerebral hemispheres. Note in **Figure 13.8B** that the lateral ventricles resemble rams' horns when viewed from the anterior side. The smaller **third ventricle** is housed within the diencephalon. It is continuous with the **fourth ventricle**, found in the brainstem, by a small canal called the **cerebral aqueduct**. The fourth ventricle is continuous with a canal that runs down the central spinal cord called the **central canal**.

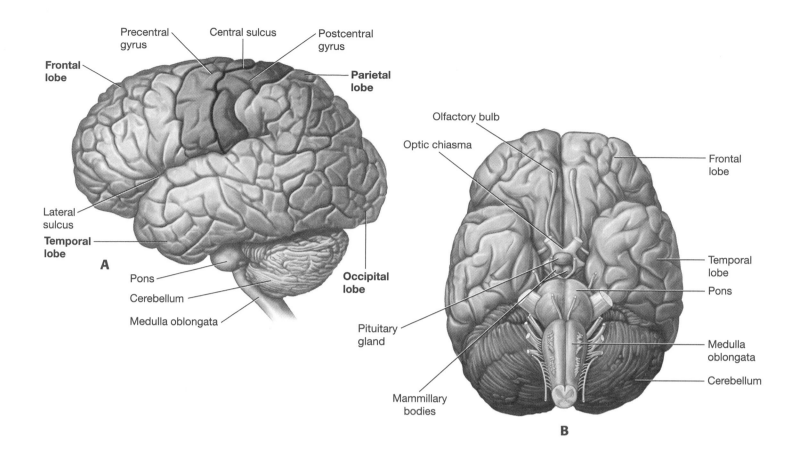

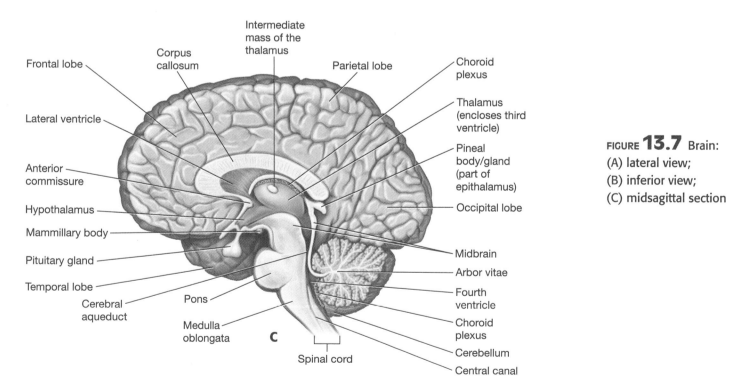

FIGURE **13.7** Brain:
(A) lateral view;
(B) inferior view;
(C) midsagittal section

Recall from Unit 12 that the ventricles are lined by neuroglial cells called ependymal cells, whose cilia beat to circulate CSF. One of the main functions of CSF is to reduce brain weight (the brain is buoyant in the CSF). Without CSF, your brain literally would crush itself under its own weight!

The cerebrum is the most superior portion of the brain. Its surface consists of elevated ridges called **gyri** and shallow grooves called **sulci**. Deep grooves, called **fissures**, separate major regions of the cerebral hemispheres. The **longitudinal**

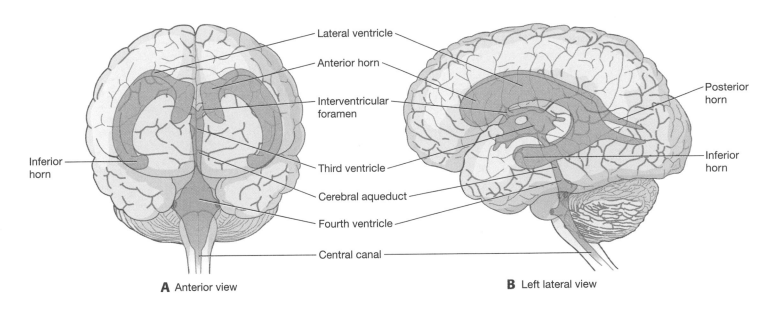

A Anterior view

B Left lateral view

FIGURE **13.8** Ventricles: (A) anterior view; (B) lateral view

fissure separates the right and left hemispheres. The cerebrum consists of five lobes: the **frontal, parietal, temporal, occipital,** and deep **insula** lobes (remember this last one by the mnemonic "the *insula* is *insula*ted"). The cerebral hemispheres are responsible for the brain's cognitive functions, including learning and language, conscious interpretation of sensory information, conscious planning of movement, and personality.

The cell bodies and unmyelinated axons and dendrites of the cerebral neurons lie in the cerebrum's outer 2 millimeters in a region called the **cerebral cortex**. These portions of the neurons are unmyelinated, which gives the cerebral cortex a gray color and for this reason it is called **gray matter**. The cell bodies and processes of the cerebral cortex communicate with other parts of the nervous system by bundles of myelinated axons called **white matter**. The largest tract of cerebral white matter is called the **corpus callosum,** and it connects the right and left cerebral hemispheres.

Gray matter isn't confined to the cerebral cortex. Clusters of cell bodies called **nuclei** are found throughout the cerebrum. An important group of nuclei, the **basal nuclei,** monitor voluntary motor functions. The neurons of these nuclei are connected to other parts of the nervous system by various tracts of white matter in the cerebrum.

Deep to the cerebral hemispheres in the central core of the brain we find the **diencephalon,** which is composed of three main parts:

1. **Thalamus**. The thalamus is a large egg-shaped mass of gray and white matter that comprises 80% of the diencephalon. It is a major integration and relay center that edits and sorts information going into and out of the cerebrum. It essentially functions as the "gateway" into and out of the cerebrum.

2. **Hypothalamus**. The hypothalamus is located on the anterior and inferior aspect of the diencephalon. It is a deceptively small structure that contains the nuclei whose neurons carry out many of the body's homeostatic functions including: helping to regulate the endocrine system; monitoring the sleep–wake cycle; controlling thirst, hunger, and body temperature; and helping to monitor the autonomic nervous system (good things do come in small packages, after all!). An endocrine organ called the **pituitary gland** is connected to the hypothalamus by a stalk called the **infundibulum.**

3. **Epithalamus**. The epithalamus is located on the posterior and superior aspect of the diencephalon. It contains an endocrine organ called the **pineal gland,** which secretes the hormone **melatonin** that helps to regulate the sleep-wake cycle.

The third major portion of the brain, the **brainstem,** influences the automatic functions of the body, such the rhythm for breathing, heart rate, blood pressure, and certain reflexes. The most superior portion of the brainstem is the **midbrain,** and inferior to it we find the rounded **pons** that bulges anteriorly. The last segment of the brain stem, the **medulla oblongata** (or simply *medulla*), is continuous inferiorly with the spinal cord.

The fourth major component of the brain is the large posterior **cerebellum**. It consists of two highly convoluted lobes connected by a piece called the **vermis**. Like the cerebral hemispheres, the cerebellum has an outer **cerebellar cortex** composed of gray matter and inner white matter. The cerebellar white matter is called the **arbor vitae** because of its resemblance to the branches of a tree. The cerebellum coordinates and plans ongoing motor activities and is critical in reducing and preventing motor error with movement.

Note in **Figure 13.9** that a set of three membranes, collectively called the **meninges** (singular – *meninx*), surrounds the brain. The meninges include the following:

1. **Dura mater.** The outermost meninx is the thick, leathery, double-layered dura mater. The superficial dural layer is fused to the skull, and the deeper layer is continuous with the dura mater of the spinal cord. The two layers of the dura are fused, but in three regions the deep layer separates from the superficial layer and dives into the brain to form three structures:

   a. the **falx cerebri**, which forms a partition between the right and left cerebral hemispheres,

   b. the **falx cerebelli**, which separates the two cerebellar hemispheres, and

   c. the **tentorium cerebelli**, which separates the cerebrum from the cerebellum.

   At these locations there are spaces between the two dural layers collectively called the **dural sinuses**. All deoxygenated blood from the brain drains into the dural sinuses, which in turn drain into veins exiting the head and neck.

2. **Arachnoid mater.** The middle meninx, the arachnoid mater, is separated from the dura by a space called the **subdural space**. This space contains little bundles of the arachnoid mater called the **arachnoid granulations**, which are vascular structures that project into the dural sinuses and allow CSF to reenter the blood.

3. **Pia mater.** The thinnest, innermost meninx is the pia mater. The pia mater clings to the surface of the cerebral hemispheres and is richly supplied with blood vessels. A space between the pia mater and the arachnoid mater, called the **subarachnoid space**, is filled with CSF.

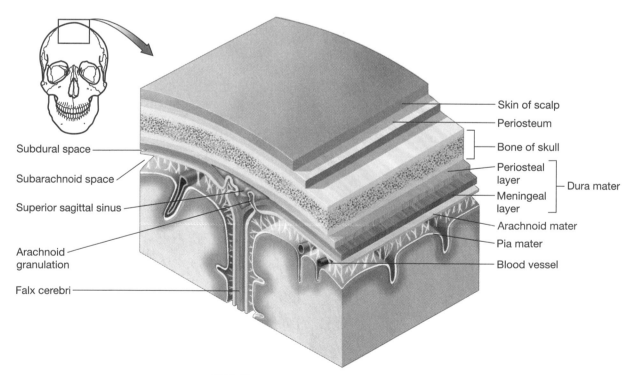

FIGURE 13.9 Brain and meninges, frontal section

EXPLORING ANATOMY & PHYSIOLOGY IN THE LABORATORY

Procedure Model Inventory for the Brain

Identify the following structures of the brain on models and diagrams, using your textbook and this unit for reference. As you examine the anatomical models and diagrams, record on the model inventory in Table 13.2 the name of the model and the structures you were able to identify. The brain's structure is fairly complex, and it's best to examine models in as many different planes of section as possible. When you have completed the activity, answer Check Your Understanding questions 1 through 3 (pp. 301–302).

1. Cerebral hemispheres
2. Corpus callosum (cerebral white matter)
3. Cerebral cortex (gray matter)
4. Basal nuclei (these will only be visible on certain sections of the brain)
5. Lobes of the cerebrum
   a. Frontal lobe
   b. Parietal lobe
   c. Occipital lobe
   d. Temporal lobe
   e. Insula lobe
6. Fissures
   a. Longitudinal fissure
   b. Transverse fissure
7. Sulci
   a. Central sulcus
      (1) Pre-central gyrus
      (2) Post-central gyrus
   b. Lateral sulcus
   c. Parieto-occipital sulcus
8. Diencephalon
   a. Thalamus
   b. Hypothalamus
      (1) Infundibulum
      (2) Pituitary gland
   c. Epithalamus
   d. Pineal gland

9. Brainstem
   a. Midbrain
   b. Pons
   c. Medulla oblongata
10. Cerebellum
   a. Vermis
   b. Arbor vitae
11. Brain coverings
   a. Dura mater
      (1) Subdural space
      (2) Falx cerebri
      (3) Falx cerebelli
      (4) Tentorium cerebelli
   b. Arachnoid mater
      (1) Subarachnoid space
   c. Pia mater
12. Ventricles
   a. Lateral ventricles
   b. Third ventricle
   c. Fourth ventricle
   d. Cerebral aqueduct
13. Vascular structures
   a. Dural sinuses
      (1) Superior sagittal sinus
      (2) Transverse sinus
      (3) Cavernous sinus
   b. Choroid plexus
   c. Arachnoid granulations

| TABLE **13.2** | Model Inventory for the Brain |

Model	Structures Identified

 Exercise 2

The Spinal Cord

MATERIALS NEEDED

> Spinal cord models: whole and sectioned

The medulla oblongata passes through the foramen magnum of the occipital bone and becomes the **spinal cord** (**Figure 13.10**). Note in **Figure 13.10A** that the spinal cord does not extend the entire length of the vertebral column; rather, it ends between the first and second lumbar vertebra. At this point it tapers to form the **conus medullaris**, which gives off a tuft of nerve roots called the **cauda equina** (meaning "horse's tail"). The cauda equina fills the remainder of the vertebral column to the sacrum and exits out of the appropriate foramina to become spinal nerves.

The cranial meninges are continuous with spinal meninges and are similar in name and structure (**Figure 13.10B** and **13.11**). The CSF that surrounds the brain and fills the spaces between the cranial meninges also fills the spaces between the spinal meninges. But there are a few notable differences between the cranial and spinal meninges. For one, the spinal dura mater consists of only *one* layer rather than two like the cranial dura. This single dural layer does not attach to the vertebral column, which creates a space between the spinal dura and the interior vertebral foramen called the **epidural space**. Because the cranial dura is fused to the interior of the skull, there is no epidural space around the brain.

Another difference between the cranial and spinal meninges is the presence of small extensions of pia mater in the spinal cord called **denticulate ligaments**. These tiny ligaments secure the spinal cord to the vertebral column. In addition, the pia mater continues long after the spinal cord ends, and forms a long, fibrous extension called the **filum terminale**, which eventually attaches to the coccyx.

Internally, the spinal cord consists of a butterfly-shaped core of gray matter, which surrounds the CSF-filled **central canal** (**Figure 13.11**). The gray matter is divided into regions, or **horns**. Anteriorly are the **anterior** (or *ventral*) **horns**, which contain the cell bodies of motor neurons. The axons of the neurons of the anterior horn exit the spinal cord and form the **ventral root**, which eventually becomes part of a spinal nerve. On the posterior side are the **posterior** (or *dorsal*) **horns**, which contain the cell bodies of sensory neurons. These neurons receive input from sensory axons in the peripheral nervous system. These sensory axons are found in bundles called the **posterior** (or *dorsal*) **root** (also part of the spinal

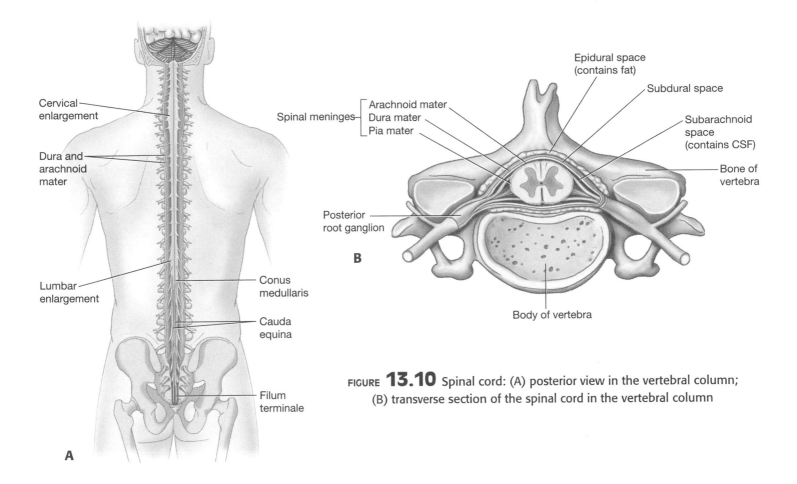

Cervical enlargement

Dura and arachnoid mater

Lumbar enlargement

Conus medullaris

Cauda equina

Filum terminale

A

Spinal meninges
- Arachnoid mater
- Dura mater
- Pia mater

Epidural space (contains fat)

Subdural space

Subarachnoid space (contains CSF)

Bone of vertebra

Posterior root ganglion

B

Body of vertebra

FIGURE **13.10** Spinal cord: (A) posterior view in the vertebral column; (B) transverse section of the spinal cord in the vertebral column

nerve). The posterior root's cell bodies are located just lateral to the spinal cord in a swollen knob called the **posterior** (or dorsal) **root ganglion.** In the thoracic and lumbar regions of the spinal cord we also find the **lateral horns,** which contain the cell bodies of autonomic neurons.

As you can see in Figure 13.11, it is possible to discern the anterior from the posterior spinal cord by looking at the shapes of the anterior and posterior horns. The anterior horns are broad and flat on the ends, whereas the posterior horns are more tapered, and they extend farther out toward the edge.

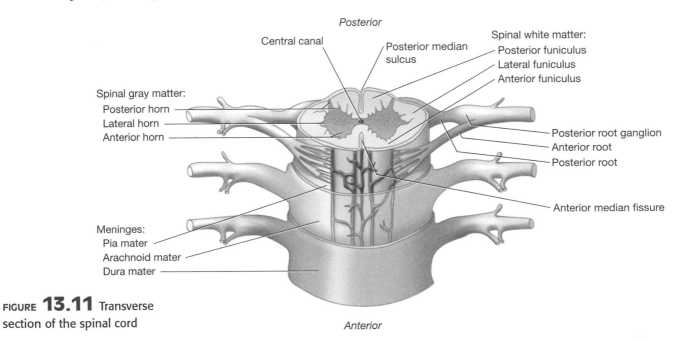

Posterior

Central canal

Posterior median sulcus

Spinal white matter:
Posterior funiculus
Lateral funiculus
Anterior funiculus

Spinal gray matter:
Posterior horn
Lateral horn
Anterior horn

Posterior root ganglion
Anterior root
Posterior root

Anterior median fissure

Meninges:
Pia mater
Arachnoid mater
Dura mater

FIGURE **13.11** Transverse section of the spinal cord

Anterior

Surrounding the spinal gray matter is the spinal white matter, which can be divided into three **funiculi** (or *columns*): the **anterior, posterior,** and **lateral funiculi**. Each funiculus contains myelinated axons that are grouped into bundles called **tracts**. Tracts contain axons that have the same beginning and end points, and the same general function. **Ascending tracts** carry sensory information from sensory neurons to the brain, and **descending tracts** carry motor information from the brain to motor neurons.

Procedure Model Inventory for the Spinal Cord

Identify the following structures of the spinal cord on models and diagrams, using your textbook and this unit for reference. As you examine the anatomical models and diagrams, record on the model inventory in Table 13.3 the name of the model and the structures you were able to identify. When you have completed the activity, answer Check Your Understanding questions 4 and 5 (p. 302).

1. Meninges
   a. Dura mater
   b. Arachnoid mater
   c. Pia mater
      (1) Denticulate ligaments

2. Spinal gray matter
   a. Anterior horn
   b. Lateral horn
   c. Posterior (dorsal) horn

3. Spinal white matter
   a. Anterior funiculus
   b. Lateral funiculus
   c. Posterior funiculus

4. Spinal nerve roots:
   a. Posterior root
      (1) Posterior root ganglion
   b. Anterior root

5. Conus medullaris

6. Filum terminale

7. Cauda equina

8. Cervical enlargement

9. Lumbar enlargement

10. Central canal

11. Anterior median fissure

12. Posterior median sulcus

TABLE **13.3**	Model Inventory for the Spinal Cord
Model	**Structures Identified**

Exercise 3

Brain and Spinal Cord Dissection

MATERIALS NEEDED

- Sheep brain and spinal cord
- Hammer and chisel (if the brain is still in the skull)
- Dissection equipment and trays

Often structures of the brain and spinal cord are difficult to see on anatomical models. This exercise will allow you to examine these structures more closely by dissecting a preserved sheep brain. You will note that certain structures, such as the frontal lobes of the cerebral hemispheres, are proportionally smaller in the sheep than in the human brain.

 Procedure Dissection

Note: Goggles and gloves are required.

---

1 If the brain is still encased in the skull, you have your work cut out for you. The best way to approach extracting it from the skull is to take a hammer and chisel and gently (at least as gently as one can with a hammer and chisel!) remove it piece by piece.

2 As you gently hack away at the skull, you will note a thick membrane holding the skull in place. This is the dura mater, and it can make removal of the skull somewhat difficult. Ideally, you would like to preserve the dura, but you may end up cutting through it as you remove the brain.

3 Once you have removed most of the skull, gently lift out the brain. (If you're careful, you may be able to get the brain out with the pituitary gland still attached.) You may have to loosen the remaining attachments of the dura with your finger.

4 Once the brain is out, note the thick part of the dura covering the longitudinal fissure. If you cut through this with scissors, you will enter the superior sagittal sinus.

5 Next remove the dura to reveal the thin membrane on top of the brain. This is the arachnoid mater.

6 Remove an area of the arachnoid mater to see the shiny inner membrane—the pia mater—directly touching the surface of the brain. Note that the pia mater follows the convolutions of the gyri and sulci.

7 Examine the surface anatomy of both the superior and the inferior surfaces of the sheep brain (**Figures 13.12** and **13.13**). Below, draw what you see, and label your drawing with all of structures from Exercise 1 that you are able to identify.

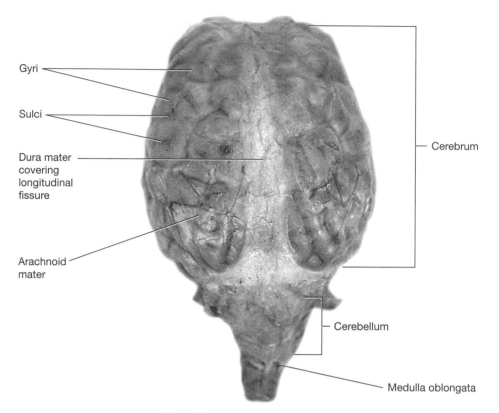

Gyri

Sulci

Dura mater
covering
longitudinal
fissure

Arachnoid
mater

Cerebrum

Cerebellum

Medulla oblongata

FIGURE 13.12 Superior view of the sheep brain

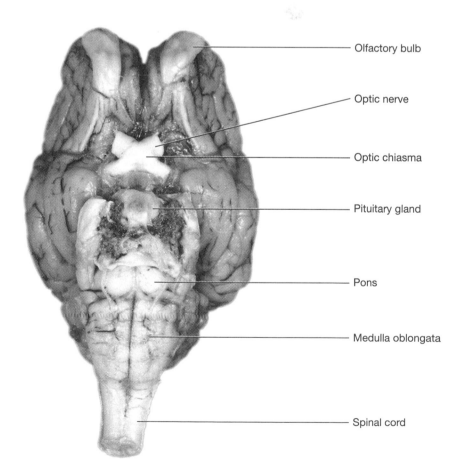

Olfactory bulb

Optic nerve

Optic chiasma

Pituitary gland

Pons

Medulla oblongata

Spinal cord

FIGURE 13.13 Inferior view of the sheep brain

EXPLORING ANATOMY & PHYSIOLOGY IN THE LABORATORY

8 Note the size of the olfactory bulbs. Are they larger or smaller than those that you observed in the human brain? Why do you think this is so?

9 Spread the two cerebral hemispheres, and identify the corpus callosum.

10 Make a cut down the brain's midsagittal plane to separate the two cerebral hemispheres.

11 Examine the brain's internal anatomy (**Figure 13.14**), and stick your finger in the lateral ventricle. You will see (or feel) that it is much larger than it appears. Draw what you see in the space below and label your drawing with all of structures from Exercise 1 that you are able to identify.

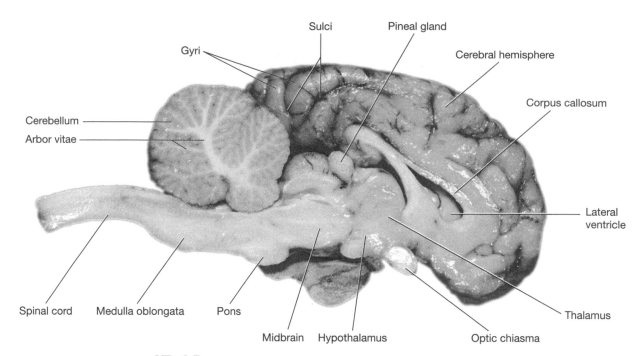

Gyri Sulci Pineal gland Cerebral hemisphere

Cerebellum

Arbor vitae

Corpus callosum

Lateral ventricle

Spinal cord Medulla oblongata Pons

Midbrain Hypothalamus

Optic chiasma

Thalamus

FIGURE 13.14 Lateral view of the sheep brain; midsagittal section

12 Section one of the halves of the brain in the frontal plane, approximately along the central sulcus. Note the outer cerebral cortex (the gray matter) and the inner white matter. From this view you also can see the lateral and third ventricles. Again draw what you see in the space below and label your drawing with all of the structures from Exercise 1 that you are able to identify.

13 If the brain on which you are working contains a segment of the spinal cord, move to it now, and identify the single-layered dura surrounding it. Peel it back to see the arachnoid mater.

14 Section the cord in the transverse plane. Note the butterfly-shaped inner gray matter and the outer white matter. Draw what you see in the space below, and label your drawing with all of structures from Exercise 2 that you are able to identify.

NAME _____

SECTION _____ DATE _____

Check Your Recall

1 Label the following parts of the brain in **Figures 13.15** and **13.16.**

- Frontal lobe
- Parietal lobe
- Temporal lobe
- Occipital lobe
- Central sulcus
- Pre-central gyrus
- Post-central gyrus
- Lateral sulcus
- Pons
- Medulla oblongata
- Cerebellum
- Thalamus
- Hypothalamus
- Pineal gland
- Midbrain
- Arbor vitae
- Lateral ventricle
- Third ventricle
- Cerebral aqueduct
- Fourth ventricle

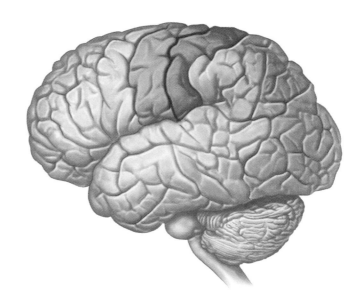

FIGURE **13.15** Brain, lateral view

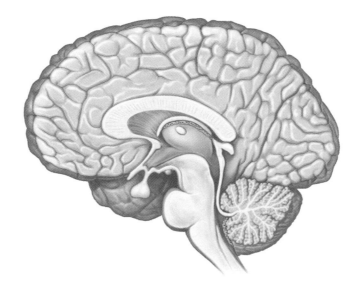

FIGURE **13.16** Brain, midsagittal section

2 Label the following parts of the spinal cord in **Figure 13.17**.

- Spinal dura mater
- Spinal arachnoid mater
- Anterior horn
- Posterior horn
- Lateral horn
- Anterior funiculus
- Posterior funiculus
- Lateral funiculus
- Anterior root
- Posterior root
- Posterior root ganglion
- Posterior median sulcus
- Anterior median fissure
- Central canal

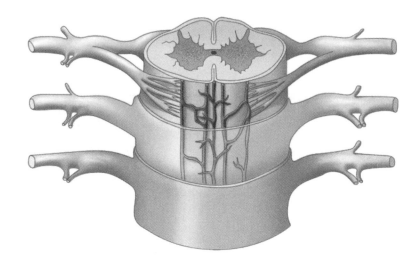

FIGURE 13.17 Spinal cord, transverse section

3 Which of the following is *not* one of the four main regions of the brain?

a. Cerebral hemispheres
b. Brainstem
c. Cerebral aqueduct
d. Cerebellum
e. Diencephalon

4 Fill in the blanks: The fluid-filled spaces within the brain are called _____, and they are filled with _____.

5 Where are the cell bodies of the cerebral neurons housed? What is the function of the cerebral neurons?

6 White matter consists of
   a. unmyelinated cell bodies, dendrites, and axons.
   b. myelinated axons.
   c. unmyelinated axons.
   d. myelinated cell bodies, dendrites, and axons.

7 Matching: Match the following terms with the correct definition.

   a. Thalamus   _____   A. Most inferior portion of the brainstem

   b. Pia mater   _____   B. Posterior part of the brain that controls and monitors ongoing movement

   c. Medulla oblongata _____   C. Innermost and thinnest meninx

   d. Epithalamus   _____   D. Most superior part of the brainstem

   e. Cerebellum   _____   E. Largest, egg-shaped component of the diencephalon; edits and sorts information
                                    coming into or out of the cerebrum
   f. Hypothalamus   _____
                                 F. Middle part of the brainstem that bulges anteriorly
   g. Dura mater   _____
                                 G. Middle meninx
   h. Pons   _____
                                 H. Inferior part of the diencephalon that controls many aspects of homeostasis
   i. Midbrain   _____
                                 I. Contains the pineal gland and secretes the hormone melatonin
   j. Arachnoid mater   _____
                                 J. Outermost and thickest meninx

8 Fill in the blanks: The spinal cord extends from the _____ of the occipital bone to the

   _____ vertebra. It terminates as the _____ and gives off a bundle

   of nerve roots called the _____.

9 Which of the following spaces is found around the spinal cord but *not* around the brain?
   a. Epidural space
   b. Subdural space
   c. Subarachnoid space
   d. Suprarachnoid space

10 The _____ of the spinal cord contains the cell bodies of motor neurons, whereas the

_____ of the spinal cord contains the cell bodies of neurons that receive information from

sensory neurons.

a. lateral horn; anterior horn
b. lateral horn; posterior horn
c. anterior horn; lateral horn
d. anterior horn; posterior horn

Check Your Understanding | **Critical Thinking and Application Questions**

1 Traumatic brain injuries may result in damage to the blood vessels in the spaces between the meninges. These damaged blood vessels leak blood into either the subdural or subarachnoid spaces—a potentially deadly condition called a *subdural* or *subarachnoid hemorrhage*. Why do you think that this condition is so dangerous?

2 Predict the effects of injuries to the following areas (you may want to use your text for this one):

a. Frontal lobes:

b. Brainstem:

c. Cerebellum:

d. Occipital lobes:

e. Basal nuclei:

3 Which of the injuries from #2 do you think would be the most damaging to survival? Why?

4 A common way to deliver anesthesia for surgery and childbirth is to inject the anesthetic agent into the epidural space (*epidural anesthesia*). A possible complication of this procedure is a tear in the dura mater that causes CSF to leak out of the central nervous system. Why would a loss of cerebrospinal fluid be problematic? What symptoms do you predict with this condition? (*Hint*: Think about the function of cerebrospinal fluid.)

5 During a procedure called a *lumbar puncture*, CSF surrounding the meninges is withdrawn with a needle. This procedure generally is performed between L3 and L5. Why do you think the fluid is withdrawn here rather than from higher up in the vertebral column (e.g., the cervical vertebrae)? Often, the sampled CSF is tested for bacteria or viruses if a brain infection is suspected. Why would CSF sampled from the spinal cord give you information about the condition of the brain?

Peripheral and Autonomic Nervous System

OBJECTIVES

Once you have completed this unit, you should be able to:

- Identify structures of a peripheral nerve.

- Identify, describe, and demonstrate functions of cranial nerves.

- Identify spinal nerves and plexuses.

- Describe a simple spinal reflex arc and test the effects of mental concentration on the patellar tendon (knee-jerk) reflex.

- Describe the effects of the two branches of the autonomic nervous system on the body systems.

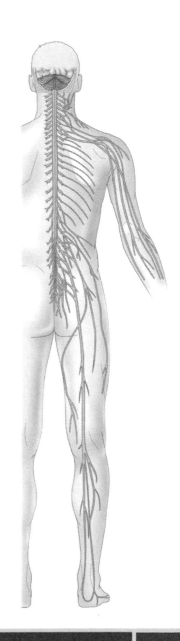

Pre-Lab Exercise 1
Key Terms

Table 14.1 lists the key terms with which you should be familiar before coming to lab. Please note that the individual cranial and spinal nerves are covered in Pre-Lab Exercises 4 and 6, respectively.

TABLE 14.1	Key Terms

Term   **Definition**

General Terms

Peripheral nerve _____

_____

Cranial nerve   _____

_____

Spinal nerve   _____

_____

Nerve plexus   _____

_____

Autonomic nervous system _____

_____

Sympathetic nervous system _____

_____

Parasympathetic nervous system _____

_____

Parts of a Peripheral Nerve

Fascicle   _____

_____

Axon   _____

_____

Epineurium _____

_____

Perineurium _____

_____

Endoneurium _____

_____

Spinal Nerve Plexuses

Cervical plexus _____

_____

Brachial plexus _____

_____

Lumbar plexus _____

_____

Sacral plexus _____

_____

✎ Pre-Lab Exercise 2
Peripheral Nerve Anatomy

Label and color the peripheral nerve diagram
depicted in Figure 14.1 with the terms from
Exercise 1. Use your text and Exercise 1 in
this unit for reference.

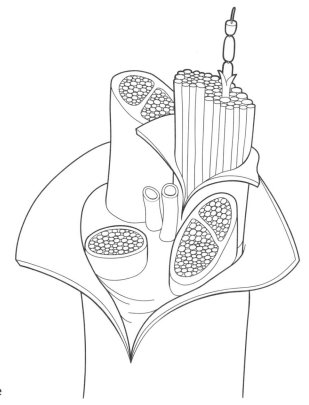

FIGURE **14.1** Anatomy of a peripheral nerve

Cranial Nerve Anatomy

Label and color the cranial nerves depicted in **Figure 14.2** with the terms from Exercise 2. Use your text and Exercise 2 in this unit for reference.

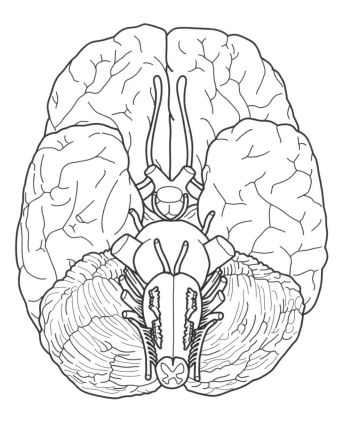

FIGURE **14.2** Inferior view of the brain with cranial nerves

Pre-Lab Exercise 4
Cranial Nerve Locations and Functions
Complete Table 14.2 with the functions of each pair of cranial nerves and whether the nerve is motor, sensory, or mixed.

TABLE **14.2** The Cranial Nerves		
Cranial Nerve	**Functions**	**Motor, Sensory, or Mixed**
CN I: Olfactory nerve		
CN II: Optic nerve		
CN III: Oculomotor nerve		
CN IV: Trochlear nerve		
CN V: Trigeminal nerve		
CN VI: Abducens nerve		
CN VII: Facial nerve		
CN VIII: Vestibulocochlear nerve		
CN IX: Glossopharyngeal nerve		
CN X: Vagus nerve		
CN XI: Accessory nerve		
CN XII: Hypoglossal nerve		

 Pre-Lab Exercise 5

Nerve Plexus and Spinal Nerve Anatomy

Label the diagram of the nerve plexuses and their spinal nerves depicted in **Figure 14.3** with the terms from Exercise 3. Use your text and Exercise 3 in this unit for reference. Note that this diagram is presented in color to facilitate identification of the nerves.

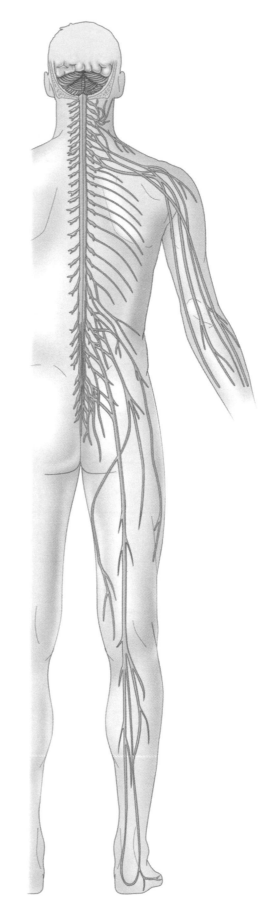

FIGURE **14.3** The nerve plexuses and the ventral rami of the spinal nerves

EXPLORING ANATOMY & PHYSIOLOGY IN THE LABORATORY

Pre-Lab Exercise 6
Spinal Nerve Locations and Functions

The spinal nerves (with the exception of the intercostal nerves) can be traced to the specific nerve plexus from which they originate. Fill in Table 14.3 with the nerve plexus from which each nerve originates and the functions of each nerve.

TABLE **14.3** The Spinal Nerves		
Spinal Nerve	**Nerve Plexus**	**Functions**
Phrenic nerve		
Radial nerve		
Musculocutaneous nerve		
Ulnar nerve		
Median nerve		
Femoral nerve		
Sciatic nerve		
Tibial nerve		
Common peroneal (fibular) nerve		

Pre-Lab Exercise 7
Branches of the Autonomic Nervous System

The autonomic nervous system (ANS) has two subdivisions: (a) the sympathetic nervous system (SNS), and (b) the parasympathetic nervous system (PSNS). Fill in Table 14.4 with the general characteristics of each branch of the ANS and the effects of each system on its target organs.

TABLE 14.4 Sympathetic and Parasympathetic Nervous Systems		
Characteristic	**SNS**	**PNS**
Main function		
Location of nerve roots		
Neurotransmitter(s)		
Effects on Target Organs and Physiologic Processes		
Heart		
Bronchioles (airway passages of the lungs)		
Blood vessels to abdominal organs and skin		
Blood vessels to skeletal muscle		
Digestive and urinary functions		
Metabolism		
Pupils		

Exercises

In Unit 13 we discussed the central nervous system (CNS), and now we turn to the other division of the nervous system— the **peripheral nervous system** (**PNS**). The PNS consists of organs called **peripheral nerves**, or simply *nerves*. There are two types of nerves classified by location:

1. nerves that originate from the brain and the brainstem, called **cranial nerves**, and

2. nerves that originate from the spinal cord, called **spinal nerves**.

The PNS is divided broadly into two divisions based upon function:

1. the **motor** or **afferent division**, which innervates muscle cells and glands, and

2. the **sensory** or **efferent division**, which innervates the skin, joints, special sensory organs, and other sensory receptors.

The motor and sensory divisions may be further subdivided based upon the structures that are served: (a) the **somatic sensory** and **somatic motor divisions**, which serve skin and skeletal muscles, respectively, and (b) the **visceral sensory** and **visceral motor divisions,** which serve smooth muscle, cardiac muscle, and glands. The visceral motor division controls most of our automatic functions such as heart rate, digestion, and urine production, and also is known as the **autonomic nervous system** (**ANS**).

In the following exercises, you will examine the structure and functions of the organs and divisions of the PNS. As you study the PNS, bear in mind that its functions are closely interwoven with the CNS and the branches of the nervous system cannot function independently of one another.

 Exercise 1

Peripheral Nerve Anatomy

MATERIALS NEEDED

▶ Peripheral nerve models

A **nerve** is an organ consisting of myelinated and unmyelinated axons (also known as *nerve fibers*), blood vessels, lymphatic vessels, and connective tissue sheaths (**Figure 14.4**). Each individual axon is covered by its axolemma (plasma membrane), superficial to which we find a connective tissue sheath called the **endoneurium**. Axons are grouped into bundles called **fascicles**, which are covered with another connective tissue sheath called the **perineurium**. Groups of fascicles, blood vessels, and lymphatic vessels are surrounded by the most superficial connective tissue sheath: the **epineurium**.

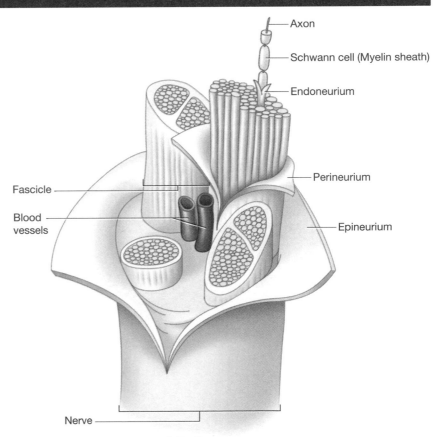

FIGURE **14.4** Anatomy of a peripheral nerve

Procedure Model Inventory

Identify the following parts of a nerve on models and diagrams, using your textbook and this unit for reference. As you examine the anatomical models and diagrams, record on the model inventory in Table 14.5 the name of the model and the structures you were able to identify. When you have completed the activity, answer Check Your Understanding questions 1 and 2 (p. 331).

1. Nerve
2. Fascicle
3. Axon (nerve fiber)
4. Blood vessels
5. Nerve sheaths
   a. Epineurium
   b. Perineurium
   c. Endoneurium
   d. Myelin sheath

TABLE **14.5**	Model Inventory for Peripheral Nerve Anatomy
Model/Diagram	**Structures Identified**

 Exercise 2

The Cranial Nerves

MATERIALS NEEDED

- Brain models
- Preserved brain specimens
- Penlight
- Snellen vision chart
- Tuning fork
- Unknown samples to smell
- PTC, thiourea, and sodium benzoate tasting papers

The 12 pairs of cranial nerves originate from the brain (Figure 14.5). Each nerve is given two names: (1) a sequential Roman numeral in order of its origin from the brain, and (2) a name that describes the nerve's location or function. For example, cranial nerve III is the third cranial nerve to arise from the brain. It is also called the oculomotor nerve because one of its functions is to provide motor fibers to some of the muscles that move the eyeball.

All cranial nerves innervate structures of the head and neck. Three cranial nerves are **sensory nerves** whose fibers have purely sensory functions; four are **mixed nerves** that contain both sensory and motor fibers; and five are **motor nerves** that contain primarily motor fibers. Following is an overview

of the main functions of each cranial nerve (note that
CN = "cranial nerve"):

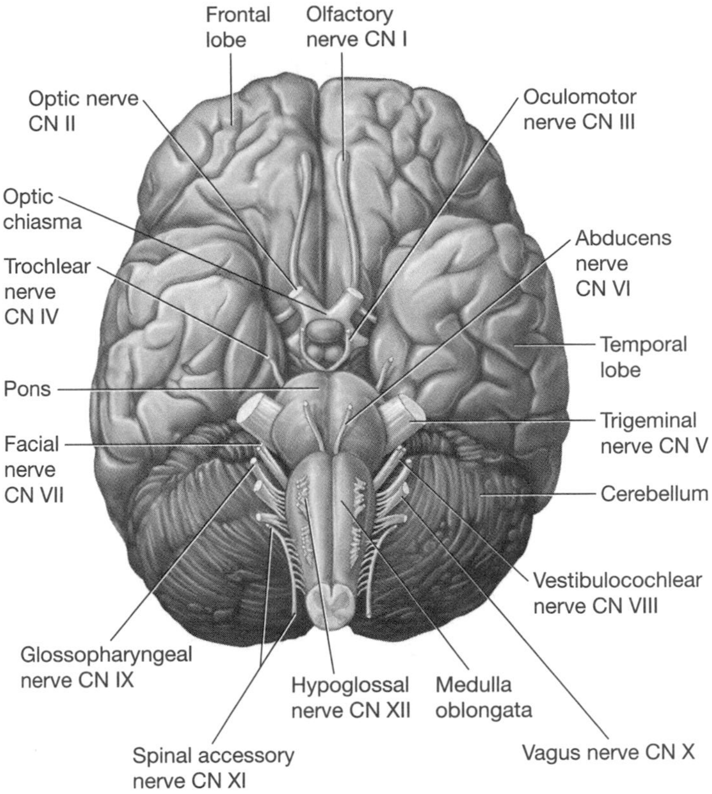

1. **CN I: Olfactory Nerve.** The olfactory nerve is
   a purely sensory nerve that innervates the olfactory mucosa in the superior nasal cavity where
   it provides for the sense of smell.

2. **CN II: Optic Nerve.** The optic nerve is also a
   purely sensory nerve that provides for the sense
   of vision. Its fibers emerge from the retina of the
   eye and meet at the **optic chiasma**, where the
   nerves partially exchange fibers before diverging
   to form the **optic tracts**.

3. **CN III: Oculomotor Nerve.** The oculomotor
   nerve is a motor cranial nerve that innervates
   four of the six extraocular muscles that move
   the eyeball, the muscle that opens the eyelid, the
   muscle that constricts the pupil, and the muscle
   that changes the shape of the lens for near
   vision, an adjustment called **accommodation**.

4. **CN IV: Trochlear Nerve.** The trochlear nerve is
   a small motor nerve that innervates one of the
   six extraocular muscles that move the eyeball
   (the *superior oblique muscle*).

5. **CN V: Trigeminal Nerve.** The trigeminal nerve
   is a large mixed nerve named for its three
   branches that together provide sensory innervation to the face and motor innervation to the muscles of mastication
   (chewing).

FIGURE **14.5** Inferior view of the brain with the cranial nerves

6. **CN VI: Abducens Nerve.** The abducens is a small motor nerve that innervates the final extraocular muscle (the
   lateral rectus muscle).

7. **CN VII: Facial Nerve.** The mixed fibers of the facial nerve innervate many structures and provide for the following:
   motor to the muscles of facial expression; taste to the anterior two-thirds of the tongue; motor to the glands that
   produce tears (the lacrimal glands), mucus, and saliva; and sensory to part of the face and mouth.

8. **CN VIII: Vestibulocochlear Nerve.** The final sensory nerve, the vestibulocochlear nerve, innervates the structures of
   the inner ear and provides for the senses of hearing and equilibrium.

9. **CN IX: Glossopharyngeal Nerve.** The small mixed glossopharyngeal nerve provides motor fibers to the muscles of
   the pharynx (throat) involved in swallowing and sensory fibers to the posterior one-third of the tongue for taste
   sensation.

10. **CN X: Vagus Nerve.** The mixed vagus nerve is the only cranial nerve that "wanders" outside of the head and neck
    ("vagus" means "wanderer"). In the head and the neck it provides some sensory fibers to the skin of the head and
    the pharynx, motor fibers to muscles involved in speech and swallowing, and motor fibers to certain salivary glands.
    Outside the head and the neck it innervates most of the thoracic and abdominal viscera as the main nerve of the
    parasympathetic nervous system.

11. **CN XI: Accessory Nerve.** The accessory nerve is the only cranial nerve that has both a cranial component originating
    from the brainstem and a spinal component originating from the spinal cord. Its motor fibers innervate the muscles
    that move the head and the neck, such as the trapezius and sternocleidomastoid muscles.

12. **CN XII: Hypoglossal Nerve.** The hypoglossal nerve is a small motor nerve that innervates the muscles that move the
    tongue. Note that the hypoglossal nerve *moves* the tongue but does not provide any taste sensation to the tongue.

HINTS & TIPS

Remembering the Order of the Cranial Nerves

Many cranial nerve mnemonics have been created over the years to help students remember their correct order. Following is one of my favorite mnemonics, but if this one doesn't stick for you, try making up your own or doing an Internet search for "cranial nerve mnemonics":

Oh (Olfactory)

Once (Optic)

One (Oculomotor)

Takes (Trochlear)

The (Trigeminal)

Anatomy (Abducens)

Final (Facial)

Very (Vestibulocochlear)

Good (Glossopharyngeal)

Vacations (Vagus)

Are (Accessory)

Happening (Hypoglossal)

You also can help yourself remember the olfactory and optic nerves by reminding yourself that you have one nose (CN I, the *olfactory* nerve) and two eyes (CN II, the *optic* nerve).

Procedure Model Inventory

Identify cranial nerves I–XII and associated structures on anatomical models and/or preserved specimens of the brain. List the nerves and the structures that you identify in Table 14.7, which is located in Exercise 3 (p. 322).

1. Olfactory nerves
   a. Olfactory bulbs
2. Optic nerves
   a. Optic chiasma
   b. Optic tract
3. Oculomotor nerves
4. Trochlear nerves
5. Trigeminal nerves
6. Abducens nerves
7. Facial nerves
8. Vestibulocochlear nerves
9. Glossopharyngeal nerves
10. Vagus nerves
11. Accessory nerves
12. Hypoglossal nerves

Procedure Testing the Cranial Nerves

A component of every complete physical examination performed by health care professionals is the cranial nerve exam. In this exercise you will put on your "doctor hat" and perform the same tests of the cranial nerves that would be done during a physical examination. Pair up with another student and take turns performing the following tests. For each test, first document your observations (in many cases, this will be "able to perform" or "unable to perform"). Then state which cranial nerve(s) you have checked with each test. Keep in mind that some tests check more than one cranial nerve, some cranial nerves are tested more than once, and each nerve is tested in this exercise at least once. When you have completed the activity, answer Check Your Understanding question 1 (p. 331).

1 Have your partner perform the following actions individually (not all at once—your partner may have difficulty smiling and frowning at the same time!): smile, frown, raise his or her eyebrows, and puff his or her cheeks.

Observations:

CN(s) tested:

2 Have your partner open and close his or her jaw and clench his or her teeth.

Observations:

CN(s) tested:

3 Have you partner elevate and depress his or her shoulders and turn his or her head to the right and the left.

Observations:

CN(s) tested:

4 Draw a large, imaginary "Z" in the air with your finger. Have your partner follow your finger with his or her eyes without moving his or her head. Repeat the procedure, this time drawing the letter "H" in the air with your finger.

Observations:

CN(s) tested:

5 Test pupillary response:

a. Dim the lights in the room about halfway.

b. Place your hand vertically against the bridge of your partner's nose as illustrated in **Figure 14.6**. This forms a light shield to separate the right and left visual fields.

c. Shine the penlight indirectly into the left eye from an angle as illustrated in Figure 14.6. Watch what happens to the pupil in the left eye.

d. Move the penlight away and watch what happens to the left pupil.

e. Shine the light into the left eye again and watch what happens to the pupil in the right eye. Move the penlight away and watch what happens to the right pupil.

f. Repeat this process with the right eye.

g. Record your results in Table 14.6.

CN(s) tested.

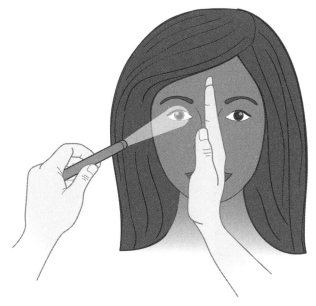

FIGURE 14.6 Method for testing the pupillary response

TABLE **14.6**	Pupillary Response Results	
Action	**Response of Left Pupil**	**Response of Right Pupil**
Light shined into left eye		
Light removed from left eye		
Light shined into right eye		
Light removed from right eye		

6 Have your partner focus on an object on the far side of the room (e.g., the chalkboard or a chart) for 1 minute. Then have your partner switch his or her focus to an object in your hand (e.g., a pencil). Watch your partner's pupils carefully as the point of focus is changed from far to near.

Observations:

CN(s) tested:

7 Place your hand lightly on your partner's throat, and have him or her swallow and speak. Feel for symmetrical movement of the larynx (throat).

Observations:

CN(s) tested:

8 Have your partner protrude his or her tongue. Check for abnormal deviation or movement (e.g., does the tongue move straight forward or does it move to one side?).

Observations:

CN(s) tested:

9 Test your partner's vision by having him or her stand 20 feet from a Snellen chart and read the chart, starting at the largest line and progressing to the smallest line that he or she is able to see clearly. Record the ratio (e.g., 20/30) next to the smallest line your partner can read.

Observations:

CN(s) tested:

10 Hold a tuning fork by its handle and strike the tines with a rubber mallet (or just tap it lightly on the lab table). Touch the stem to the top of your partner's head along the midsagittal line. Ask your partner if he or she hears the vibration better in one ear or if he or she hears the sound equally well in both ears.

Observations:

CN(s) tested:

11 Have your partner stand with his or her eyes closed and arms at his or her sides for several seconds. Evaluate his or her ability to remain balanced.

Observations:

CN(s) tested:

12 Hold an unknown sample under your partner's nose, and have him or her identify the substance by its scent.

Observations:

CN(s) tested:

13 Evaluate your partner's ability to taste using tasting papers. Have your partner place a piece of PTC paper on his or her tongue and determine if he or she can taste it (the ability to taste PTC is genetically determined, about half of the population can taste it). If your partner cannot taste the PTC, try the thiourea paper instead (a word of warning—thiourea tastes bad). If your partner cannot taste either of these papers, try the sodium benzoate paper.

Observations:

CN(s) tested:

Exercise 3

Spinal Nerves and Reflexes

MATERIALS NEEDED

- Model of the spinal nerves
- Reflex hammer

Each of the 31 pairs of **spinal nerves** forms from the fusion of the dorsal and ventral roots of the spinal cord. Recall that the ventral roots carry motor fibers emerging from the spinal cord and the dorsal roots carry sensory fibers to the spinal cord. Because each spinal nerve carries both motor and sensory fibers, all spinal nerves are mixed nerves.

Shortly after the dorsal and ventral roots fuse to form the spinal nerve, it splits into three branches: a **dorsal ramus**, a **ventral ramus**, and a small **meningeal branch**. The dorsal rami serve the skin, joints, and musculature of the posterior trunk. The meningeal branches reenter the vertebral canal to innervate spinal structures. The larger ventral rami travel anteriorly to supply the muscles of the upper and lower limbs, the anterior thorax and abdomen, and part of the back. The distribution of the ventral rami is illustrated in **Figure 14.7**.

The ventral rami of the thoracic spinal nerves travel anteriorly as 11 separate pairs of **intercostal nerves** that innervate the intercostal muscles, the abdominal muscles, and the skin of the chest and abdomen. The ventral rami of the cervical, lumbar, and sacral nerves combine to form four large **plexuses**, or networks, of nerves: the **cervical**, **brachial**, **lumbar**, and **sacral plexuses**. The major nerves of the cervical, brachial, thoracic, lumbar, and sacral plexuses are as follows:

1. **Cervical plexus**. The cervical plexus consists of the ventral rami of C1–C4 with a small contribution from C5. Its branches serve the skin of the head and the neck and certain neck muscles. Its major branch is the **phrenic nerve** (C3–C5), which serves the diaphragm.

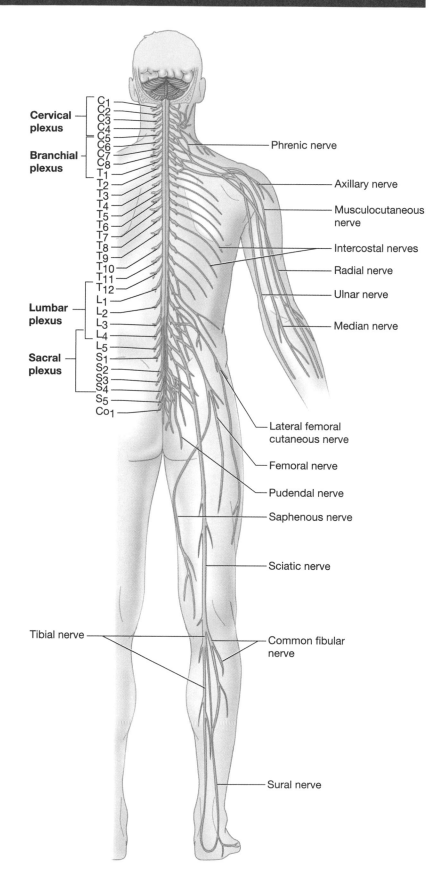

FIGURE **14.7** The nerve plexuses and the ventral rami of the spinal nerves

2. **Brachial plexus**. The complicated-looking brachial plexus consists of the ventral rami of C5–T1. Following are the major nerves that stem from the brachial plexus:

   a. **Axillary nerve**. The axillary nerve serves structures near the axilla, including the deltoid and teres minor muscles and the skin around this region.

   b. **Musculocutaneous nerve**. The musculocutaneous nerve is located in the lateral arm and serves the anterior arm muscles (such as the biceps brachii) and the skin of the lateral forearm.

   c. **Radial nerve**. The radial nerve is located in the posterior arm. It serves the posterior arm muscles and the forearm extensors, as well as the skin in the lateral hand.

   d. **Ulnar nerve**. The ulnar nerve, which you likely know as the "funny bone nerve," begins posteriorly but then crosses over to the anterior side of the arm as it curves around the medial epicondyle of the humerus. At this point the nerve is superficial and is easily injured when you smack your elbow on something. The ulnar nerve supplies certain forearm flexors, most of the intrinsic muscles of the hand, and the skin over the medial hand.

   e. **Median nerve**. The median nerve is named as such because it travels approximately down the middle of the arm and forearm. It supplies most of the forearm flexors, certain intrinsic hand muscles, and the skin over the anterior and lateral hand. As the median nerve enters the wrist, it travels under a band of connective tissue called the *flexor retinaculum*. Occasionally the median nerve becomes trapped and inflamed under the flexor retinaculum, which results in **carpal tunnel syndrome**.

3. **Lumbar plexus**. Each lumbar plexus consists of the ventral rami of L1–L4 with a small contribution from T12. The largest nerve of this plexus is the **femoral nerve**, which provides motor innervation to most of the anterior thigh muscles and sensory innervation to the skin of the anterior and medial thigh, the leg, and the foot. Another smaller branch is the **lateral femoral cutaneous nerve**, which provides mainly sensory innervation to the anterolateral thigh.

4. **Sacral plexus**. The sacral plexus forms from ventral rami of L4–S4. Its largest nerve, and indeed the largest nerve in the body, is the **sciatic nerve**. The sciatic nerve travels in the posterior thigh, where it splits into two branches: the **tibial nerve** and the **common peroneal (fibular) nerve**. The tibial nerve provides motor innervation to the posterior muscles of the thigh, posterior leg, and foot, and sensory innervation to the posterior leg and foot. The common peroneal nerve provides motor and sensory innervation to the anterolateral leg and the foot. A smaller branch of the sacral plexus is the **pudendal nerve**, which innervates the muscles of the pelvic floor and anogenital sphincters.

 Procedure Model Inventory

Identify the following nerves and nerve plexuses on models and diagrams, using your textbook and this unit for reference. As you examine the anatomical models and diagrams, record on the model inventory in Table 14.7 the name of the model and the structures that you were able to identify. Note that you also used this model inventory for the cranial nerves in Exercise 2. When you have completed the activity, answer Check Your Understanding question 2 (p. 331).

1. Cervical plexus
   a. Phrenic nerve

2. Brachial plexus
   a. Axillary nerve
   b. Radial nerve
   c. Musculocutaneous nerve
   d. Ulnar nerve
   e. Median nerve

3. Thoracic (intercostal) nerves

4. Lumbar plexus
   a. Femoral nerve
   b. Lateral femoral cutaneous nerve

5. Sacral plexus
   a. Sciatic nerve
     (1) Tibial nerve
     (2) Common peroneal (fibular) nerve
   b. Pudendal nerve

TABLE **14.7**	Model Inventory for Cranial Nerve and Spinal Nerve Anatomy
Model/Diagram	**Structures Identified**

Procedure Testing Spinal Reflexes

A **reflex** is an involuntary, predictable motor response to a stimulus. The pathway through which information travels, shown below and in **Figure 14.8**, is called a **reflex arc**:

> sensory receptor detects the stimulus ⟶ sensory neurons bring the stimulus to the CNS ⟶ the CNS processes and integrates the information ⟶ the CNS sends its output via motor neurons to an effector ⟶ the muscle contracts

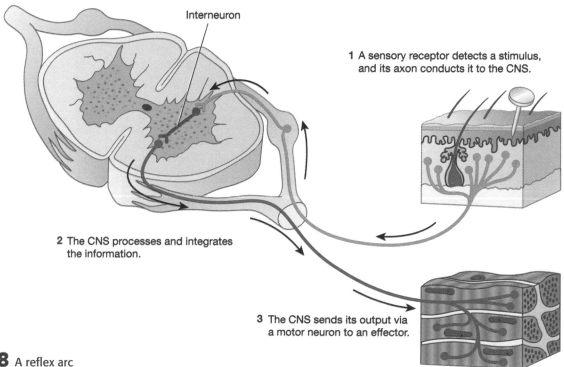

Interneuron

1 A sensory receptor detects a stimulus, and its axon conducts it to the CNS.

2 The CNS processes and integrates the information.

3 The CNS sends its output via a motor neuron to an effector.

FIGURE 14.8 A reflex arc

EXPLORING ANATOMY & PHYSIOLOGY IN THE LABORATORY

The human body has many different reflex arcs, one of the simplest of which is the **stretch reflex**. Stretch reflexes are important in maintaining posture and equilibrium and are initiated when a muscle is stretched. The stretch is detected by **muscle spindles**, which are specialized stretch receptors in the muscles, and this information is sent via sensory neurons to the CNS. The CNS then sends impulses down the motor neurons to the muscle that trigger a muscle contraction that counters the stretch.

You can demonstrate the stretch reflex easily: Sit down, with your knees bent and relaxed, and palpate (feel) the musculature of your posterior thigh. How do the muscles feel (taut or soft)? Now stand up and bend over to touch your toes. Palpate the muscles of your posterior thigh again. How do they feel? The reason they feel different in this position is that you stretched the hamstring muscles when you bent over to touch your toes. This triggered a stretch reflex that resulted in shortening (or tightening) of those muscles.

Technically, this reflex can be carried out without the help of the cerebral cortex and can be mediated solely by the spinal cord. We will see shortly, however, that the cortex is involved in even the simplest example of a stretch reflex—the patellar tendon (knee-jerk) reflex. When you have completed the following activity, answer Check Your Understanding question 3 (p. 332).

1. Have your lab partner sit in a chair with his or her legs dangling freely.

2. Palpate your partner's patellar tendon between the tibial tuberosity and the patella.

3. Tap this area with the flat end of a reflex hammer (sometimes a few taps are necessary to hit the right spot). What is the result?

4. Now give your partner a difficult math problem to work (long division with decimals usually does the trick). As your partner works the problem, tap the tendon again. Is this response different from the original response? If yes, how, and why?

 # Exercise 4

The Autonomic Nervous System

MATERIALS NEEDED

- Automated or manual sphygmomanometer
- Stethoscope

The **autonomic nervous system** (ANS) is the largely involuntary branch of the peripheral nervous system charged with maintaining homeostasis in the face of changing conditions. The ANS has two branches—the **sympathetic nervous system** (SNS) and the **parasympathetic nervous system** (PSNS).

The cells bodies of the neurons of the SNS are located in the thoracic and lumbar regions of the spinal cord. Functionally, the SNS is often described in terms of the "fight or flight" response, in which the body prepares to respond to an emergency situation (e.g., running from a hungry alligator). This gives rise to a useful mnemonic: Your body is *sympathetic* to your emergency situation, so your *sympathetic* nervous system comes to the rescue. Although reaction to an emergency is a primary function of the SNS, it isn't the whole story. The SNS is activated by *any* excitement, emotion, or exercise, even something as simple as standing up. The axons of the SNS release the neurotransmitters **epinephrine**, **norepinephrine**, and **acetylcholine**. These neurotransmitters trigger the following effects in their target tissues:

- increased heart rate, which increases the blood pressure;
- constriction of the blood vessels serving the abdominal viscera and the skin, which also increases the blood pressure;
- dilation of the blood vessels serving the skeletal muscles;
- dilation of the airway passages in the lungs;
- decreased digestive and urinary functions;
- increased metabolic rate;
- increased release of fatty acids and glucose; and
- dilation of the pupils.

The cell bodies of the neurons of the PSNS are located in cranial nerve nuclei and in the sacral portion of the spinal cord. Functionally, the PSNS is discussed in terms of its "rest and recovery" role, in which it promotes functions associated with digestion, defecation, and diuresis (urine production). The axons of the PSNS release the neurotransmitter acetylcholine, which has the following effects on its target tissues:

- decreased heart rate, which decreases blood pressure;
- dilation of the blood vessels serving the skin and abdominal viscera, which decreases blood pressure (note that this is *not* a direct effect of the PSNS on the blood vessels. Instead, it is an indirect effect resulting from an inhibition of sympathetic nerves);
- mild constriction of the airway passages in the lungs;
- increased digestion and secretion of digestive products;
- increased urine production;
- decreased metabolic rate;
- storage of fats and glucose;
- constriction of the pupils; and
- adjustment of the lens for near vision.

Note that the effects of the SNS and PSNS are exactly opposite. Generally, the PSNS is subordinate to the SNS (digesting that doughnut isn't all that important when you're being chased by a hungry alligator); however neither branch of the ANS is ever completely quiet. The divisions work together constantly to provide the proper balance for any given situation.

 Procedure **Testing Blood Pressure and Heart Rate Response to Exercise**

One of the simplest ways to activate the SNS is with exercise. In this activity you will be measuring the effects of exercise on the heart rate and the blood pressure to see the SNS in action. When you have completed this activity, answer Check Your Understanding questions 4 and 5 (p. 332).

1 Have your partner remain seated quietly for 2 minutes.

2 After 2 minutes, measure your partner's resting heart rate by placing two fingers on his or her radial pulse (on the lateral side of the wrist). Count the number of beats for 1 minute. Record the rate in Table 14.8.

3 Obtain a stethoscope and listen to your partner's heart. How does the heartbeat sound (quiet, loud, etc.)?

4 Measure your partner's resting blood pressure using a sphygmomanometer (blood pressure cuff) and a stethoscope (instructions for taking blood pressures are found in Unit 18 on p. 425). You may also use an automated sphygmomanometer if one is available. Record the blood pressure in Table 14.8.

5 Have your partner perform vigorous exercise for several minutes (running up and down stairs works well). Note that even minor exercise will produce changes in blood pressure and pulse, but the changes are more apparent with vigorous exercise (especially for those that are novices at taking blood pressure).

6 Immediately after the vigorous exercise, repeat the heart rate and blood pressure measurements, then listen to your partner's heart with a stethoscope. How does the heartbeat sound compared to the sounds you heard at rest?

7 Have your partner rest while seated for 5 minutes. Repeat the heart rate and blood pressure measurements, and again listen to the heartbeat. Does the heart sound different after rest?

TABLE **14.8** Results of Exercise and ANS Experiment		
Situation	**Blood Pressure**	**Heart Rate**
Seated (at rest)		
Immediately after exercise		
After resting for 5 minutes		

8 Interpret your results:

a. In which situation(s) did the SNS dominate? How do you know?

b. In which situation(s) did the PSNS dominate? How do you know?

NAME _____

SECTION _____ DATE _____

 Check Your Recall

1 Label the following structures on **Figure 14.9**.

- Nerve
- Epineurium
- Fascicle
- Perineurium
- Axon
- Myelin sheath
- Endoneurium

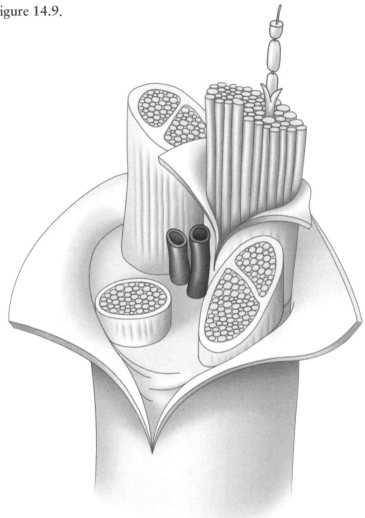

FIGURE **14.9** A peripheral nerve

2 Another name for an axon is

a. myelin sheath.
b. endoneurium.
c. nerve fiber.
d. peripheral nerve.

3 Fill in the blanks: Write the name of each cranial nerve next to its Roman numeral below.

CN I _____

CN II _____

CN III _____

CN IV _____

CN V _____

CN VI _____

CN VII _____

CN VIII _____

CN IX _____

CN X _____

CN XI _____

CN XII _____

4 Matching: Match the cranial nerve with its main functions.

CN I _____ A. Sensory to the face, motor to the muscles of mastication

CN II _____ B. Motor to the trapezius and sternocleidomastoid muscles

CN III _____ C. Hearing and equilibrium

CN IV _____ D. Olfaction (smell)

CN V _____ E. Motor to the muscles of swallowing; taste to the posterior one-third of the tongue

CN VI _____ F. Motor to 1/6 extraocular muscles (superior oblique muscle)

CN VII _____ G. Motor to the tongue

CN VIII _____ H. Vision

CN IX _____ I. Motor to the muscles of facial expression, taste to the anterior 2/3 of the tongue

CN X _____ J. Motor to 4/6 extraocular muscles, dilates the pupil, opens the eye, changes the shape of the lens

CN XI _____ K. Motor to the muscles of swallowing and speaking, motor to the thoracic and abdominal viscera

CN XII _____ L. Motor to 1/6 extraocular muscles (lateral rectus muscle)

5 Which of the following cranial nerves is *not* a purely sensory nerve?
a. Optic nerve
b. Vestibulocochlear nerve
c. Hypoglossal nerve
d. Olfactory nerve

6 Label the following nerves and plexuses on **Figure 14.10.**

- Femoral nerve
- Sciatic nerve
- Median nerve
- Intercostal nerves
- Ulnar nerve
- Radial nerve
- Cervical plexus
- Brachial plexus
- Lumbar plexus
- Sacral plexus

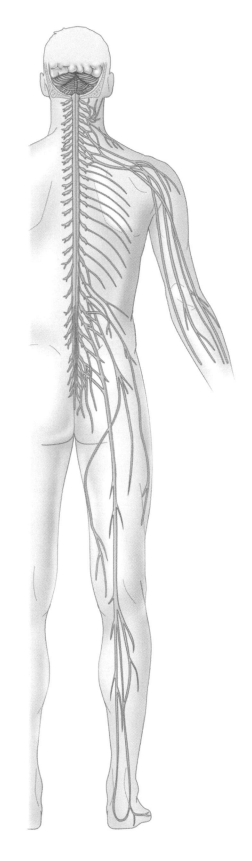

FIGURE 14.10 Nerve plexuses and the ventral rami of the spinal nerves

7 Matching: Match each spinal nerve with the main structures that it supplies.

a. Phrenic nerve _____   A. Anterior and lateral leg muscles and skin

b. Median nerve _____   B. Posterior thigh and leg muscles, foot

c. Tibial nerve _____   C. Diaphragm

d. Radial nerve _____   D. Some forearm flexors, most intrinsic hand muscles, skin on medial hand

e. Femoral nerve _____   E. Posterior arm muscles, forearm extensors, skin on lateral hand

f. Ulnar nerve _____   F. Anterior arm muscles, skin on lateral forearm

g. Common fibular nerve _____   G. Most forearm flexors, skin on anterior and lateral hand

h. Musculocutaneous nerve _____   H. Anterior thigh muscles, skin on anterior and medial thigh and leg

8 Number the events of a reflex arc from 1 (first event) through 5 (last event).

_____ CNS sends output via motor neurons to an effector.

_____ Sensory neurons bring the stimulus to the CNS.

_____ The muscle contracts.

_____ CNS processes and integrates the information.

_____ Sensory receptor detects the stimulus.

9 The receptor that detects the stretch in a stretch reflex is called a(an)
a. mitotic spindle.
b. muscle spindle.
c. capsular receptor.
d. efferent neuron.

10 Mark each of the following as an effect of the sympathetic nervous system (SNS) or the parasympathetic nervous system (PSNS):

_____ Increased blood pressure

_____ Dilation of the pupils

_____ Changing the shape of the lens for near vision

_____ Dilation of the bronchioles

_____ Decreased metabolic rate

_____ Increased digestion

_____ Storage of fats and glucose

_____ Decreased urinary and digestive functions

_____ Increased heart rate

_____ Constriction of blood vessels serving skin and abdominal viscera

NAME _____

SECTION _____ DATE _____

 Check Your Understanding | **Critical Thinking and Application Questions**

1 Damage to which cranial nerve(s) might produce the following results?

a. Inability to move the tongue

b. Inability to taste

c. Inability to move the eyes in any direction

d. Inability to shrug the shoulders

e. Inability to swallow

2 Damage to which spinal nerve(s) might produce the following physical findings?

a. Inability to flex the forearm (with the biceps brachii muscle)

b. Inability to breathe

c. Inability to flex the hip and extend the knee with the anterior thigh muscles

d. Inability to extend the hip and flex the knee with the posterior thigh muscles

e. Inability to move the hand or feel the skin over the medial hand

3 Sometimes the reflex response is diminished or absent—a phenomenon termed *hyporeflexia*. Do you think hyporeflexia would be caused by disorders of the central nervous system or of the peripheral nervous system? Explain your reasoning.

4 Drugs used to treat hypertension (high blood pressure) often work by blocking different components of the SNS. Explain how inhibiting the SNS may act to lower blood pressure.

5 The disease *asthma* is characterized by increased airway resistance secondary to constriction of the airways, airway inflammation, and excessive secretion of mucus. One of the main classes of drugs used to treat asthma mimics the effects of the SNS on the airways. Explain how mimicking the SNS might help alleviate an asthma attack.

General and Special Senses

OBJECTIVES

Once you have completed this unit, you should be able to:

- Identify structures of the eye.

- Describe the extraocular muscles that move the eyeball.

- Compare the functions of the rods and cones.

- Identify structures of the ear.

- Perform tests of hearing and equilibrium.

- Identify structures of the olfactory and taste senses.

- Determine the relative concentration of cutaneous sensory receptors in different regions of the body.

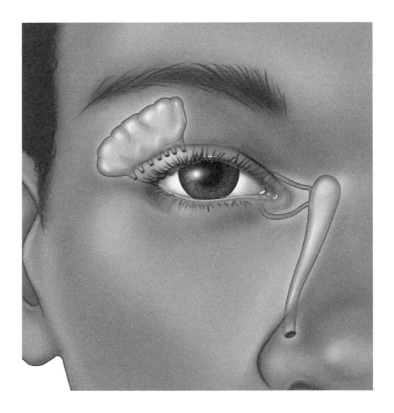

Pre-Lab Exercises | Complete the following exercises prior to coming to lab, using your textbook and lab manual for reference.

Pre-Lab Exercise 1
Key Terms

Table 15.1 lists the key terms with which you should be familiar before coming to lab.

TABLE 15.1	Key Terms

Term	Definition
Structures of the Eye	
Conjunctiva	_____
	_____
Lacrimal gland	_____
	_____
Sclera	_____
	_____
Cornea	_____
	_____
Iris	_____
	_____
Pupil	_____
	_____
Lens	_____
	_____
Ciliary body	_____
	_____
Choroid	_____
	_____
Retina	_____
	_____

Rods _____

_____

Cones _____

_____

Structures of the Ear

Auricle _____

_____

External auditory canal _____

_____

Tympanic membrane _____

_____

Auditory ossicles _____

_____

Pharyngotympanic tube _____

_____

Vestibule _____

_____

Semicircular canals _____

_____

Cochlea _____

_____

Structures of Taste and Smell

Chemosenses _____

_____

Olfactory epithelium _____

_____

Tongue papillae _____

_____

Pre-Lab Exercise 2

Anatomy of the Eye

Label and color the structures of the eye depicted in Figures 15.1 and 15.2 with the terms from Exercise 1. Use your text and Exercise 1 in this unit for reference.

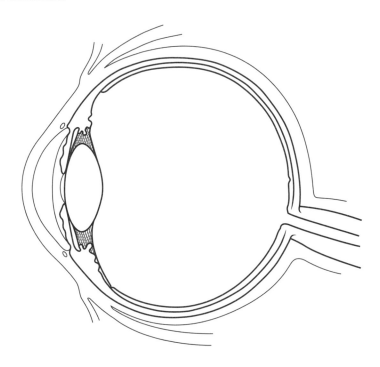

FIGURE **15.1** Eyeball, sagittal section

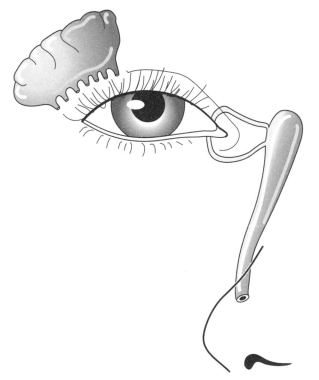

FIGURE **15.2** External and accessory structures of the eye

EXPLORING ANATOMY & PHYSIOLOGY IN THE LABORATORY

Pre-Lab Exercise 3
Extraocular Muscles

The six muscles that move the eyeball are called **extraocular muscles**. First, label and color the extraocular muscles illustrated in Figure 15.3. Then, fill in Table 15.2 with the location, action, and cranial nerve innervation for each of the extraocular muscles.

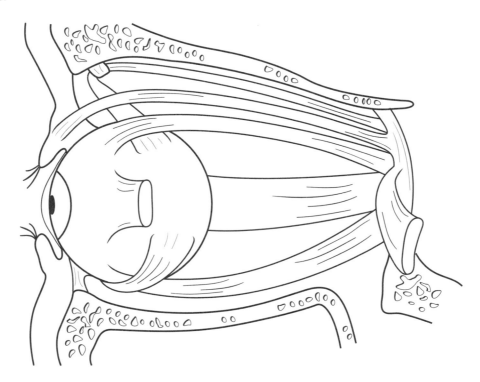

FIGURE **15.3** The extraocular muscles

TABLE **15.2**	The Extraocular Muscles		
Muscle	**Location**	**Action**	**Cranial Nerve Innervation**
Superior rectus muscle			
Inferior rectus muscle			
Medial rectus muscle			
Lateral rectus muscle			
Superior oblique muscle			
Inferior oblique muscle			

Pre-Lab Exercise 4
Anatomy of the Ear

Label and color the structures of the ear in **Figure 15.4** with the terms from Exercise 2.

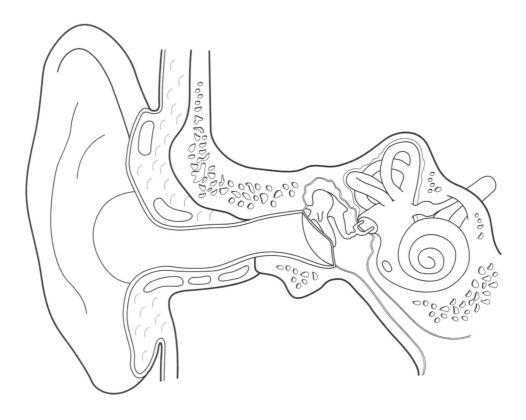

FIGURE **15.4** The ear

Exercises

Sensation is broadly defined as of the detection of changes in the internal and external environments. Sensation may be conscious or subconscious, depending on the destination of the sensory information. For example, certain blood vessels have receptors that detect blood pressure. This information is taken to the brainstem, which makes changes as necessary to ensure that blood pressure remains relatively constant. This information never makes it to the cerebral cortex, so you are not consciously aware of it. However, information that is eventually taken to the cerebral cortex (e.g., the taste of your food or the level of light in a room) is something of which you are consciously aware. This type of sensation is called **perception,** and it is the focus of this unit.

The following exercises ask you to examine the anatomy and physiology of the organs of perception, commonly called the **special senses:** vision, hearing and equilibrium, hearing, taste, and smell. You also will examine the **general senses** in this unit: touch and stretch.

EXPLORING ANATOMY & PHYSIOLOGY IN THE LABORATORY

Anatomy of the Eye and Vision

MATERIALS NEEDED

- Eye models
- Preserved eyeballs
- Dissection equipment
- Dissection trays
- Snellen vision chart
- Dark green or blue paper

The eye is a complex organ consisting of three components:

1. external structures such as the **eyelids** (**Figure 15.5**),

2. accessory structures such as the **lacrimal gland** and the extraocular muscles (**Figures 15.5** and **15.6**), and

3. the **eyeball** (**Figure 15.7**).

Many of the external and accessory structures of the eye protect the delicate eyeball.

Anteriorly, the eye is covered by the eyelids, or **palpebrae**. The eyelids meet medially and laterally at the **medial** and **lateral canthi**, respectively. The internal surface of the eyelids and much of the anterior eyeball is covered with a thin mucous membrane called the **conjunctiva**. Within the eyelids are the **meibomian glands**, whose oily secretions help to keep the eye lubricated.

One of the most prominent accessory structures of the eye is the **lacrimal apparatus**, which produces and drains tears. The lacrimal apparatus consists of the **lacrimal gland**, located in the superolateral orbit, and the ducts that drain the tears it produces. The other major accessory structures are the **extraocular muscles**, which move the eyeball. As you learned in Unit 14, there are six extraocular muscles (**Figure 15.6**):

1. **Lateral rectus:** moves the eyeball laterally

2. **Medial rectus:** moves the eyeball medially

3. **Superior rectus:** moves the eyeball superiorly

4. **Inferior rectus:** moves the eyeball inferiorly

5. **Superior oblique:** moves the eyeball inferiorly and laterally

6. **Inferior oblique:** moves the eyeball superiorly and laterally

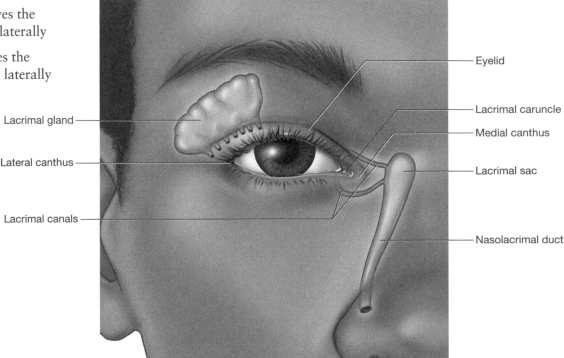

FIGURE **15.5** External and accessory structures of the eye

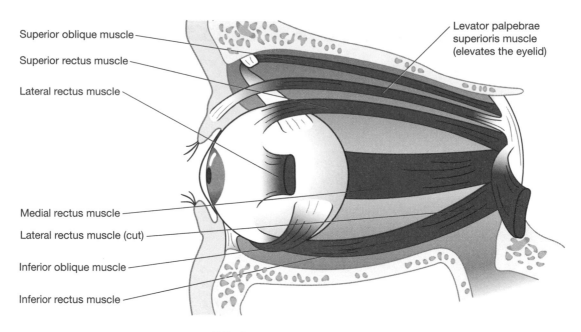

Superior oblique muscle

Superior rectus muscle

Lateral rectus muscle

Medial rectus muscle

Lateral rectus muscle (cut)

Inferior oblique muscle

Inferior rectus muscle

Levator palpebrae superioris muscle (elevates the eyelid)

FIGURE **15.6** The extraocular muscles

The eyeball itself is a hollow organ with three distinct tissue layers, or tunics (Figure 15.7):

1. **Fibrous tunic.** This outermost layer of the eyeball consists mostly of dense irregular connective tissue. It is avascular (lacks a blood supply), and consists of two parts:

    a. **Sclera.** The sclera is the white part of the eyeball, which makes up the posterior 5/6 of the fibrous tunic. It is white because of numerous collagen fibers that contribute to its thickness and toughness (in the same way that a joint capsule or a ligament is tough and white).

    b. **Cornea.** The clear cornea makes up the anterior 1/6 of the fibrous tunic and is one of the refractory media of the eyeball (it bends light coming into the eye).

2. **Vascular tunic.** Also called the **uvea,** the vascular tunic carries most of the blood supply to the tissues of the eye. It is composed of three main parts:

    a. **Choroid.** The highly vascular choroid makes up the posterior part of the vascular tunic. The choroid is brown in color to prevent light scattering in the eye.

    b. **Ciliary body.** The ciliary body is located at the anterior aspect of the eye. It is made chiefly of the **ciliary muscle,** which controls the shape of the lens. The muscle attaches to the lens via small **suspensory ligaments.**

    c. **Iris.** The pigmented iris is the most anterior portion of the uvea. It consists of muscle fibers arranged around an opening called the **pupil.** As the fibers contract, the pupil either constricts or dilates.

3. **Sensory tunic.** This layer consists of the **retina** and the **optic nerve.** The retina is a thin, delicate structure that contains **photoreceptors** called **rods** and **cones.**

    a. Rods are scattered throughout the retina and are responsible for vision in dim light and for peripheral vision.

    b. Cones are concentrated at the posterior portion of the retina and are found in highest numbers in an area called the **macula lutea.** At the center of the macula lutea is the **fovea centralis,** which contains only cones. Cones are responsible for color and high-acuity (sharp) vision in bright light.

        Note that there are no rods or cones at the posteriormost aspect of the eyeball where the optic nerve leaves the eyeball. This location is called the **optic disc,** or blind spot.

Another component of the eyeball is the **lens,** which allows for precise focusing of light on the retina. The lens divides the eyeball into the **anterior** and **posterior cavities** (sometimes called the anterior and posterior segments). The anterior cavity is filled with a watery fluid called **aqueous humor,** and the posterior cavity contains a thicker fluid called **vitreous humor.** Both help to refract light onto the retina.

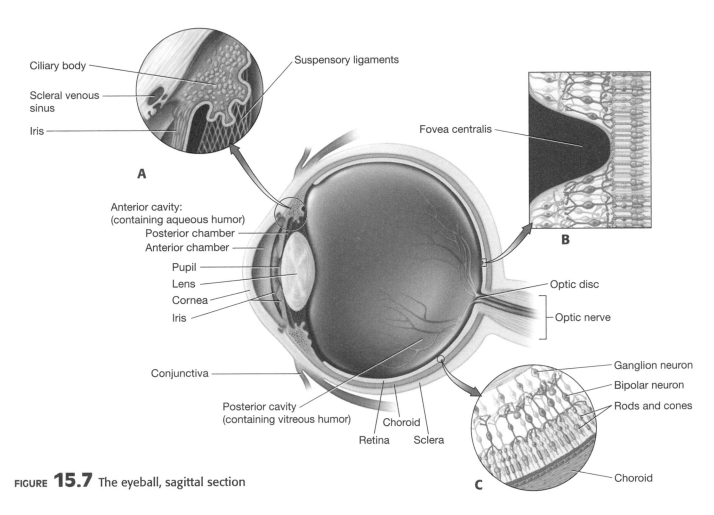

A

Ciliary body

Scleral venous sinus

Iris

Suspensory ligaments

Fovea centralis

B

Anterior cavity: (containing aqueous humor)
Posterior chamber
Anterior chamber
Pupil
Lens
Cornea
Iris

Conjunctiva

Posterior cavity (containing vitreous humor)

Choroid

Retina   Sclera

Optic disc

Optic nerve

Ganglion neuron

Bipolar neuron

Rods and cones

Choroid

C

FIGURE **15.7** The eyeball, sagittal section

Procedure Model Inventory

Identify the following structures of the eye and the eyeball on models and diagrams using your textbook and this unit for reference. As you examine the anatomical models and diagrams, record on the model inventory in Table 15.3 the name of the model and the structures you were able to identify.

Eyeball

1. Fibrous tunic
   a. Sclera
   b. Cornea
2. Vascular tunic (uvea)
   a. Choroid
   b. Ciliary body
   c. Suspensory ligaments
   d. Iris
   e. Pupil

3. Lens
4. Sensory tunic
   a. Retina
   b. Optic disc
   c. Macula lutea
   d. Fovea centralis
5. Optic nerve
6. Optic chiasma

7. Anterior cavity
   a. Anterior chamber
   b. Posterior chamber
   c. Aqueous humor
   d. Scleral venous sinus
8. Posterior cavity
   a. Vitreous humor
   b. Hyaloid canal

Accessory Structures

1. Palpebrae
2. Palpebral fissures
3. Medial and lateral canthi

4. Lacrimal apparatus
   a. Lacrimal gland
   b. Lacrimal caruncle
   c. Lacrimal canal
   d. Nasolacrimal duct

5. Conjunctiva
6. Extraocular muscles
   a. Superior oblique
   b. Inferior oblique

   c. Superior rectus
   d. Inferior rectus
   e. Medial rectus
   f. Lateral rectus

TABLE **15.3**	Model Inventory for the Eye
Model/Diagram	**Structures Identified**

Procedure Eyeball Dissection

In this exercise you will examine the structures of the eyeball on a fresh or preserved eyeball. I promise that eyeball dissection isn't as gross as it sounds!

Note: Goggles and gloves are required.

---

1 Examine the external anatomy of the eyeball (**Figure 15.8**), and record below the structures you can identify.

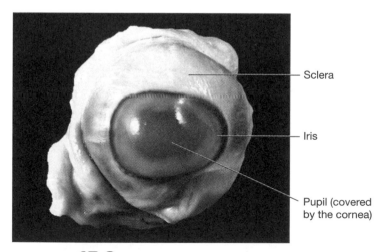

Sclera

Iris

Pupil (covered by the cornea)

FIGURE 15.8 Anterior view of an eyeball

EXPLORING ANATOMY & PHYSIOLOGY IN THE LABORATORY

2 Use scissors to remove the adipose tissue surrounding the eyeball. Identify the optic nerve.

3 Hold the eyeball at its anterior and posterior poles, and use a sharp scalpel or scissors to make an incision in the frontal plane. Watch out, as aqueous humor and vitreous humor are likely to spill everywhere.

4 Complete the incision, and separate the anterior and posterior portions of the eyeball (**Figure 15.9**). Take care to preserve the fragile retina—the thin, delicate yellow-tinted inner layer.

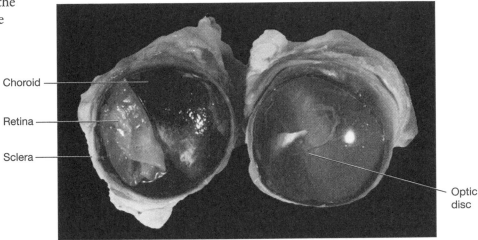

Choroid

Retina

Sclera

Optic disc

FIGURE **15.9** Frontal section of an eyeball showing the tunics

5 List the structures that you can identify in the anterior half of the eyeball (**Figure 15.10**):

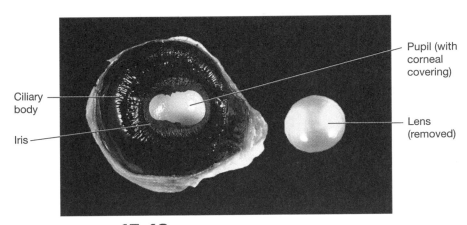

Ciliary body

Iris

Pupil (with corneal covering)

Lens (removed)

FIGURE **15.10** Posterior view of the anterior portion of the eyeball and lens

6 List the structures that you can identify in the posterior half of the eyeball:

Procedure Comparing the Distribution of Rods and Cones

We discussed earlier the unequal distribution of the photoreceptors in the retina. In this exercise you will see (no pun intended) firsthand the differences in vision produced by the rods and the vision produced by the cones. After you have completed the activity, answer Check Your Understanding question 1 (p. 359).

1 On a small sheet of paper, write the phrase "Anatomy is fun" in your regular-size print.

2 Hold this piece of paper about 10 inches directly in front of your lab partner's eyes, and have your lab partner read the phrase.

Can your partner read the phrase clearly?

Which photoreceptors are producing the image?

3 Now write a second phrase on the paper and don't tell your partner what the phrase says. Hold the paper about 10 inches from your partner's peripheral vision field. Have him or her continue to stare forward and attempt to read what you have written.

Can your partner read the phrase clearly?

Which photoreceptors are producing the image?

4 For the next test, dim the lights in the room. Have your partner stand 20 feet in front of a Snellen eye chart and read the chart. Record the number of the smallest line that he or she can read (e.g., 20/40).

Visual acuity:

5 With the lights still dimmed and your partner standing in the same place, hold a piece of dark green or dark blue paper over the Snellen chart. Ask your partner to identify the color of the paper you are holding:

Paper color:

6 Repeat the above processes with the lights illuminated:

Visual acuity:

Paper color:

7 In which scenario was visual acuity and color vision better? Explain your findings.

Procedure Testing the Extraocular Muscles

In this procedure, you will determine which extraocular muscles are responsible for moving the eyeballs in each direction.

1 Trace a horizontal line in the air about one foot in front of your partner's eyes, moving from right to left. Have your partner follow your finger without moving his or her head. Which extraocular muscles produce the movements you see for each eye?

Right eye _____

Left eye _____

2 Now trace a diagonal line, starting at the upper right corner and moving to the lower left corner. Have your partner follow your finger again. Which extraocular muscles produce the movements you see for each eye?

Right eye _____

Left eye _____

3 Again have your partner follow your finger, but trace a horizontal line from left to right. Which extraocular muscles produce the movements you see for each eye?

Right eye _____

Left eye _____

4 Finally, trace another diagonal, this time from the lower left to the upper right, and have your partner follow along (you should have traced out an hourglass shape overall). Which extraocular muscles produce the movements you see for each eye?

Right eye _____

Left eye _____

Exercise 2

Anatomy of the Ear, Hearing, and Equilibrium

MATERIALS NEEDED

- Ear models
- Tuning fork (500–1000 Hz)
- Chalk

The ear contains structures for both hearing and equilibrium. It is divided into three regions: the outer, middle, and inner ear (**Figure 15.11**).

1. **Outer ear.** The outer ear begins with the **auricle,** or *pinna*, a shell-shaped structure composed of elastic cartilage that surrounds the opening to the **external auditory canal.** The external auditory canal extends about 2.5 cm into the temporal bone, where it ends in the **tympanic membrane,** which separates the outer ear from the middle ear.

2. **Middle ear.** The middle ear is a small air-filled cavity within the temporal bone that houses tiny bones called the **auditory ossicles**—the **malleus** (hammer), **incus** (anvil), and **stapes** (stirrup). The ossicles transmit vibrations from the tympanic membrane to the inner ear through a structure called the **oval window.** An additional structure in the middle ear is the **pharyngotympanic tube** (also called the Eustachian or auditory tube), which connects the middle ear to the pharynx (throat) and equalizes pressure in the middle ear.

3. **Inner ear.** The inner ear contains the sense organs for hearing and equilibrium. It consists of cavities collectively called the **bony labyrinth** that are filled with a fluid called **perilymph.** Within the perilymph is a series of membranes called the **membranous labyrinth,** which contains a thicker fluid called **endolymph.** The bony labyrinth has three regions:

   a. **Vestibule.** The vestibule is an egg-shaped bony cavity that houses two structures responsible for equilibrium—the **saccule** and the **utricle.** Both structures transmit impulses down the vestibular portion of the vestibulocochlear nerve.

   b. **Semicircular canals.** Situated at right angles to one another, the semicircular canals house the **semicircular ducts** and the **ampulla,** which work together with the organs of the vestibule to maintain equilibrium. Their orientation allows them to sense rotational movements of the head and body. Like the saccule and utricle, the semicircular ducts and the ampulla transmit impulses down the vestibular portion of the vestibulocochlear nerve.

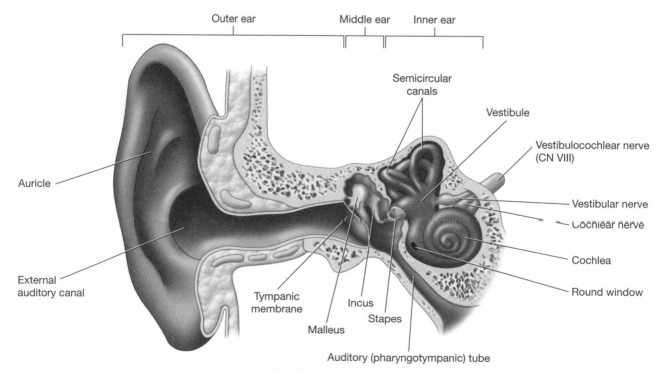

FIGURE 15.11 The anatomy of the ear

c. **Cochlea.** The cochlea is a spiral bony canal that contains the **organ of Corti**, whose specialized **hair cells** transmit sound impulses to the cochlear portion of the vestibulocochlear nerve.

 Procedure Model Inventory

Identify the following structures of the ear on models and diagrams, using your textbook and this unit for reference. As you examine the anatomical models and diagrams, record on the model inventory in Table 15.4 the name of the model and the structures you were able to identify. After you have completed the activity, answer Check Your Understanding questions 2 and 3 (p. 359).

1. Outer ear
   a. Auricle (pinna)
   b. External auditory canal
2. Middle ear
   a. Tympanic membrane
   b. Ossicles
      (1) Malleus
      (2) Incus
      (3) Stapes
   c. Oval window
   d. Round window
   e. Pharyngotympanic (auditory or Eustachian) tube

3. Inner ear
   a. Vestibule
      (1) Saccule
      (2) Utricle
   b. Semicircular canals
      (1) Semicircular duct
      (2) Ampulla
   c. Cochlea
      (1) Organ of Corti
      (2) Cochlear duct
   d. Vestibulocochlear nerve
      (1) Vestibular nerve
      (2) Cochlear nerve

TABLE **15.4**	Model Inventory for the Ear
Model/Diagram	**Structures Identified**

Hearing Acuity

There are two possible types of hearing loss:

1. **Conductive hearing loss** results from interference of sound conduction through the outer and/or middle ear.

2. **Sensorineural hearing loss** results from damage to the inner ear or the vestibulocochlear nerve.

Two clinical tests can help a healthcare professional determine if hearing loss is conductive or sensorineural—the Weber test and the Rinne test. Both tests use tuning forks that vibrate at specific frequencies when struck. The tuning forks are placed directly on the bones of the skull to evaluate bone conduction—the ability to hear the vibrations transmitted through the bone. The forks are then held near the ear, not touching bone, to evaluate air conduction—the ability to hear the vibrations transmitted through the air. After you have completed both tests, answer Check Your Understanding question 4 (p. 360).

 Procedure Weber Test

1 Obtain a tuning fork with a frequency of 500–1000 Hz (cycles per second).

2 Hold the tuning fork by the base and strike it lightly with a mallet or tap it on the edge of the table. The fork should begin ringing softly. If it is ringing too loudly, grasp the tines to stop it from ringing, and try again.

3 Place the base of the vibrating tuning fork on the midline of your partner's head.

4 Ask your partner if the sound is heard better in one ear or if the sound is heard equally in both ears. If the sound is heard better in one ear, this is called lateralization.

Was the sound lateralized? If yes, to which ear?

5 To illustrate what it would sound like if the sound were lateralized, have your partner place his or her finger in one ear. Repeat the test.

a. In which ear was the sound heard better?

b. If a patient has conduction deafness, in which ear do you think the sound will be heard most clearly (the deaf ear or the good ear)?_____ Why? (If you are confused, think about your results when one ear was plugged.)

c. If a patient has sensorineural deafness, in which ear do you think the sound will be best heard?

Procedure Rinne Test

1 Strike the tuning fork lightly to start it ringing.

2 Place the base of the tuning fork on your partner's mastoid process.

3 Time the interval during which your partner can hear the sound. Your partner will have to tell you when he or she can no longer hear the ringing.

Time interval in seconds:

4 Once your partner cannot hear the ringing, quickly move the still-vibrating tuning fork 1–2 cm lateral to the external auditory canal (the fork should not be touching your partner at this point).

5 Time the interval from the point when you moved the tuning fork in front of the external auditory canal to when your partner can no longer hear the sound.

Time interval in seconds:

Which situation tested bone conduction?

Which situation tested air conduction?

6 Typically, the air-conducted sound is heard twice as long as the bone-conducted sound. For example, if the bone-conducted sound was heard for 15 seconds, the air-conducted sound should be heard for 30 seconds.

Were your results normal?

What type of deafness is present if the bone-conducted sound is heard longer than the air-conducted sound?

Equilibrium

A common and simple test of equilibrium is the **Romberg test,** in which the person is asked to stand still, first with the eyes open and then with the eyes closed. Under normal conditions, the vestibular apparatus should be able to maintain equilibrium in the absence of visual input. If the vestibular apparatus is impaired, however, the brain relies on visual cues to maintain balance.

Procedure Romberg Test

1 Have your partner stand erect with the feet together and the arms at the sides in front of a chalkboard.

2 Use chalk to draw lines on the board on either side of your partner's torso. These lines are for your reference in the next part.

3 Have your partner stand in front of the chalkboard for 1 minute, staring forward with his or her eyes open. Use the lines on either side of his or her torso to note how much he or she sways as she stands. Below, record the amount of side-to-side swaying (i.e., minimal or significant):

4 Now have your partner stand in the same position for 1 minute with his or her eyes closed. Again note the amount of side-to-side swaying, using the chalklines for reference.

Was the amount of swaying more or less with his/her eyes closed?

Why do you think this is so?

What do you predict would be the result for a person with an impaired vestibular apparatus? Explain.

 Exercise 3

Olfactory and Taste Senses

MATERIALS NEEDED

- Head and neck models
- Tongue model

Both olfaction and taste are sometimes referred to as the **chemosenses,** because they both rely on chemoreceptors to relay information about the environment to the brain. The chemoreceptors of the olfactory sense are located in a small patch in the roof of the nasal cavity called the **olfactory epithelium (Figure 15.12).** The olfactory epithelium contains bipolar neurons called **olfactory receptor cells.** Their axons penetrate the holes in the cribriform plate to synapse on the olfactory bulb, which then sends the impulses down the axons of the olfactory nerves (cranial nerve I) to the olfactory cortex.

Taste receptors are located on taste buds housed on projections from the tongue called **papillae (Figure 15.13).** Of the four types of papillae—**filiform, fungiform, foliate,** and **circumvallate**—all but filiform papillae house taste buds. Fungiform papillae are scattered over the surface of the tongue, whereas the large circumvallate papillae are located at the posterior aspect of the tongue, arranged in a V-shape.

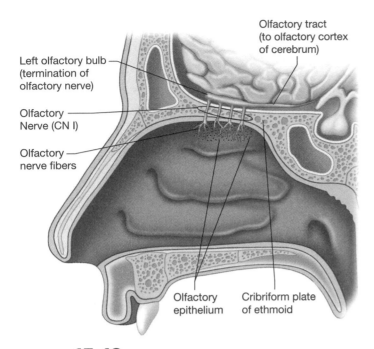

FIGURE **15.12** The nasal cavity and olfactory epithelium

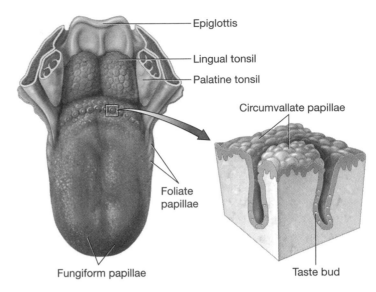

FIGURE **15.13** The surface of the tongue

 Procedure Model Inventory

Identify the following structures of the olfactory and taste senses on anatomical models and charts. As you examine the anatomical models and diagrams, record on the model inventory in Table 15.5 the name of the model and the structures you were able to identify. After you have completed the activity, answer Check Your Understanding question 5 (p. 360).

Olfaction

1. Nasal cavity
   a. Nasal conchae
   b. Nasal septum
2. Olfactory epithelium
   a. Olfactory receptor cells
3. Olfactory bulbs
4. Olfactory nerves
5. Cribriform plate

Taste

1. Papillae
   a. Fungiform papillae
   b. Circumvallate papillae
   c. Foliate papillae
2. Taste buds

TABLE **15.5**	Model Inventory for Olfaction and Taste
Model	**Structures Identified**

Exercise 4

The General Senses: Cutaneous Sensation

MATERIALS NEEDED

- Water-soluble marking pen (two colors)
- Ruler
- 2 wooden applicator sticks

Sensory receptors in the skin respond to different stimuli, including temperature, touch, and pain. These receptors are not distributed throughout the skin equally but, instead, are concentrated in certain regions of the body. The following experiments will allow you to determine the relative distribution of the receptors for touch in the skin by performing two tests: the error of localization and two-point discrimination. After you complete the activities, answer Check Your Understanding question 6 (p. 360).

Error of Localization

Every region of the skin corresponds to an area of the somatosensory association area in the cerebral cortex. Some regions are better represented than others and, therefore, are capable of localizing stimuli with greater precision than are less well-represented areas. The **error of localization** (also called tactile localization) tests the ability to determine the location of the skin that has been touched and demonstrates how well-represented each region of the skin is in the cerebral cortex.

 Procedure Testing Error of Localization

1 Have your partner sit with his or her eyes closed.

2 Use a water-soluble marking pen to place a mark on your partner's anterior forearm.

3 Using a different color of marker, have your partner, still with his or her eyes closed, place a mark as close as possible to where he or she believes the original spot is located.

4 Use a ruler to measure the distance between the two points in millimeters. This is your error of localization.

5 Repeat this procedure for each of the following locations:
  a. Anterior thigh
  b. Face
  c. Palm of hand
  d. Fingertip

6 Record your data in Table 15.6.

TABLE 15.6	Error of Localization
Location	**Error of Localization (mm)**
Anterior forearm	
Anterior thigh	
Face	
Palm of hand	
Fingertip	

Two-Point Discrimination

The **two-point discrimination test** assesses the ability to perceive the number of stimuli ("points") that are placed on the skin. Areas that have a higher density of touch receptors are better able to distinguish between multiple stimuli than those with fewer touch receptors.

Procedure Testing Two-Point Discrimination

1 Have your partner close his or her eyes.

2 Place the ends of two wooden applicator sticks close together (they should be nearly touching) on your partner's skin on the anterior forearm. Ask your partner how many points he or she can discriminate—one or two.

3 If he or she can sense only one point, move the sticks farther apart. Repeat this procedure until your partner can distinguish two separate points touching his or her skin.

4 Use a ruler to measure the distance between the two sticks in millimeters. This is your two-point discrimination.

5 Repeat this procedure for each of the following locations:

a. Anterior thigh
b. Face (around the lips and/or eyes)
c. Vertebral region
d. Fingertip

6 Record your data in Table 15.7.

7 What results did you expect for each test? Explain.

TABLE 15.7	Two-Point Discrimination
Location	**Two-Point Discrimination (mm)**
Anterior forearm	
Anterior thigh	
Face	
Vertebral region	
Fingertip	

8 Did your observations agree with your expectations? Interpret your results.

NAME _____

SECTION _____ DATE _____

Check Your Recall

1 Label the following parts of the eyeball on **Figure 15.14**.

- Sclera
- Ciliary body
- Optic nerve

- Cornea
- Suspensory ligaments
- Anterior cavity

- Iris
- Choroid
- Posterior cavity

- Lens
- Retina

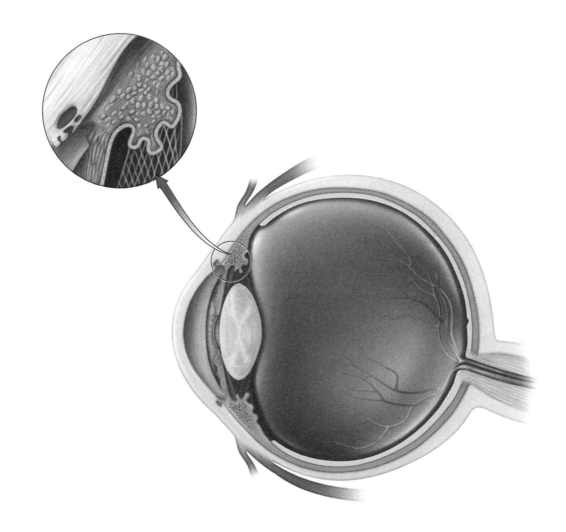

FIGURE **15.14** The eyeball, sagittal section

2 The lacrimal gland is located in the _____ and produces _____.
   a. superolateral orbit; mucus
   b. inferomedial orbit; tears
   c. superolateral orbit; tears
   d. inferomedial orbit; mucus

3 The rods are responsible for _____, whereas the cones are responsible for _____.
   a. peripheral and dim light vision; high-acuity color vision
   b. high-acuity color vision; peripheral and dim light vision
   c. peripheral and color vision; high-acuity and dim light vision
   d. high-acuity and dim light vision; peripheral and color vision

4 Label the following parts of the ear on **Figure 15.15.**
   - Auricle
   - External auditory canal
   - Tympanic membrane
   - Malleus
   - Incus
   - Stapes
   - Pharyngotympanic tube
   - Vestibule
   - Semicircular canals
   - Cochlea
   - Vestibulocochlear nerve

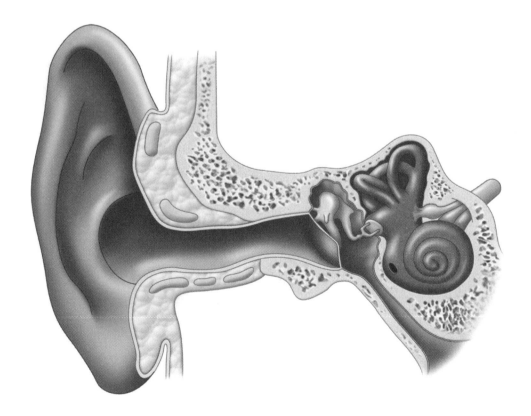

FIGURE **15.15** The anatomy of the ear

EXPLORING ANATOMY & PHYSIOLOGY IN THE LABORATORY

5 The auditory ossicles transmit vibrations from the _____ to the _____.
   a. auricle; malleus
   b. tympanic membrane; inner ear
   c. tympanic membrane; middle ear
   d. auricle; inner ear

6 The structures of the cochlea are responsible for _____, whereas the structures of the vestibule and semicircular canals are responsible for _____.
   a. equilibrium; balance
   b. equilibrium; hearing
   c. hearing; equilibrium
   d. hearing; audition

7 Mark the following statements as true (T) or false (F). If the statement is false, correct it so it is a true statement.

   _____ a. Conductive hearing loss results from damage to the inner ear or the vestibulocochlear nerve.

   _____ b. The ciliary muscle controls the diameter of the iris.

   _____ c. The Weber and Rinne tests assess balance and equilibrium.

   _____ d. The six extraocular muscles move the eyeball.

   _____ e. Smell and taste are considered chemosenses.

   _____ f. The photoreceptors are located in the vascular tunic of the eyeball.

8 The receptors for smell are located in the
   a. gustatory mucosa.
   b. olfactory epithelium.
   c. olfactory fossa.
   d. squamous epithelium.

9 Taste buds are located on the
   a. filiform, fungiform, and circumvallate papillae.
   b. filiform, fungiform, and foliate papillae.
   c. filiform and circumvallate papillae only.
   d. fungiform and foliate papillae only.

10 You would expect the error of localization and the two-point discrimination threshold to be lowest on the

    a. back.

    b. forearm.

    c. fingertip.

    d. thigh.

Check Your Understanding | **Critical Thinking and Application Questions**

1 The disease *macular degeneration* is characterized by a gradual loss of vision as a result of degeneration of the macula lutea. Considering the type of cells located in the macula lutea, which type of vision do you think a sufferer of macular degeneration would lose? Why?

2 What signs and symptoms would you expect to see from *otitis interna* (inner ear infection)? Why?

3 Explain why infectious *otitis media* (inflammation of the middle ear) may result in a simultaneous *pharyngitis* (inflammation of the throat).

4 *Otosclerosis* is a condition that results in irregular ossification (bone formation) around the stapes bone. Would you expect this to result in conductive or in sensorineural hearing loss? What results would you expect from the Rinne and Weber tests?

5 The chemoreceptors of the tongue respond to monosaccharides (sugars), metal ions (salty tastes), alkaloids (bitter tastes), and acids (sour tastes). Which type of receptor do you think is most sensitive? Why? (*Hint:* One of the four types of chemicals is present in many plant poisons.)

6 The *sensory homunculus* is a small figure of a man whose body parts are represented in proportion to the amount of the somatosensory cortex dedicated to them. Parts of the body that are well represented in the somatosensory cortex are drawn as larger, and those that are poorly represented in the somatosensory cortex are drawn as smaller. Which areas of the body do you think would be larger on the homunculus? Which areas do you think would be smaller? Explain your reasoning.

The Cardiovascular System—Part I: The Heart

OBJECTIVES

Once you have completed this unit, you should be able to:

- Identify structures of the heart.

- Trace the pathway of blood flow through the heart.

- Describe the histology of cardiac tissue.

- Identify components of the cardiac conduction system and trace an action potential through the pacemaker and nonpacemaker cells of the heart.

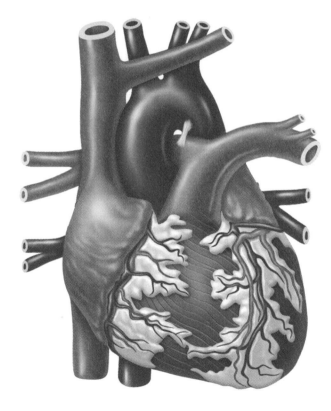

Pre-Lab Exercise 1
Key Terms

Table 16.1 lists the key terms with which you should be familiar before coming to lab.

TABLE 16.1	Key Terms

Term   **Definition**

Layers of the Heart Wall

Fibrous pericardium _____

_____

Serous pericardium (parietal and visceral layers) _____

_____

Myocardium _____

_____

Endocardium _____

_____

Structures of the Heart

Atria (right and left) _____

_____

Ventricles (right and left) _____

_____

Tricuspid valve _____

_____

Mitral (bicuspid) valve _____

_____

Pulmonary valve _____

_____

Aortic valve _____

_____

Chordae tendineae _____

_____

Papillary muscles _____

_____

Great Vessels

Superior vena cava _____

_____

Inferior vena cava _____

_____

Pulmonary trunk (artery) _____

_____

Pulmonary veins _____

_____

Aorta _____

_____

Structures of the Cardiac Conduction System

Sinoatrial (SA) node _____

_____

Atrioventricular (AV) node _____

_____

Atrioventricular bundle (bundle of His) _____

_____

Bundle branches _____

_____

Purkinje fibers _____

_____

Pre-Lab Exercise 2
Anatomy of the Heart

Label and color the three views of the heart in **Figure 16.1** with the terms from Exercise 1. Use your text and Exercise 1 in this unit for reference.

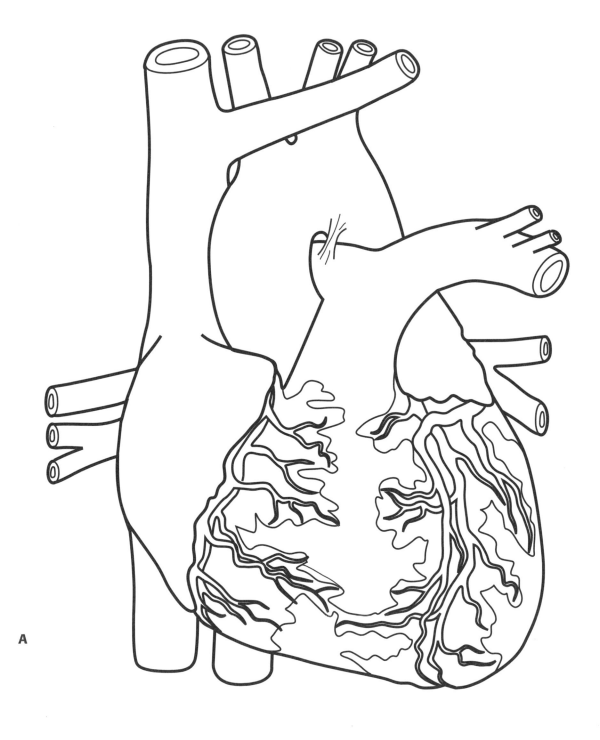

A

FIGURE 16.1 The heart: (A) anterior view

EXPLORING ANATOMY & PHYSIOLOGY IN THE LABORATORY

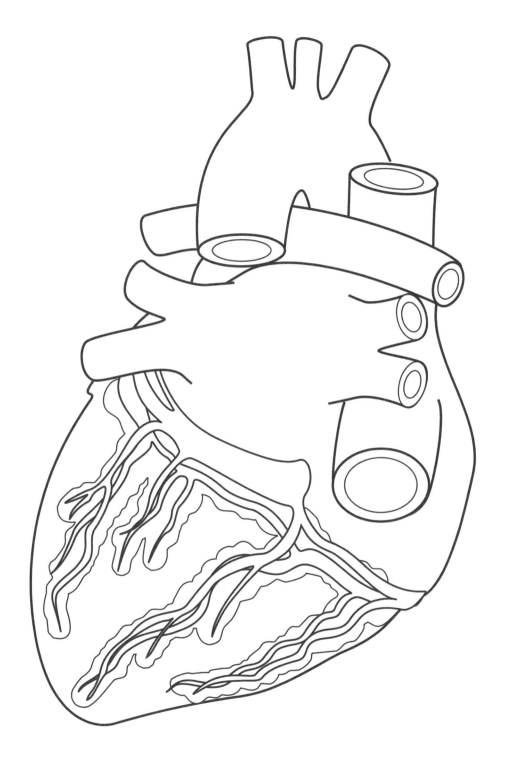

B

FIGURE **16.1** The heart (cont.): (B) posterior view

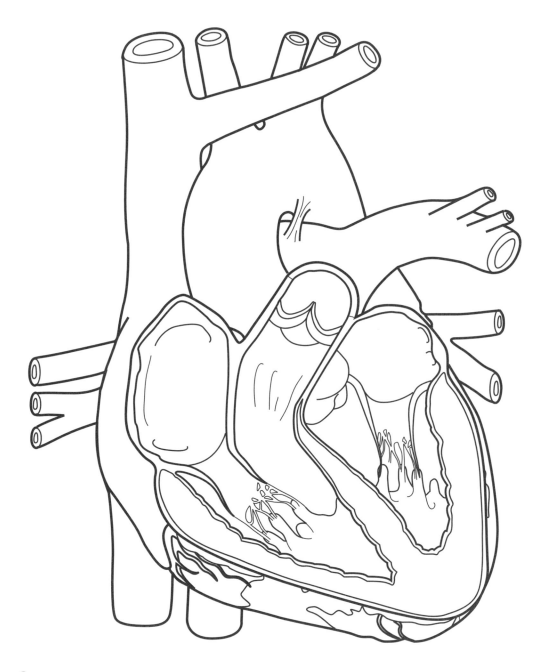

C

FIGURE **16.1** The heart (cont.): (C) frontal section

EXPLORING ANATOMY & PHYSIOLOGY IN THE LABORATORY

Pre-Lab Exercise 3
Pathway of Blood Flow Through the Heart

Answer the following questions about the pathway of blood flow through the heart. Use your textbook and Exercise 1 in this unit for reference.

1. Regarding veins:
   a. Where do veins carry blood?

   b. Is this blood generally oxygenated or deoxygenated?

   c. Does this rule have any exceptions? If yes, where?

2. Regarding arteries:
   a. Where do arteries carry blood?

   b. Is this blood generally oxygenated or deoxygenated?

   c. Does this rule have any exceptions? If yes, where?

3. Where does each atrium pump blood when it contracts?
   a. Right atrium:

   b. Left atrium:

4. Where does each ventricle pump blood when it contracts?
   a. Right ventricle:

   b. Left ventricle:

Exercises

The cardiovascular system transports oxygen, nutrients, wastes, other solutes, and cells throughout the body. In this unit we begin our exploration of the cardiovascular system with the pump that drives it—the heart. The heart is a remarkable organ, tirelessly beating more than 100,000 times per day to pump more than 8,000 liters of blood around the body. In this unit we examine the anatomy of this remarkable organ, including the blood flow through the heart and the histology of cardiac muscle. We also examine the electrical events of cardiac action potentials and the group of cells called the *cardiac conduction system*, which allows the heart to function as a unit.

Exercise 1

Anatomy of the Heart

MATERIALS NEEDED

- Heart models
- Preserved heart
- Dissection equipment
- Dissection tray
- Water-soluble marking pen
- Laminated outline of the heart and lungs

The heart lies in the mediastinum and is, on average, about the size of a fist. The heart is surrounded by a double-layered membrane called the **pericardium**. The outermost layer of the pericardium, called the **fibrous pericardium,** anchors the heart to surrounding structures. It is made of dense irregular collagenous connective tissue that is not very distensible, which helps to prevent the heart from overfilling.

The inner layer, called the **serous pericardium**, is itself composed of two layers. The outer portion, called the **parietal pericardium,** is functionally fused to the fibrous pericardium. The inner portion is attached to the heart muscle and is called the **visceral pericardium,** also known as the **epicardium.** Between the parietal and visceral layers we find a thin layer of serous fluid that occupies a narrow space called the **pericardial cavity.** The fluid within the pericardial cavity helps the heart to beat without friction.

The heart itself is an organ that consists of three tissue layers:

1. **Epicardium.** The epicardium or visceral pericardium is considered the outermost layer of the heart wall. It consists of a layer of epithelial tissue and loose connective tissue.

2. **Myocardium.** The middle myocardium is the actual muscle of the heart. It consists of cardiac muscle tissue and its fibrous skeleton.

3. **Endocardium.** The innermost endocardium is a type of simple squamous epithelium called **endothelium.** It is continuous with the endothelium lining all blood vessels in the body.

As you can see in **Figure 16.2,** the heart consists of four hollow chambers: two **atria** that receive blood from the body's **veins,** and two **ventricles** that eject blood into the body's **arteries.** In between the atria is a thin wall called the **interatrial septum.** The much thicker **interventricular septum** separates the two ventricles. The heart's four chambers include the:

1. **Right atrium.** The right atrium is the superior right chamber. It receives deoxygenated blood from the body's main veins—the **superior vena cava,** the **inferior vena cava,** and the **coronary sinus.** These veins drain a series of blood vessels collectively called the **systemic circuit,** in which gases and nutrients are exchanged between the blood and the tissue cells.

2. **Right ventricle.** The right ventricle is a large chamber inferior to the right atrium, from which it receives deoxygenated blood. It ejects blood into a vessel called the **pulmonary trunk,** or the **pulmonary artery.** The pulmonary trunk branches into right and left pulmonary arteries, which deliver deoxygenated blood to the lungs through a series of vessels collectively called the **pulmonary circuit.** Within the pulmonary circuit, gases are exchanged and the blood becomes oxygenated.

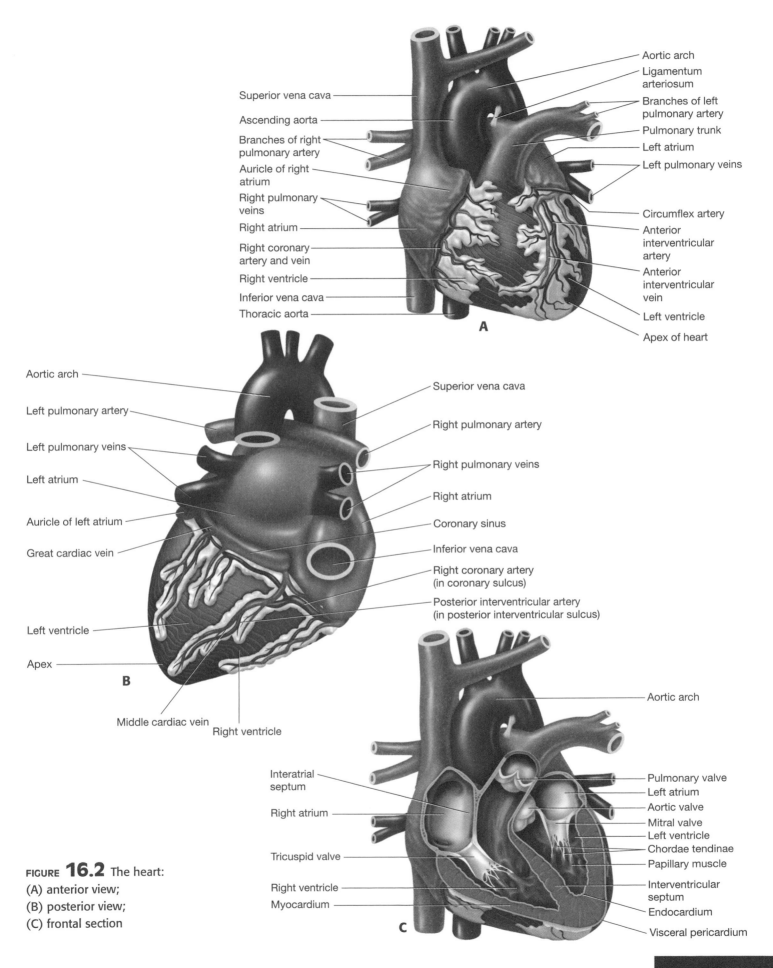

FIGURE 16.2 The heart:
(A) anterior view;
(B) posterior view;
(C) frontal section

A

- Aortic arch
- Ligamentum arteriosum
- Branches of left pulmonary artery
- Pulmonary trunk
- Left atrium
- Left pulmonary veins
- Circumflex artery
- Anterior interventricular artery
- Anterior interventricular vein
- Left ventricle
- Apex of heart

- Superior vena cava
- Ascending aorta
- Branches of right pulmonary artery
- Auricle of right atrium
- Right pulmonary veins
- Right atrium
- Right coronary artery and vein
- Right ventricle
- Inferior vena cava
- Thoracic aorta

B

- Aortic arch
- Left pulmonary artery
- Left pulmonary veins
- Left atrium
- Auricle of left atrium
- Great cardiac vein
- Left ventricle
- Apex

- Superior vena cava
- Right pulmonary artery
- Right pulmonary veins
- Right atrium
- Coronary sinus
- Inferior vena cava
- Right coronary artery (in coronary sulcus)
- Posterior interventricular artery (in posterior interventricular sulcus)

- Middle cardiac vein
- Right ventricle

C

- Interatrial septum
- Right atrium
- Tricuspid valve
- Right ventricle
- Myocardium

- Aortic arch
- Pulmonary valve
- Left atrium
- Aortic valve
- Mitral valve
- Left ventricle
- Chordae tendinae
- Papillary muscle
- Interventricular septum
- Endocardium
- Visceral pericardium

3. **Left atrium.** The superior left chamber is the left atrium. It receives oxygenated blood returning from the pulmonary circuit via four **pulmonary veins.**

4. **Left ventricle.** The left ventricle receives oxygenated blood from the left atrium and pumps it into the largest artery in the systemic circuit, the **aorta.** The aorta then branches repeatedly to deliver the oxygenated blood to the body's cells. Note that the left ventricle is considerably thicker than the right ventricle, reflecting the fact that the pressure is much higher in the systemic circuit than it is in the pulmonary circuit. The higher pressure requires the left ventricle to pump harder, and thus it is thicker.

HINTS & TIPS

Red or Blue?

On anatomical models, vessels that carry oxygenated blood are red, whereas those that carry deoxygenated blood are blue. Systemic arteries carry oxygenated blood to the body's cells, and so are red on anatomical models. Systemic veins, on the other hand, carry deoxygenated blood back to the right atrium and so are blue. But be sure to remember that the reverse is true in the pulmonary circuit: The pulmonary arteries carry deoxygenated blood to the lungs and the pulmonary veins carry oxygenated blood to the heart. So, in the pulmonary circuit, the arteries are blue and the veins are red.

In between the chambers of the heart and the ventricles and their vessels are **valves** that prevent the blood from flowing backward in the heart. The valves between the chambers are called **atrioventricular valves.** The three-cusped **tricuspid valve** is between the right atrium and right ventricle, and the two-cusped **mitral** or **bicuspid valve** is between the left atrium and left ventricle. Each cusp of the atrioventricular valves is attached to collagenous "strings" called **chordae tendineae,** which are attached to muscles within the ventricular wall called **papillary muscles.** When the ventricles contract, the papillary muscles pull the chordae tendineae taut, which puts tension on the cusps and prevents them from everting into the atria, a condition called **prolapse.**

The valves between the ventricles and their arteries are called **semilunar valves.** The **pulmonary valve** lies between the right ventricle and the pulmonary trunk, and the **aortic valve** lies between the left ventricle and the aorta. Note that there are no chordae tendineae or papillary muscles attached to the semilunar valves.

The final structures you will examine in this lab period are the vessels of the **coronary circulation.** The **coronary arteries** branch off the base of the aorta and bring oxygenated blood to the cells of the myocardium, and they are drained by a set of **cardiac veins.** The first coronary artery, the **right coronary artery,** travels in the right coronary sulcus. It branches into the **marginal artery,** which serves the lateral part of the right atrium and right ventricle, and the **posterior interventricular artery,** which serves the posterior heart. The other coronary artery is the **left coronary artery,** which branches shortly after it forms into the **anterior interventricular artery** (also known as the **left anterior descending artery**), which travels along the interventricular septum to supply the anterior heart, and the **circumflex artery,** which travels in the left coronary sulcus to supply the left atrium and posterior left ventricle. When a coronary artery is blocked, the reduced blood flow to the myocardium can result in hypoxic injury and death to the tissue, a condition termed **myocardial infarction** (commonly called a heart attack).

The anatomy of the cardiac veins often varies from person to person, but the following three main veins generally are present:

1. **small cardiac vein,** which drains the inferolateral heart,

2. **middle cardiac vein,** which drains the posterior heart, and

3. **great cardiac vein,** which drains most of the left side of the heart.

All three veins drain into the large **coronary sinus** located on the posterior right atrium. The coronary sinus drains into the right atrium.

Procedure Model Inventory

Identify the following structures of the heart on models and diagrams, using your textbook and this unit for reference. As you examine the anatomical models and diagrams, record on the model inventory in Table 16.2 the name of the model and the structures you were able to identify. When you have completed the activity, answer Check Your Understanding questions 1 through 3 (p. 383).

1. General structures
   a. Mediastinum
      (1) Pericardial cavity
   b. Pericardium
      (1) Fibrous pericardium
      (2) Parietal pericardium
      (3) Visceral pericardium (epicardium)
   c. Myocardium
   d. Endocardium
   e. Apex of the heart
   f. Base of the heart
   g. Anterior interventricular sulcus
   h. Posterior interventricular sulcus
   i. Coronary sulcus

2. Right atrium
   a. Opening of the superior vena cava
   b. Opening of the inferior vena cava
   c. Opening of the coronary sinus
   d. Right auricle
   e. Interatrial septum
   f. Fossa ovalis
   g. Pectinate muscles

3. Left atrium
   a. Opening of the pulmonary veins
   b. Left auricle
   c. Interatrial septum

4. Structures of the ventricles
   a. Right ventricle
   b. Left ventricle
   c. Interventricular septum
   d. Chordae tendineae
   e. Papillary muscles
   f. Trabeculae carneae

5. Atrioventricular valves
   a. Tricuspid valve
   b. Mitral valve

6. Semilunar valves
   a. Pulmonary valve
   b. Aortic valve

7. Great vessels
   a. Superior vena cava
   b. Inferior vena cava
   c. Pulmonary trunk (artery)
   d. Right and left pulmonary arteries
   e. Pulmonary veins
   f. Aorta
   g. Ligamentum arteriosum

8. Coronary arteries
   a. Right coronary artery
   b. Marginal artery
   c. Posterior interventricular artery
   d. Left coronary artery
   e. Anterior interventricular artery
   f. Circumflex artery

9. Cardiac veins
   a. Small cardiac vein
   b. Middle cardiac vein
   c. Great cardiac vein
   d. Coronary sinus

TABLE **16.2**	Model Inventory for the Heart
Model	**Structures Identified**

Procedure Heart Dissection

You will now examine a preserved heart or a fresh heart, which is likely from a sheep or a cow. Follow the procedure below to find the structures indicated.

Note: Safety glasses and gloves are required.

---

1 Orient yourself by first determining the superior aspect and the inferior aspect of the heart. The superior aspect of the heart is the broad end, and the inferior aspect (apex) is the pointy tip. Now orient yourself to the anterior and posterior sides. The easiest way to do this is to locate the pulmonary trunk—the vessel directly in the middle of the anterior side. Find the side from which the pulmonary trunk originates, and you will be on the anterior side. Structures to locate at this time are the

  a. parietal pericardium (may not be attached),
  b. visceral pericardium (shiny layer over the surface of the heart),
  c. aorta,
  d. pulmonary trunk,
  e. superior vena cava,
  f. inferior vena cava,
  g. pulmonary veins,
  h. ventricles, and
  i. atria.

Finding the coronary vessels tends to be difficult because the superficial surface of the heart is covered with adipose tissue. To see the coronary vessels, carefully dissect the adipose tissue.

2 Locate the superior vena cava. Insert scissors or a scalpel into the superior vena cava and cut down into the right atrium. Before moving on to Step 3, note the structure of the tricuspid valve and draw it below.

3 Once the right atrium is exposed, continue the cut down into the right ventricle. Structures to locate at this time include the

    a. tricuspid valve,
    b. chordae tendineae,
    c. papillary muscles,
    d. myocardium, and
    e. endocardium (shiny layer on the inside of the heart).

4 Insert the scissors into the pulmonary trunk. Note the structure of the pulmonary valve and draw it below.

5 Insert the scissors into a pulmonary vein. Cut down into the left atrium. Note the structure of the mitral valve, and draw it below.

6 Continue the cut into the left ventricle. Note the thickness of the left ventricle. How does it compare to the thickness of the right ventricle?

7 Insert the scissors into the aorta. Extend the cut until you can see the aortic valve. Draw the aortic valve below.

EXPLORING ANATOMY & PHYSIOLOGY IN THE LABORATORY

8 Your instructor may wish you to identify other structures on the heart. List any additional structures below.

Procedure Tracing Blood Through the Heart

Use water-soluble markers and a laminated outline of the heart to trace the pathway of blood as it flows through the heart and pulmonary circulation. Use a blue marker to indicate areas that contain deoxygenated blood and a red marker to indicate areas that contain oxygenated blood. If no laminated outline is available, use **Figure 16.3**.

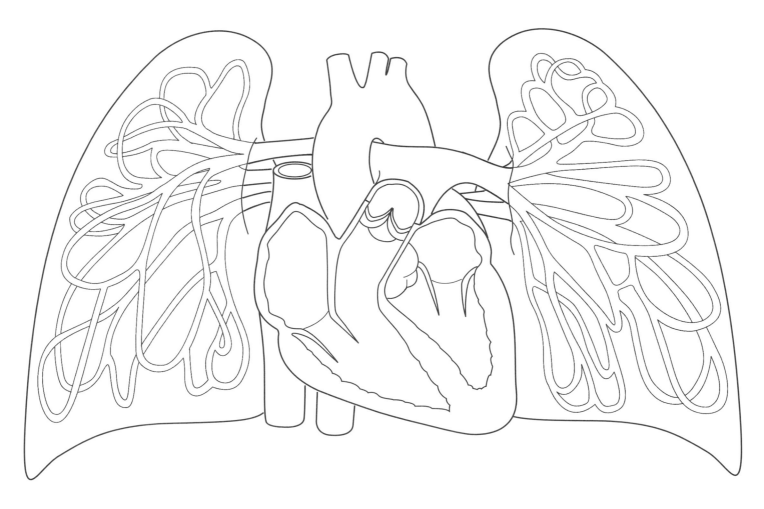

FIGURE **16.3** Heart, lungs, and pulmonary circulation

Exercise 2

Cardiac Muscle Histology

MATERIALS NEEDED

- Cardiac muscle tissue slide
- Light microscope
- Colored pencils

Recall from Unit 5 (p. 98) that cardiac muscle tissue is striated like skeletal muscle tissue but otherwise is quite different (**Figure 16.4**). Following are some important differences:

  - ▶ Cells of cardiac muscle, known as **cardiac myocytes,** are shorter, wider, and branched, whereas skeletal muscle fibers are long, thin, and unbranched.

▶ Cardiac muscle cells typically are uninucleate. The nucleus is generally located in the center of the cell.

▶ These cells contain specialized adaptations called **intercalated discs,** which appear as dark lines that run parallel to the striations. They contain desmosomes and gap junctions that hold adjacent cardiac cells tightly together and to allow the cells to communicate chemically and electrically. The intercalated discs enable the heart to contract as a unit.

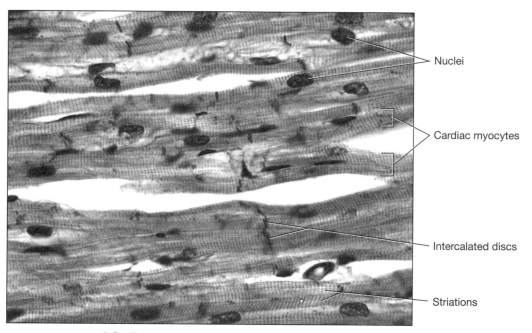

Nuclei

Cardiac myocytes

Intercalated discs

Striations

FIGURE **16.4** Cardiac muscle tissue, photomicrograph

Procedure Microscopy of Cardiac Muscle Tissue

Examine a prepared slide of cardiac muscle tissue on high power. Use colored pencils to draw what you see, and label the structures indicated. When you have completed the activity, answer Check Your Understanding question 4 (p. 384).

Label the following structures:

1. Cardiac myocyte
2. Intercalated discs
3. Striations
4. Nucleus

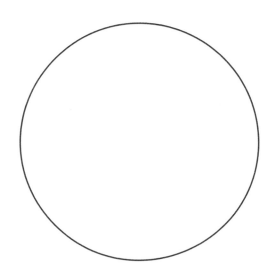

The Cardiac Conduction System

The heart has two populations of cardiac myocytes. The majority of the cells (about 99%) are known as **nonpacemaker cells,** which are regular, contractile myocytes. The remaining 1% are specialized cells known as **pacemaker cells.** Pacemaker cells have the unique property of spontaneously depolarizing and generating action potentials. The action potentials they initiate trigger action potentials of the nonpacemaker cells, which is the stimulus for these cells to contract.

The heart's pacemaker cells are located at specific places throughout the heart in clusters. Together, these clusters of pacemaker cells make up the **cardiac conduction system.** The parts of the cardiac conduction system, shown in **Figure 16.5,** include the following:

1. **Sinoatrial (SA) node.** The SA node is located in the upper right atrium, where it acts as the main pacemaker of the heart. It depolarizes spontaneously around 60 times per minute. (Note that the heart rate is generally higher than 60 due to the influence of the autonomic nervous system.)

2. **Atrioventricular (AV) node.** The AV node is located in the lower right atrium just medial to the tricuspid valve. It acts as a backup pacemaker and is capable of pacing the heart at about 40 beats per minute. Conduction through the AV node is slow, producing what is known as the **AV node delay.** This delay allows the atria to depolarize and contract before the ventricles, which allows the ventricles to fill with blood.

3. **Purkinje system.** The impulses are transmitted from the AV node to the ventricles by the group of pacemaker cells collectively called the Purkinje system. The myocytes of the Purkinje system pace the heart slowly but conduct impulses more rapidly than any part of the conduction system. The Purkinje system has three components:

   a. **Atrioventricular (AV) bundle.** The AV bundle is the small group of fibers in the lower interatrial septum and upper interventricular septum that transmits impulses from the AV node to the ventricles.

   b. **Right and left bundle branches.** Impulses are transmitted from the AV bundle down either side of the interventricular septum by the right and left bundle branches.

   c. **Purkinje fibers.** At the end of the interventricular septum, the right and left bundle branches fan out through the myocardium as the Purkinje fibers. These fibers extend about one-third of the way into the heart muscle, after which they blend with regular nonpacemaker cardiac myocytes.

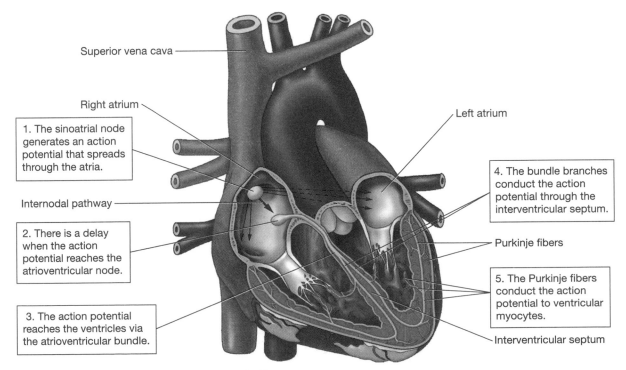

FIGURE 16.5 The cardiac conduction system

The general pathway of impulse conduction through the heart starts with the cells of the SA node generating an action potential. This pacemaker action potential spreads to the nonpacemaker cells of the atria and triggers them to have action potentials and contract. The impulses are carried through the atria until they hit the AV node, where they encounter a delay. Once the AV nodal cells are depolarized, this spreads to the Purkinje system and finally to the non-pacemaker cardiac myocytes of the ventricles. These cells are then triggered to depolarize and contract. Note that while ventricular depolarization is occurring, the atria are repolarizing. As the ventricles are repolarizing, the cells of the SA node are slowly depolarizing, which begins the cycle again.

Procedure Tracing the Electrical Events of the Heart

Let's put all of these electrophysiological events together and trace the electrical events that occur within the heart, including

- ▶ the steps of the pacemaker action potentials that initiate the cycle;
- ▶ the spread of electrical activity through the atria;
- ▶ the steps of the nonpacemaker action potentials of the atria;
- ▶ the spread of electrical activity through the ventricles;
- ▶ the steps of the nonpacemaker action potentials of the ventricles through repolarization.

You will want to refer to your textbook for parts of this exercise—particularly the steps of the pacemaker and non-pacemaker cell action potentials. You will find hints and blank spaces throughout this tracing exercise, to ensure that you stay on the right track. When you have completed the exercise, answer Check Your Understanding question 5 (p. 384).

Start ⟶ SA nodal cells slowly depolarize to threshold ⟶

_____ ⟶

_____ ⟶

nonpacemaker atrial cells depolarize to threshold ⟶

_____ ⟶

_____ ⟶

calcium channels open and the nonpacemaker atrial cells enter the plateau phase ⟶

_____ ⟶

_____ ⟶

impulse delayed at the AV node ⟶

_____ ⟶

_____ ⟶

_____ ⟶

nonpacemaker ventricular cells depolarize to threshold ⟶

_____ ⟶

_____ ⟶

_____ ⟶

nonpacemaker ventricular cells repolarize as potassium ions exit the cells and returns them to their resting membrane potential ⟶ **End**

NAME _____

SECTION _____ DATE _____

 Check Your Recall

1 Label the following parts of the heart on **Figure 16.6.**

- Superior vena cava
- Inferior vena cava
- Pulmonary trunk
- Pulmonary veins
- Aorta
- Right coronary artery
- Marginal artery
- Anterior interventricular artery
- Circumflex artery

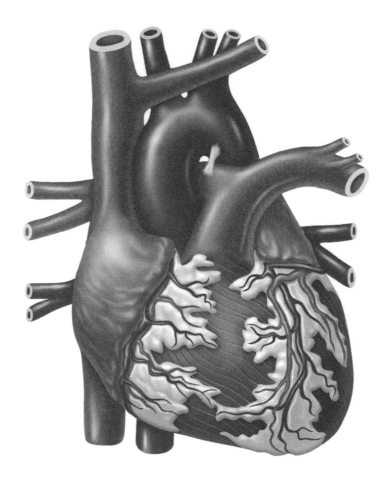

FIGURE **16.6** Heart, anterior view

2 Label the following parts
of the heart on **Figure 16.7.**

- Right atrium
- Left atrium
- Right ventricle
- Left ventricle
- Tricuspid valve
- Mitral valve
- Pulmonary valve
- Aortic valve
- Chordae tendineae
- Papillary muscles
- Interventricular septum
- Interatrial septum

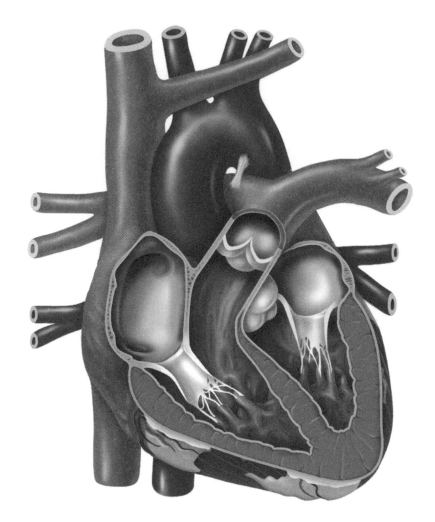

FIGURE 16.7 Heart,
frontal section

3 Matching: Match the following terms with the correct definition.

a. Myocardium _____   A. Located between the left ventricle and the aorta

b. Parietal pericardium _____   B. Located between the left atrium and left ventricle

c. Tricuspid valve _____   C. Largest artery of the systemic circuit

d. Aortic valve _____   D. Bring(s) oxygenated blood to the left atrium from the lungs

e. Papillary muscles _____   E. Layer of the heart composed of cardiac muscle tissue

f. Pulmonary veins _____   F. Bring(s) deoxygenated blood to the lungs

g. Mitral valve _____   G. Located between the right atrium and the right ventricle

h. Visceral pericardium _____   H. Largest vein of the coronary circulation

i. Pulmonary trunk _____   I. Outer layer of the serous pericardium

j. Coronary sinus _____   J. Fibrous cord(s) that attach(es) to valves

k. Chordae tendineae _____   K. Inner layer of the serous pericardium

l. Aorta _____   L. Fingerlike muscular projections from the ventricles

4 The veins of the systemic circuit carry _____ blood and the veins of the pulmonary circuit carry _____ blood.

a. oxygenated; deoxygenated
b. oxygenated; oxygenated
c. deoxygenated; deoxygenated
d. deoxygenated; oxygenated

5 The pulmonary and aortic valves are known as the

a. atrioventricular (AV) valves.
b. semilunar valves.
c. coronary valves.
d. chordae tendineae.

6 The right and left coronary arteries are the first branches off the

a. aorta.
b. superior vena cava.
c. pulmonary trunk.
d. pulmonary veins.

7 Fill in the blanks: Cardiac muscle cells are also known as _____. Adjacent cells are joined together

by _____, which allow the heart to _____.

8 Label the following components of the cardiac conduction system on **Figure 16.8**.

- SA node
- AV node
- AV bundle
- Right bundle branch
- Left bundle branch
- Purkinje fibers

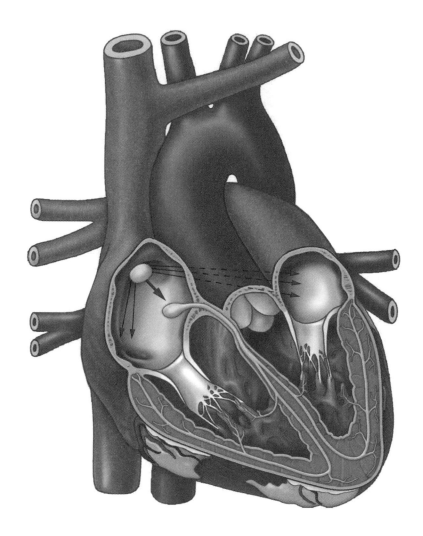

FIGURE **16.8** The cardiac conduction system

9 Which of the following is the normal pacemaker of the heart?

a. AV node
b. AV bundle
c. Purkinje fibers
d. SA node

10 The purpose of the AV node delay is to

a. allow the atria and ventricles to depolarize and contract simultaneously.
b. allow the ventricles to depolarize and contract before the atria.
c. allow the atria to depolarize and contract before the ventricles and allow the ventricles to fill.
d. allow the right side of the heart to depolarize and contract before the left side of the heart.

EXPLORING ANATOMY & PHYSIOLOGY IN THE LABORATORY

Check Your Understanding | **Critical Thinking and Application Questions**

1 When the pericardium fills with blood, it produces a condition called *cardiac tamponade*, which can be rapidly lethal. Why is this condition so dangerous? (*Hint:* Consider the structure of the fibrous pericardium.)

2 High pressure in the systemic and pulmonary circuits often results in *ventricular hypertrophy*, in which the ventricle enlarges to pump against greater force. Which side(s) of the heart would be affected by high pressure in the pulmonary circuit? Which side(s) of the heart would be affected by high pressure in the systemic circuit? Explain.

3 Note the smooth texture of the valves and how the leaflets fit together. How do you think the function of the valves would be affected if the valves were tough and filled with calcium deposits? Explain.

4 Skeletal muscle cells exhibit a phenomenon known as *recruitment*, in which the number of muscle cells recruited to contract is proportional to the strength of muscle contraction needed. In this way, we activate a few fibers to produce a small contraction to pick up a piece of paper and we activate many fibers to produce a larger contraction to pick up a textbook. Would you expect to see recruitment in cardiac muscle tissue? Why or why not?

5 One of the more common electrical abnormalities seen in the heart is a *heart block*, in which conduction is blocked or delayed along some part of the cardiac conduction system. Predict what you would see if a heart block were present in each of the following locations:

a. AV node:

b. Right or left bundle branches:

c. Purkinje fibers:

The Cardiovascular System—Part II: Blood Vessel Anatomy

OBJECTIVES

Once you have completed this unit, you should be able to:

- Identify selected arteries and veins.

- Describe the organs and regions supplied and drained by each artery and vein.

- Describe the unique blood flow patterns through the brain and the hepatic portal system.

- Trace the pathway of blood flow through various arterial and venous circuits.

- Describe the histological differences between arteries and veins.

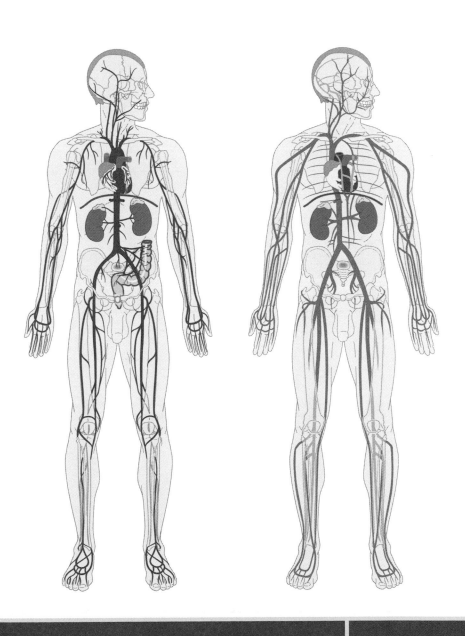

Pre-Lab Exercise 1

Key Terms: Arteries

Describe the location and the organ or region of the body supplied by each of the arteries in Table 17.1.

TABLE 17.1	Major Arteries of the Body	
Artery	**Location**	**Organ/Region Supplied**
Arteries of the Trunk		
Brachiocephalic artery		
Celiac trunk		
Common iliac artery		
Hepatic artery		
Inferior mesenteric artery		
Internal iliac artery		
Left gastric artery		
Renal artery		
Splenic artery		
Superior mesenteric artery		
Arteries of the Head and Neck		
Anterior cerebral artery		
Basilar artery		
Circle of Willis		
Common carotid artery		

Artery	Location	Organ/Region Supplied
External carotid artery		
Internal carotid artery		
Middle cerebral artery		
Posterior cerebral artery		
Temporal artery		
Vertebral artery		
Arteries of the Upper Limb		
Axillary artery		
Brachial artery		
Radial artery		
Subclavian artery		
Ulnar artery		
Arteries of the Lower Limb		
Anterior tibial artery		
Dorsalis pedis artery		
External iliac artery		
Femoral artery		
Popliteal artery		
Posterior tibial artery		

Pre-Lab Exercise 2
Key Terms: Veins

Describe the location and the organ or region of the body drained by each of the veins in Table 17.2.

TABLE 17.2	Major Veins of the Body	
Vein	**Location**	**Organ/Region Drained**
Veins of the Trunk		
Brachiocephalic vein		
Common iliac vein		
Hepatic portal vein		
Hepatic vein		
Inferior mesenteric vein		
Internal iliac vein		
Renal vein		
Splenic vein		
Superior mesenteric vein		
Veins of the Head and Neck		
Dural sinuses		
External jugular vein		

Vein	Location	Organ/Region Drained
Internal jugular vein		
Veins of the Upper Limb		
Basilic vein		
Brachial vein		
Cephalic vein		
Median cubital vein		
Radial vein		
Subclavian vein		
Ulnar vein		
Veins of the Lower Limb		
Anterior tibial vein		
External iliac vein		
Femoral vein		
Popliteal vein		
Posterior tibial vein		

Pre-Lab Exercise 3

Arterial Anatomy

Label the arterial diagrams depicted in **Figures 17.1, 17.2,** and **17.3** with the terms from Exercise 1. Use your text and Exercise 1 in this unit for reference. Note that these diagrams are presented in color to facilitate identification of the vessels.

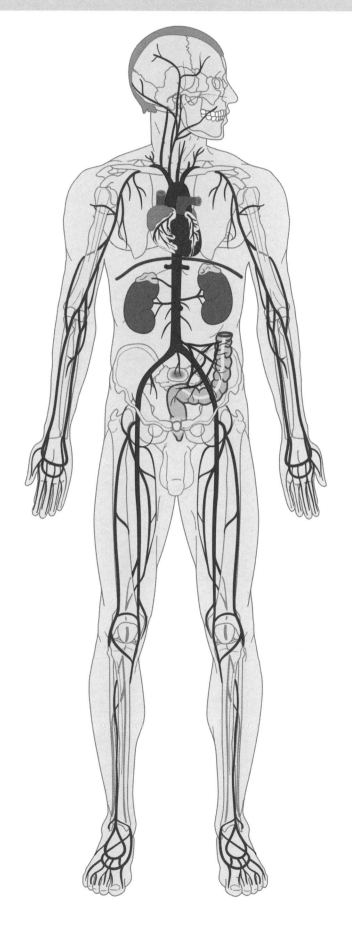

FIGURE 17.1 Major arteries of the body

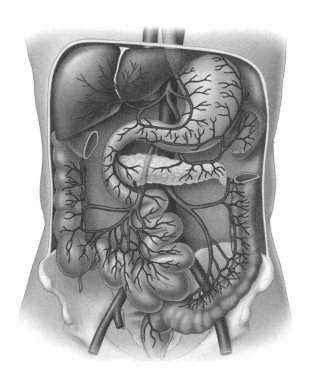

FIGURE **17.2** Arteries of the abdomen

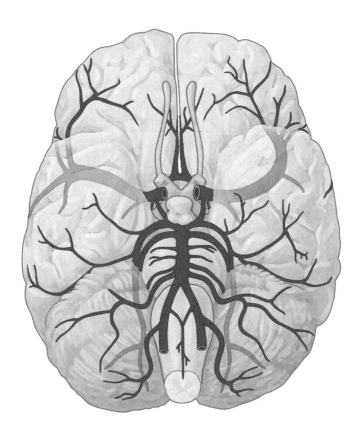

FIGURE **17.3** Arteries of the brain, inferior view

Pre-Lab Exercise 4
Venous Anatomy

Label the venous diagrams depicted in **Figures 17.4** and **17.5** with the terms from Exercise 2. Use your text and Exercise 1 in this unit for reference. Note that these diagrams are presented in color to facilitate identification of the vessels.

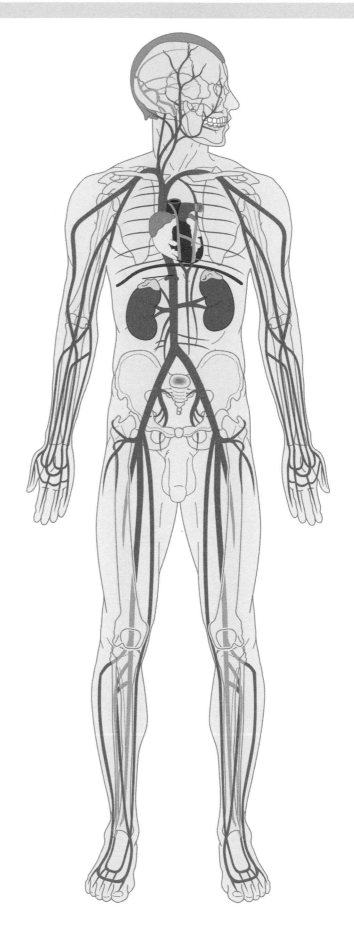

FIGURE 17.4 Major veins of the body

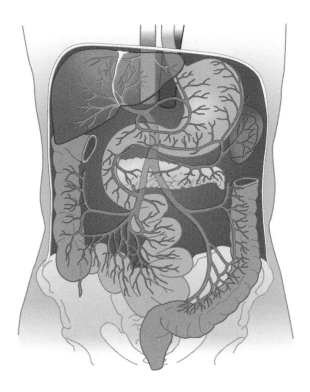

FIGURE **17.5** Veins of the abdomen and the hepatic portal system

Exercises

The blood vessels are a closed system of tubes that carry blood around the body. The heart pumps blood away from the heart through a series of **arteries.** Arteries branch as they pass through organs and tissues to form progressively smaller vessels until they branch into tiny **capillary beds,** where gas, nutrient, and waste exchange take place. The blood is drained from the capillaries via a series of **veins,** which return the blood to the heart.

The three major circuits of blood flow in the body are

1. the **systemic circuit,** which delivers oxygenated blood to most organs and tissues in the body,

2. the **coronary circuit,** which delivers oxygenated blood to the heart, and

3. the **pulmonary circuit,** which delivers deoxygenated blood to the lungs.

In this unit we will address primarily the systemic circuit, as the coronary and pulmonary circuits were discussed in Unit 16. In the upcoming exercises, you will identify the systemic circuit's major blood vessels and trace various pathways of blood flow through the body. In the final exercise you will investigate the histology of the blood vessel wall.

 **Exercise 1**

Major Arteries of the Body

MATERIALS NEEDED

▶ Blood vessel models: human torsos, brain, head and neck, abdomen, upper limb, lower limb

The systemic arterial circuit begins with the largest artery in the body, the **aorta** (**Figure 17.6**). The aorta originates from the left ventricle as the **ascending aorta** which ascends until it curves around to form the **aortic arch**. The aortic arch has three major branches:

1. **Brachiocephalic artery**. The first branch is the brachiocephalic artery, which travels superiorly and to the right. Near the clavicle it splits into two branches—the **right subclavian artery** and the **right common carotid artery**. The right subclavian artery supplies the upper limb and becomes the **axillary artery** near the axilla. In the arm, the axillary artery becomes the **brachial artery**, which branches into the **radial artery** and the **ulnar artery**. The right common carotid artery travels to the neck, where it splits into the **internal carotid artery**, which supplies the brain, and the **external carotid artery**, which supplies the face and the scalp.

2. **Left common carotid artery**. The second branch is the left common carotid artery. Like the right common carotid artery, the left common carotid artery splits into internal and external carotid arteries in the neck.

3. **Left subclavian artery**. The final branch off of the aortic arch is the left subclavian artery, which supplies the left upper limb. The branches and names of the left subclavian artery are the same as those of the right subclavian artery.

The arterial supply of the brain comes primarily from the internal carotid arteries and the **basilar artery** (**Figure 17.7**). The basilar artery is located on the base of the brainstem and is derived from the fusion of two branches off of the subclavian arteries called the **vertebral arteries**. Once the internal carotid arteries and basilar artery enter the brain, they contribute to a structure called the **circle of Willis** (also called the **cerebral arterial circle**). This circle is composed of branches of these vessels and a set of **anterior** and **posterior communicating arteries**. These vessels are connected to provide alternate routes of circulation to the brain if one of the arteries supplying the brain becomes blocked.

After these three vessels branch off of the aortic arch, it turns downward to become the **thoracic aorta**. The thoracic aorta descends through the thoracic cavity posterior to the heart, after which it passes through the diaphragm to become the **abdominal aorta**. The major branches of the abdominal aorta, shown in **Figure 17.8**, include the following:

1. **Celiac trunk**. The short, stubby celiac trunk is the first branch off of the abdominal aorta. It splits almost immediately into the **common hepatic artery**, the **splenic artery**, and the **left gastric artery**.

2. **Renal arteries**. Inferior to the celiac trunk we find the two **renal arteries** that serve the kidneys. Note that the renal arteries are not illustrated in Figure 17.8 but they are visible in **Figure 17.6**.

3. **Superior mesenteric artery**. Right around the renal arteries is another branch called the superior mesenteric artery. As its name implies, it travels through the membranes of the intestines (called the *mesentery*) and supplies the small and much of the large intestine.

4. **Inferior mesenteric artery**. The last large branch off of the abdominal aorta is the inferior mesenteric artery which supplies the remainder of the large intestine.

The abdominal aorta terminates by bifurcating into two **common iliac arteries**, which themselves bifurcate into an **internal iliac artery** and an **external iliac artery**. The internal iliac artery supplies structures of the pelvis, and the external iliac artery passes deep to the inguinal ligament to enter the thigh, where it becomes the **femoral artery**. The femoral artery continues until the area around the popliteal fossa (the posterior knee), where it becomes the **popliteal artery**. Shortly thereafter, the popliteal artery divides into its two main branches: the **anterior tibial artery**, which continues in the foot as the **dorsalis pedis artery**, and the **posterior tibial artery**.

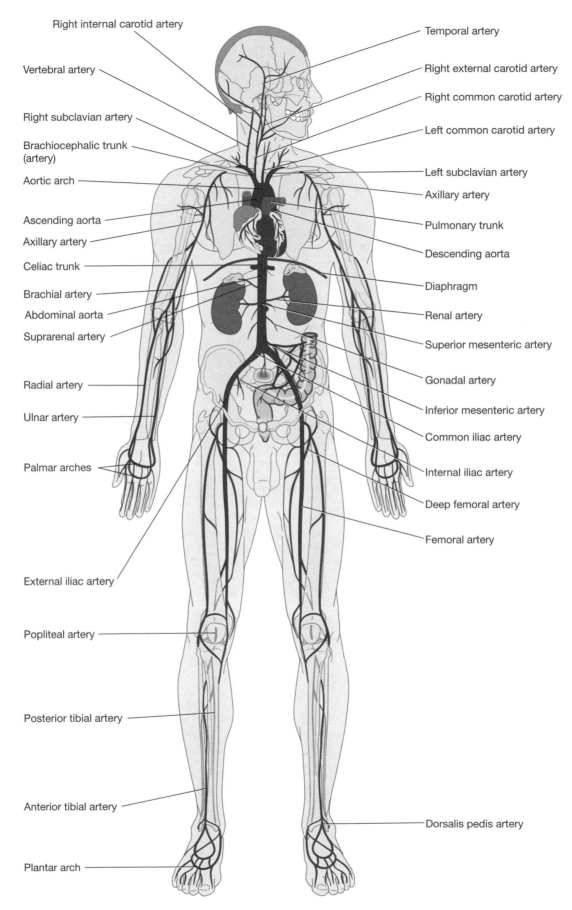

Right internal carotid artery

Vertebral artery

Right subclavian artery

Brachiocephalic trunk (artery)

Aortic arch

Ascending aorta

Axillary artery

Celiac trunk

Brachial artery

Abdominal aorta

Suprarenal artery

Radial artery

Ulnar artery

Palmar arches

External iliac artery

Popliteal artery

Posterior tibial artery

Anterior tibial artery

Plantar arch

Temporal artery

Right external carotid artery

Right common carotid artery

Left common carotid artery

Left subclavian artery

Axillary artery

Pulmonary trunk

Descending aorta

Diaphragm

Renal artery

Superior mesenteric artery

Gonadal artery

Inferior mesenteric artery

Common iliac artery

Internal iliac artery

Deep femoral artery

Femoral artery

Dorsalis pedis artery

FIGURE **17.6** Major arteries of the body

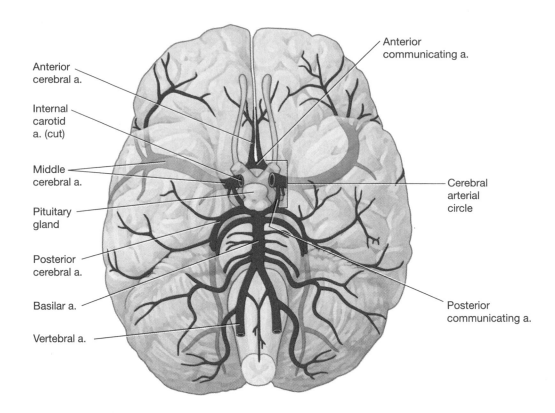

FIGURE 17.7 Arteries of the brain, inferior view

Anterior cerebral a.

Internal carotid a. (cut)

Middle cerebral a.

Pituitary gland

Posterior cerebral a.

Basilar a.

Vertebral a.

Anterior communicating a.

Cerebral arterial circle

Posterior communicating a.

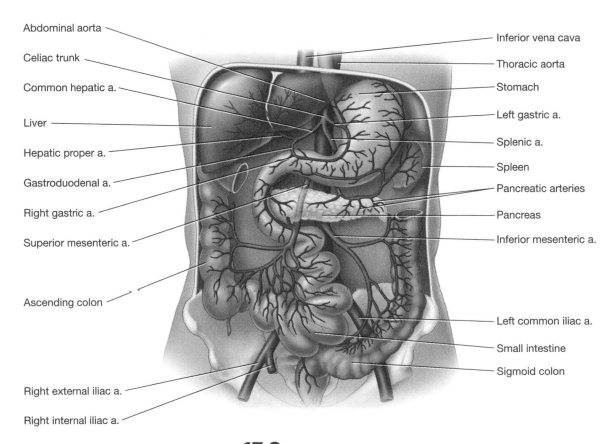

FIGURE 17.8 Arteries of the abdomen

Abdominal aorta

Celiac trunk

Common hepatic a.

Liver

Hepatic proper a.

Gastroduodenal a.

Right gastric a.

Superior mesenteric a.

Ascending colon

Right external iliac a.

Right internal iliac a.

Inferior vena cava

Thoracic aorta

Stomach

Left gastric a.

Splenic a.

Spleen

Pancreatic arteries

Pancreas

Inferior mesenteric a.

Left common iliac a.

Small intestine

Sigmoid colon

EXPLORING ANATOMY & PHYSIOLOGY IN THE LABORATORY

Procedure Model Inventory

Identify the following arteries on models and diagrams, using your textbook and this unit for reference. As you examine the anatomical models and diagrams, record on the model inventory in Table 17.3 the name of the model and the structures you were able to identify. When you have completed the activity, answer Check Your Understanding question 1 (p. 415).

Arteries of the Trunk

1. Aorta
   a. Ascending aorta
   b. Aortic arch
   c. Thoracic (descending) aorta
   d. Abdominal aorta
2. Brachiocephalic artery
3. Celiac trunk
   a. Splenic artery
   b. Left gastric artery
   c. Common hepatic artery
4. Superior mesenteric artery
5. Suprarenal artery
6. Renal artery
7. Inferior mesenteric artery
8. Gonadal artery
9. Common iliac artery
   a. Internal iliac artery

Arteries of the Head and Neck

1. Common carotid artery
   a. External carotid artery
      (1) Temporal artery
   b. Internal carotid artery
      (1) Carotid sinus
2. Vertebral artery
3. Basilar artery
4. Cerebral arterial circle
   a. Anterior communicating arteries
   b. Posterior communicating arteries
5. Anterior cerebral artery
6. Middle cerebral artery
7. Posterior cerebral artery

Arteries of the Upper Limbs

1. Subclavian artery
2. Axillary artery
3. Brachial artery
4. Radial artery
5. Ulnar artery

Arteries of the Lower Limbs

1. External iliac artery
2. Femoral artery
3. Popliteal artery
   a. Anterior tibial artery
      (1) Dorsalis pedis artery
   b. Posterior tibial artery

TABLE **17.3**	Model Inventory for Arteries
Model/Diagram	**Structures Identified**

 Exercise 2

Major Veins of the Body

MATERIALS NEEDED

> Blood vessel models: human torsos, brain, head and neck, abdomen, upper limb, lower limb, dural sinuses

Arteries of the systemic circuit deliver oxygenated, nutrient-rich blood to capillary beds, where gases, nutrients, and wastes are exchanged. The deoxygenated, carbon dioxide-rich blood is then drained from the capillary beds by a series of veins (**Figure 17.9**). The two largest veins in the body are the **superior vena cava**, which drains the structures superior to the diaphragm, and the **inferior vena cava**, which drains the structures inferior to the diaphragm.

The head and the neck are drained primarily by the **internal** and **external jugular** veins. The much smaller external jugular vein drains the face and the scalp, and the larger internal jugular vein, which travels in a sheath with the common carotid artery, drains the brain. Note, however, that venous blood from the brain does not simply drain into one vein and exit the head. Instead, it drains into spaces between the two layers of the dura mater called the **dural sinuses**. Blood from the brain capillaries drains into the dural sinuses; the inferior dural sinus, called the **transverse sinus**, drains into the internal jugular vein.

EXPLORING ANATOMY & PHYSIOLOGY IN THE LABORATORY

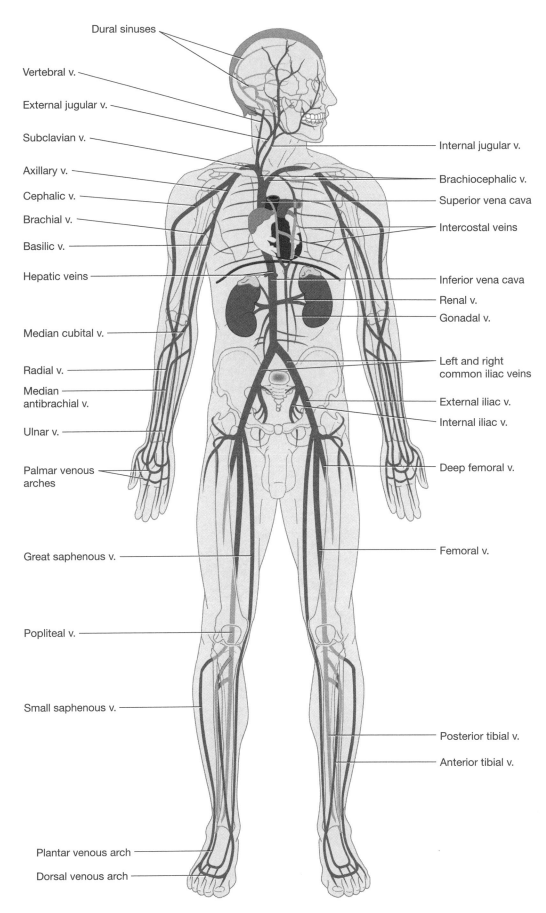

Dural sinuses

Vertebral v.

External jugular v.

Subclavian v.

Axillary v.

Cephalic v.

Brachial v.

Basilic v.

Hepatic veins

Median cubital v.

Radial v.

Median antibrachial v.

Ulnar v.

Palmar venous arches

Great saphenous v.

Popliteal v.

Small saphenous v.

Plantar venous arch

Dorsal venous arch

Internal jugular v.

Brachiocephalic v.

Superior vena cava

Intercostal veins

Inferior vena cava

Renal v.

Gonadal v.

Left and right common iliac veins

External iliac v.

Internal iliac v.

Deep femoral v.

Femoral v.

Posterior tibial v.

Anterior tibial v.

FIGURE **17.9** Major veins of the body

Veins draining blood from the organs of the abdomen are named largely in parallel with the arteries that serve the organs: The **renal veins** drain the kidneys, the **splenic vein** drains the spleen, the **gastric veins** drain the stomach, the **superior mesenteric vein** drains the small intestine and much of the large intestine, and the **inferior mesenteric vein** drains the remainder of the large intestine. Although the renal vein empties into the inferior vena cava, the blood from the latter four veins does not drain into the inferior vena cava directly. Instead, note in **Figure 17.10** that each vein drains into a common vein called the **hepatic portal vein**. Here, the nutrient-rich blood percolates through the liver, where it is processed and detoxified. In this way, everything that we ingest (except lipids, which we discuss in Unit 25) must travel through the liver before entering the systemic circulation. Once the blood has filtered through the hepatic portal system, it exits via **hepatic veins** and drains into the inferior vena cava.

Blood from the deep structures of the upper limb is drained by the **radial** and **ulnar veins**, both of which parallel the bones for which they are named. These two veins merge in the arm to form the **brachial vein**, which becomes the **axillary vein** in the axilla. Near the clavicle, the axillary vein becomes the **subclavian vein**, which drains into the **brachiocephalic vein** and finally into the superior vena cava. The superficial structures of the upper limb are drained by the **cephalic vein** on the lateral side and the **basilic vein** on the medial side. Note in **Figure 17.9** that the cephalic vein and the basilic vein are united in the antecubital fossa by the **median antecubital vein**. This is a frequent site for drawing blood with a syringe. The basilic vein joins the brachial vein to form the axillary vein, and the cephalic vein joins the axillary vein to form the subclavian vein.

The deep structures of the lower limb are drained by the **anterior** and **posterior tibial veins**, which unite in the popliteal fossa to form the **popliteal vein**. In the distal thigh, the popliteal vein becomes the **femoral vein**, which becomes the **external iliac vein** after it passes under the inguinal ligament. The external iliac vein merges with the **internal iliac vein**, which drains pelvic structures, forming the **common iliac vein**. The two common iliac veins unite to form the inferior vena cava near the superior part of the pelvis. The largest superficial vein of the lower limb is the **greater saphenous vein**, which drains the medial leg and thigh and empties into the femoral vein.

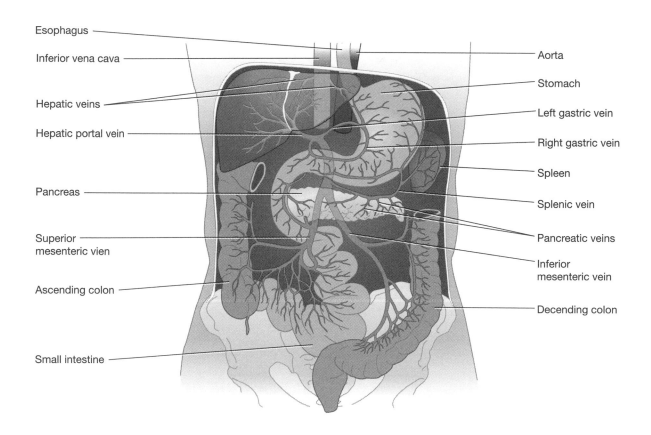

FIGURE 17.10 Veins of the abdomen and the hepatic portal system

EXPLORING ANATOMY & PHYSIOLOGY IN THE LABORATORY

Procedure Model Inventory

Identify the following veins on models and diagrams, using your textbook and this unit for reference. As you examine the anatomical models and diagrams, record on the model inventory in Table 17.4 the name of the model and the structures that you were able to identify. When you have completed the activity, answer Check Your Understanding question 2 (p. 415).

Veins of the Trunk

1. Superior vena cava
2. Inferior vena cava
3. Brachiocephalic vein
4. Hepatic veins
5. Hepatic portal vein
6. Splenic vein
7. Superior mesenteric vein
8. Inferior mesenteric vein
9. Gastric vein
10. Renal vein
11. Suprarenal vein
12. Gonadal vein
13. Internal iliac vein
14. Common iliac vein

Veins of the Head and Neck

1. Dural sinuses
   a. Superior sagittal sinus
   b. Cavernous sinus
   c. Transverse sinus
2. Internal jugular vein
3. External jugular vein
4. Vertebral vein

Veins of the Upper Limbs

1. Ulnar vein
2. Radial vein
3. Median cubital vein
4. Brachial vein
5. Basilic vein
6. Cephalic vein
7. Axillary vein
8. Subclavian vein

Veins of the Lower Limbs

1. Anterior tibial vein
2. Posterior tibial vein
3. Popliteal vein
4. Great saphenous vein
5. Femoral vein
6. External iliac vein

TABLE **17.4**	Model Inventory for Veins
Model	**Structures Identified**

EXPLORING ANATOMY & PHYSIOLOGY IN THE LABORATORY

 **Exercise 3**

Time to Trace!

MATERIALS NEEDED

▶ Laminated outline of the human body
▶ Water-soluble marking pen

In this exercise you will trace the blood flow through various places in the body. As you trace, keep the following hints in mind:

▶ Don't forget about the hepatic portal system! Remember that most venous blood coming from the abdominal organs has to go through the hepatic portal vein and the hepatic portal system before it can enter the general circulation.

▶ Don't forget that the venous blood in the brain drains first into the dural sinuses, then into a vein.

▶ If you start in a vein, you have to go through the venous system and through the heart before you can get back to the arterial system.

▶ If you start in an artery, you have to go through the arterial system and then through a *capillary bed* before you can go through the venous system. You can't go backward through the arterial system—that's cheating!

▶ If you start in an artery and end in an artery, you likely will have to go through the arterial circuit, through a capillary bed, through the venous circuit, back to the heart and lungs, and *then* reenter the arterial circuit. Whew!

▶ Sometimes there is more than one right answer, because you may take multiple paths.

▶ Following is an example, where we have started in the right popliteal vein and ended in the left internal carotid artery:

Start: right popliteal vein $\longrightarrow$ right femoral vein $\longrightarrow$ right external iliac vein $\longrightarrow$ right common iliac vein $\longrightarrow$ inferior vena cava $\longrightarrow$ right atrium $\longrightarrow$ tricuspid valve $\longrightarrow$ right ventricle $\longrightarrow$ pulmonary valve $\longrightarrow$ pulmonary artery $\longrightarrow$ lungs $\longrightarrow$ pulmonary veins $\longrightarrow$ left atrium $\longrightarrow$ mitral valve $\longrightarrow$ left ventricle $\longrightarrow$ aortic valve $\longrightarrow$ ascending aorta $\longrightarrow$ aortic arch $\longrightarrow$ left common carotid artery $\longrightarrow$ left internal carotid artery $\longrightarrow$ **End**

Wasn't that easy?

Procedure Tracing Blood Flow Patterns

Trace the path of blood flow through the following circuits, using the example on page 403 for reference. It is helpful to draw the pathway out on a laminated outline of the human body as you trace. If a laminated outline is not available, use Figures 17.11–17.14 instead.

1. *Start:* Right radial vein

   End: Right renal artery

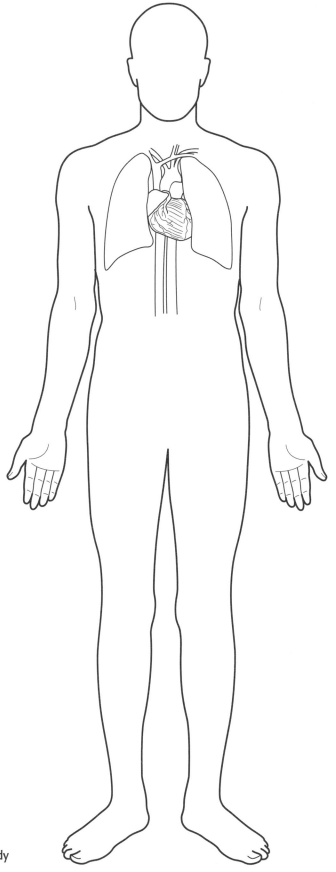

FIGURE 17.11 Outline of the human body

EXPLORING ANATOMY & PHYSIOLOGY IN THE LABORATORY

2. *Start:* Left coronary artery

   End: Dorsalis pedis artery

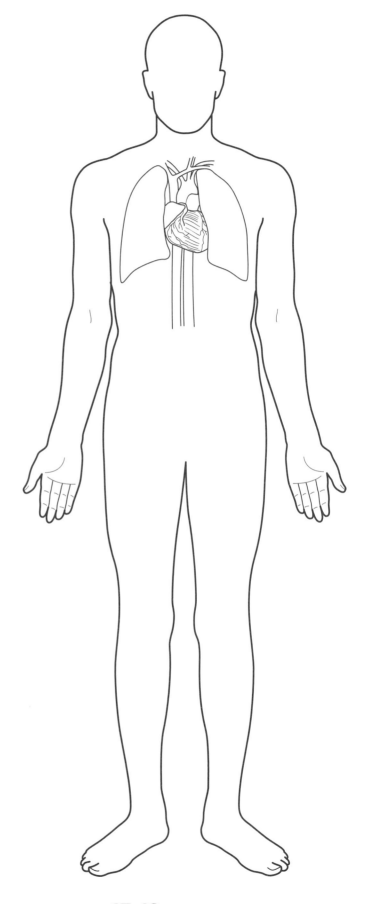

FIGURE **17.12** Outline of the human body

3. *Start:* Superior mesenteric vein

   End: Superior mesenteric artery

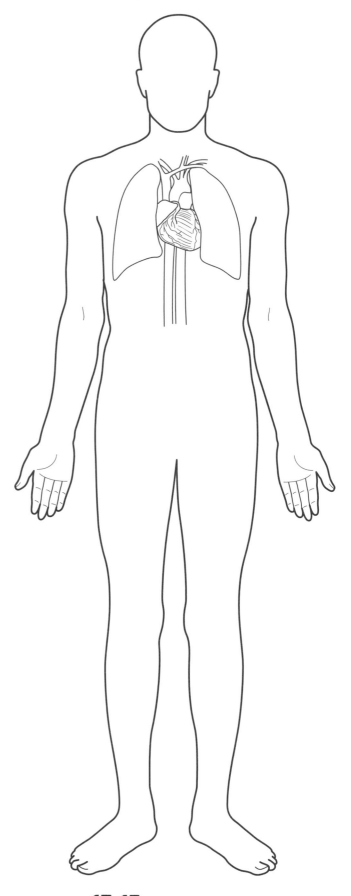

FIGURE **17.13** Outline of the human body

EXPLORING ANATOMY & PHYSIOLOGY IN THE LABORATORY

4. *Start:* Renal artery

   End: Internal jugular vein

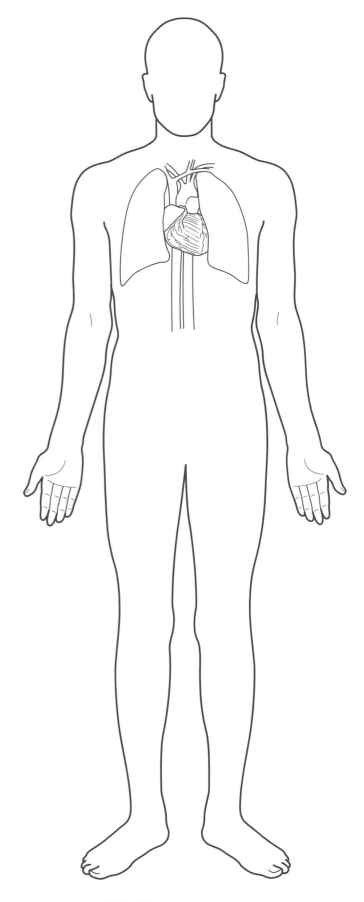

FIGURE **17.14** Outline of the human body

Histology of the Blood Vessel Wall

MATERIALS NEEDED

- Blood vessel slide
- Light microscope
- Colored pencils

Three distinct tissue layers make up the walls of arteries and veins. The three layers, shown in **Figure 17.15**, include the following:

1. **Tunica interna.** The innermost lining of the blood vessel is called the tunica interna. It consists of a specialized type of simple squamous epithelium called **endothelium**. It rests on top of a thin layer of connective tissue.

2. **Tunica media.** The middle layer of the blood vessel wall is called the tunica media and consists of smooth muscle and elastic fibers. The smooth muscle, which is innervated by the sympathetic nervous system, controls the diameter of the vessel and plays an important role in tissue perfusion and blood pressure. The elastic fibers allow the vessel to expand with changing pressure and return to its original shape and diameter.

3. **Tunica externa.** The outermost layer of the blood vessel wall is the tunica externa (or **tunica adventitia**), which consists of dense irregular collagenous connective tissue with abundant collagen fibers. The collagen fibers reinforce the blood vessel and prevent it from rupturing when the pressure in the vessel increases.

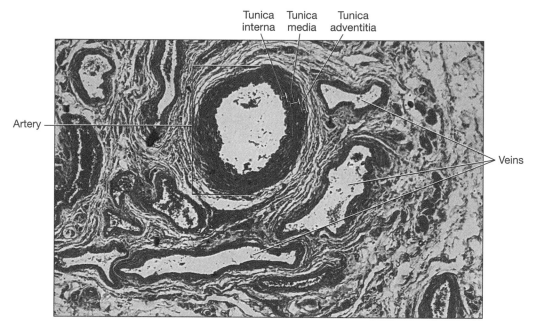

FIGURE 17.15 Photomicrograph, artery and veins

Like any other organ, the structure of each type of blood vessel follows its function. For this reason, the characteristics of the three layers of the blood vessel wall are considerably different in arteries, capillaries, and veins. They are easy to identify if you bear the following in mind:

▶ Arteries have a much thicker tunica media, with prominent elastic fibers that typically appear as a wavy purple line. They appear circular in cross-sections due to their thick walls.

▶ Veins have a thin tunica media with few elastic fibers. Because the wall is so much thinner, the lumen is wider and the vein is typically collapsed on the slide.

▶ Capillaries are extremely thin-walled and consist only of a thin tunica interna. The smallest capillaries are large enough for only one red blood cell to fit through at a time.

Procedure Microscopy

Examine prepared microscope slides of an artery, a capillary, and a vein. The capillary may be on a separate slide. Use colored pencils to draw what you see, and label your diagrams with the terms indicated. When you have completed the activity, answer Check Your Understanding questions 3 and 4 (p. 416).

1. Artery
   a. Tunica interna (endothelium)
   b. Tunica media
      (1) Smooth muscle
      (2) Elastic fibers
   c. Tunica externa
   d. Lumen

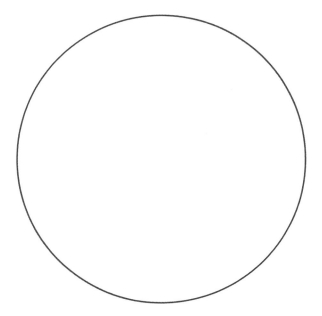

2. Capillary
   a. Tunica interna
   b. Blood cell(s)

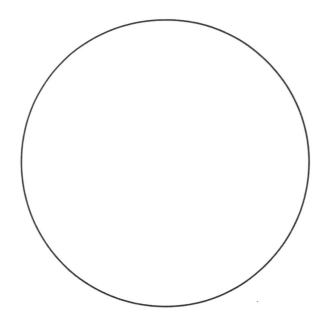

3. Vein
   a. Tunica interna
   b. Tunica media
      (1) Smooth muscle
   c. Tunica externa
   d. Lumen

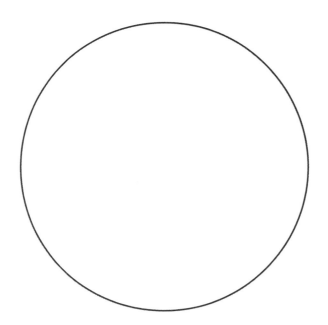

NAME _____

SECTION _____ DATE _____

 Check Your Recall

1 Label the following arteries in **Figures 17.16** and **17.17** (on the next page).

Figure 17.16:

- Dorsalis pedis artery
- Posterior tibial artery
- Femoral artery
- Common iliac artery
- Renal artery
- Left common carotid artery
- Left subclavian artery
- Right common carotid artery
- Brachiocephalic artery
- Brachial artery
- Radial artery
- Ulnar artery

Figure 17.17:

- Celiac trunk
- Common hepatic artery
- Splenic artery
- Left gastric artery
- Superior mesenteric artery
- Inferior mesenteric artery

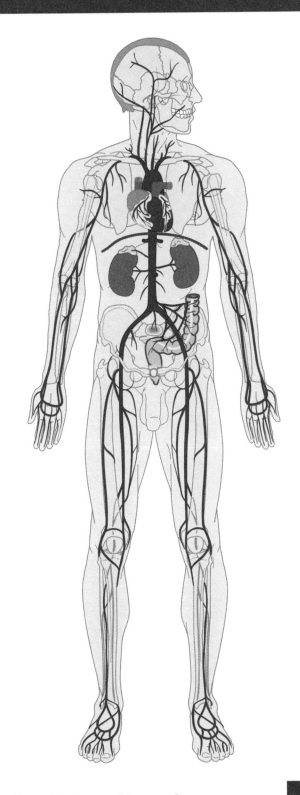

FIGURE **17.16** Major arteries of the body

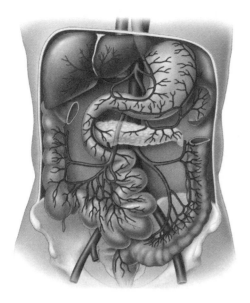

FIGURE **17.17** Arteries of the abdomen

2 Label the following veins on **Figures 17.18** and **17.19** (on the next page).

Figure 17.18:

- Great saphenous vein
- Femoral vein
- Common iliac vein
- Renal vein
- Brachiocephalic vein
- Subclavian vein
- Axillary vein
- Cephalic vein
- Brachial vein
- Internal jugular vein

Figure 17.19:
- Hepatic portal vein
- Superior mesenteric vein
- Inferior mesenteric vein
- Hepatic veins
- Splenic vein

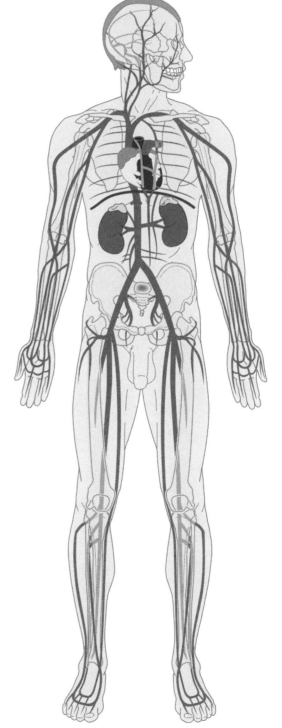

FIGURE **17.18** Major veins of the body

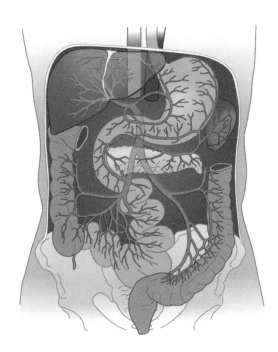

FIGURE **17.19** Veins of the abdomen and the hepatic portal system

3 Which of the following is *not* a major circuit of blood flow in the body?
   a. Coronary circuit
   b. Cerebral circuit
   c. Pulmonary circuit
   d. Systemic circuit

4 How do the right and left common carotid arteries differ?

5 The cerebral arterial circle (circle of Willis):
   a. provides alternate routes of blood flow in the brain.
   b. supplies the face and the scalp.
   c. provides alternate routes of blood flow in the liver.
   d. supplies the myocardium.

6 The venous blood of the brain drains into a set of _____ before draining into a vein.

   a. coronary arteries
   b. cerebral veins
   c. dural sinuses
   d. paranasal sinuses

7 Which of the following veins drain into the hepatic portal vein? (Circle all that apply.)

   a. Renal vein
   b. Splenic vein
   c. Superior mesenteric vein
   d. Inferior mesenteric vein
   e. Hepatic veins
   f. Gastric veins

8 Which of the following describes the purpose of the hepatic portal system?

   a. It provides the liver with oxygenated blood.
   b. It allows the liver to process and detoxify nutrient-rich blood from most of the abdominal organs.
   c. It drains blood from the liver to the inferior vena cava.
   d. All of the above.

9 From superficial to deep, the layers of the blood vessel wall include the:

   _____

   _____

   _____

10 Which of the following correctly describes the role of the smooth muscle cells in the blood vessel wall?

   a. It plays an important role in determining the amount of blood flowing to a tissue (tissue perfusion).
   b. It plays an important role in blood pressure.
   c. It controls the diameter of the blood vessel.
   d. All of the above.
   e. None of the above

Check Your Understanding | **Critical Thinking and Application Questions**

1 Would a blood clot lodged in one of the anterior or posterior communicating arteries of the cerebral arterial circle be likely to cause significant symptoms? Why or why not?

2 Certain drugs cannot be taken by mouth because the entire dose of the drug is destroyed in the liver before it ever reaches the general circulation. Explain why these same drugs can be given by injection, either intravenously or intramuscularly. (*Hint*: Consider the hepatic portal system.)

3 Why do arteries require more elastic fibers than veins?

4 The diseases known as *collagen vascular diseases* are characterized by defects in the collagen in the tunica externa. What effects would you expect such diseases to have on the blood vessels?

The Cardiovascular System—Part III: Cardiovascular Physiology

OBJECTIVES

Once you have completed this unit, you should be able to:

- Auscultate the sounds of the heart.

- Describe and demonstrate common physical examination tests of the blood vessels.

- Measure blood pressure using a stethoscope and a sphygmomanometer.

- Describe the effects of the autonomic nervous system and caffeine on blood pressure and pulse rate.

- Measure and interpret the ankle-brachial index using a Doppler ultrasound device.

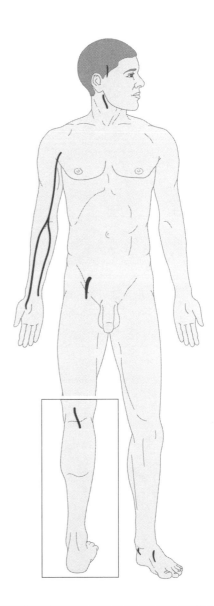

Pre-Lab Exercises

Complete the following exercises prior to coming to lab, using your textbook and lab manual for reference.

Pre-Lab Exercise 1
Key Terms

Table 18.1 lists the key terms with which you should be familiar before coming to lab.

TABLE 18.1	Key Terms

Term	Definition
Heart Sounds and Pulses	
Auscultation	
S1	
S2	
Murmur	
Tachycardia	
Bradycardia	
Pulse point	
Blood Pressure	
Systolic pressure	
Diastolic pressure	
Sounds of Korotkoff	

✎ Pre-Lab Exercise 2

Autonomic Nervous System and the Cardiovascular System

As you learned in Unit 14, the autonomic nervous system (ANS) is an important body system because of its regulation of other body systems. The cardiovascular system is no exception, which you will see in this unit's exercises. Fill in Table 18.2 with the basic effects of the two divisions of the ANS to refresh your memory.

TABLE 18.2	Effects of the Divisions of the ANS	
Organ	**Sympathetic Nervous System Effects**	**Parasympathetic Nervous System Effects**
Heart		
Blood vessels		
Net effect on blood pressure		
Lungs (bronchial smooth muscle)		
Metabolism		
Urinary functions		
Digestive functions		

Exercises

Every physical examination includes an assessment of the cardiovascular system, in which the heart rate is counted, heart sounds are auscultated, and the blood pressure is measured. In this unit you will learn how to perform these and other common procedures and the basis for their interpretation. You will also determine the effects of the autonomic nervous system and caffeine on these measurements.

Exercise 1

Heart Auscultation

MATERIALS NEEDED

▸ Stethoscope
▸ Alcohol and cotton ball

Heart sounds are produced by the closing of valves at certain points during the cardiac cycle. The first heart sound is known as **S1**, and it is caused by closure of the mitral and tricuspid valves at the beginning of ventricular systole and isovolumetric contraction. The second heart sound, **S2**, is caused by closure of the aortic and pulmonary valves at the beginning of ventricular diastole and isovolumetric relaxation.

The process of listening to heart sounds is known as **auscultation**. Heart sounds are typically auscultated in four areas, each of which is named for the valve that is heard best at that specific location. The position of each area is described relative to the sternum and the spaces between the ribs, known as *intercostal spaces*. The first intercostal space is located between the first and second rib, which is roughly below the clavicle. From the clavicle you can count down to consecutive spaces to auscultate in the appropriate areas. The four areas are as follows (Figure 18.1):

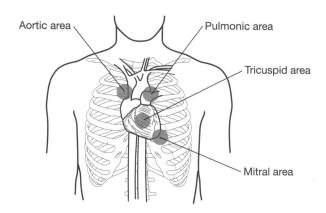

FIGURE **18.1** Areas of auscultation

1. **Aortic area.** This is the area where the sounds of the aortic valve are best heard. It is located in the second intercostal space at the right sternal border.

2. **Pulmonic area.** The pulmonic valve is best heard over the second intercostal space at the left sternal border.

3. **Tricuspid area.** The sounds produced by the tricuspid valve are best heard over the fourth intercostal space at the left sternal border.

4. **Mitral area.** The mitral area is located in the fifth intercostal space at the left midclavicular line (draw an imaginary line down the middle of the clavicle, roughly in line with the nipple).

The following variables are checked during heart auscultation:

▶ **Heart rate.** The heart rate refers to the number of heartbeats per minute. If the rate is more than 100, it is termed **tachycardia**. If the rate is below 60 beats per minute, it is termed **bradycardia**.

▶ **Heart rhythm.** The heart's rhythm refers to the pattern and regularity with which it beats. Some rhythms are *regularly irregular*, in which the rhythm is irregular but still follows a defined pattern. Others are *irregularly irregular*, in which the rhythm follows no set pattern.

▶ **Additional heart sounds.** Sometimes sounds in addition to S1 and S2 are heard, which could be a sign of pathology. These sounds are called **S3**, which occurs just after S2, and **S4**, which occurs just prior to S1.

▶ **Heart murmur.** A heart murmur is a clicking or "swooshing" noise heard between the heart sounds. Murmurs are caused by a valve leaking, called **regurgitation**, or by a valve that has lost its pliability, called **stenosis**.

Heart sounds are auscultated with a stethoscope, which you will use in this unit. Most stethoscopes contain the following parts (Figure 18.2):

▶ **Earpieces** are gently inserted into the external auditory canal and allow you to auscultate the heart sounds.

▶ The **diaphragm** is the broad, flat side of the end of the stethoscope. It is used to auscultate higher-pitched sounds and is the side used most often in auscultation of heart sounds.

▶ The **bell** is the concave, smaller side of the end of the stethoscope. It is used to auscultate lower-pitched sounds.

Note that sounds are not audible through both the bell and the diaphragm at the same time. Typically, the end can be flipped from one side to the next. Before auscultating with either side, lightly tap the end to ensure that you can hear sound through it. Note also that the ends of some stethoscopes have only one side—the diaphragm. In this case, placing light pressure on the end as you are auscultating yields sounds associated with the diaphragm while placing heavier pressure yields sounds associated with the bell.

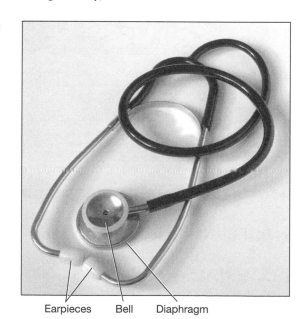

FIGURE **18.2** A stethoscope

Procedure Heart Auscultation

1 Obtain a stethoscope and clean the earpieces and diaphragm with alcohol and cotton balls.

2 Place the earpieces in your ears and gently tap the diaphragm to ensure that it is on the proper side. If it is not, flip it to the other side.

3 Lightly place the diaphragm on your partner's chest in the aortic area. (*Note:* You may wish to have your lab partner place the stethoscope on his or her chest under the shirt, as the sounds are heard best on bare skin.)

4 Auscultate several cardiac cycles and determine which sound is S1 and which sound is S2. Measure your partner's heart rate and determine if the rhythm is regular. In addition, check for heart murmurs and extra heart sounds.

5 Record your results in Table 18.3.

6 Move on to the next area, and repeat.

TABLE **18.3**	Heart Auscultation Results			
Area	**Rate**	**Rhythm**	**Extra Heart Sounds? (yes/no)**	**Murmurs Present? (yes/no)**
Aortic area				
Pulmonic area				
Tricuspid area				
Mitral area				

Exercise 2

Vascular Examination

MATERIALS NEEDED

- Blood vessel model or diagram
- Stethoscope
- Alcohol and cotton swab

A vascular examination is the portion of a physical examination that assesses the health of the arteries and veins. Commonly performed exams include *pulse palpation*, auscultation of vessels to check for noises called *bruits*, and measuring the time it takes for capillary beds to refill. Each of those procedures is outlined in the sections that follow.

Please note before you begin that you should never assess both of your lab partner's carotid arteries at the same time. This might initiate the **baroreceptor reflex**, in which the parasympathetic nervous system triggers a reflexive and often dramatic drop in blood pressure and heart rate. This could cause your lab partner to momentarily lose consciousness.

Procedure Pulse Palpation

Pulse palpation is the process of feeling the pulse with the fingertips. It is performed to assess rate, rhythm, and regularity of the heartbeat, and to assess the arterial circulation to different parts of the body. The pulses commonly measured are those found at the radial, ulnar, brachial, carotid, temporal, femoral, popliteal, posterior tibial, and dorsalis pedis arteries, shown in **Figure 18.3**.

When pulses are palpated, they are **graded** according to a standard scale. This allows healthcare professionals to communicate about a patient unambiguously and to assess the progress or deterioration of a patient's condition. The scale utilizes the following four grades:

Grade 0/4: The pulse is absent.

Grade 1/4: The pulse is barely or only lightly palpable.

Grade 2/4: The pulse is normal.

Grade 3/4: The pulse is quite strong.

Grade 4/4: The pulse is bounding, and visible through the skin.

Note that this scale has no negative numbers or decimal numbers (e.g., you would not use –1/4 or 2.5/4). In a healthy person most pulses are grade 2/4, although occasionally a pulse is weak or absent. This is simply normal anatomical variation and does not signify pathology. Students often mistakenly grade any strong pulse (such as the carotid pulse) as 4/4. If a pulse were truly 4/4, however, this would be a sign of extremely high blood pressure in that artery.

The following procedure asks you to palpate certain pulses on your lab partner. Note that we will not be palpating each pulse illustrated in **Figure 18.3**. When you have completed this activity, answer Check Your Understanding questions 1 through 3 (p. 431).

1 Wash your hands prior to palpating your lab partner's pulses.

2 On a model or diagram, locate the artery you are palpating on a model or diagram.

3 Lightly place your index finger and middle finger over the artery. You may increase the pressure slightly, but be careful not to press too hard, as you could cut off blood flow through the artery and also could mistake the pulse in your fingertips for your partner's pulse. If you are unsure if the pulse is yours or your partner's, feel the lab table. If the lab table "has a pulse," you are feeling the pulse in your own fingertips.

4 Palpate *only* one side (right or left) at a time, especially in the carotid artery.

5 Grade your partner's pulses according to the 0/4 to 4/4 scale, and record the results in Table 18.4.

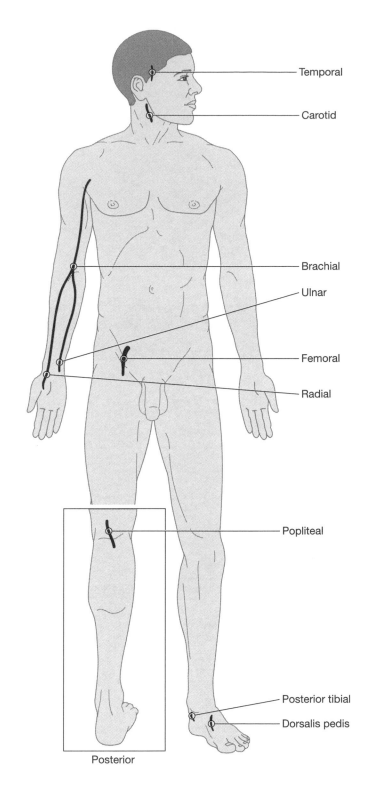

FIGURE 18.3 Common pulse points

EXPLORING ANATOMY & PHYSIOLOGY IN THE LABORATORY

TABLE 18.4	Pulse Points Grades	
Artery	**Right Side Grade**	**Left Side Grade**
Carotid		
Temporal		
Brachial		
Radial		
Ulnar		
Dorsalis pedis		
Posterior tibial		

Procedure Auscultate for Carotid Bruits

Vascular disease in large vessels may lead to turbulent blood flow through the vessel. This produces a sound that you can auscultate with a stethoscope called a **bruit**. The following procedure outlines the process for auscultating for bruits in the carotid arteries.

1 Obtain a stethoscope and switch from the diaphragm to the bell. Clean the earpieces and bell with alcohol and cotton balls.

2 Place the earpieces in your ears and gently tap the bell to ensure that it is on the proper side. If it is not, flip it to the other side. If your stethoscope has only one side, remember that you will have to use slightly heavier pressure to hear the sounds associated with the bell.

3 Palpate the carotid pulse (remember to palpate only one carotid artery at a time) and place the bell over it. Auscultate for bruits, which sound like a "swooshing" sound with each heartbeat.

4 Repeat the process for the other carotid artery.

5 Were bruits present in either side?

Procedure Measuring Capillary Refill Time

The **capillary refill time** is the time it takes for capillary beds to refill after they have been forcibly emptied by pressure. This standard physical examination test is done to evaluate a patient for a variety of vascular diseases, particularly arterial disease. A normal capillary refill time measures 1–3 seconds; a value greater than 3 seconds may signify some sort of pathology. Measure the capillary refill time for each of your lab partner's digits in the procedure outlined below. When you have completed the activity, answer Check Your Understanding question 4 (p. 432).

1 Squeeze the end of the fingertip over the fingernail until the fingernail bed blanches (turns white).

2 Release the fingernail and count the number of seconds it takes for the fingernail bed to return to a pinkish color (note that this may occur immediately, in which case simply record it as 1 second).

3 Repeat for each fingertip of each hand, and record your results in Table 18.5.

TABLE 18.5	Capillary Refill Times			
Digits: Right Hand	**Refill Time**	**Digits: Left Hand**	**Refill Time**
1 (thumb)		1 (thumb)	
2		2	
3		3	
4		4	
5		5	

 Exercise 3

Blood Pressure

MATERIALS NEEDED

- Sphygmomanometer
- Stethoscope
- Alcohol and cotton ball
- Bucket of ice water
- Caffeinated beverage
- Caffeine-free beverage

Blood pressure is defined as the pressure exerted by the blood on the walls of the blood vessels. It is determined by the following three factors:

1. **Cardiac output.** Cardiac output is the amount of blood each ventricle pumps in 1 minute. It is a product of **stroke volume,** or the amount pumped with each beat, and heart rate.

2. **Peripheral resistance.** Resistance is defined as any impedance to blood flow encountered in the blood vessels. It is determined largely by the degree of **vasoconstriction** or **vasodilation** in the systemic circulation. Vasoconstriction increases peripheral resistance, and vasodilation has the opposite effect. Other factors that influence resistance include obstructions such as atheromatous plaques within the arteries.

3. **Blood volume.** The amount of blood found in the blood vessels at any given time is known as the blood volume. It is greatly influenced by overall fluid volume and is largely controlled by the kidneys and hormones of the endocrine system.

Note that cardiac output and peripheral resistance are factors that can be altered quickly to change blood pressure. Alterations to blood volume, however, occur relatively slowly and generally require 2 to 3 days to have a noticeable effect.

Arterial blood pressure is measured clinically and experimentally using an instrument called a **sphygmomanometer** and a stethoscope. This procedure yields two pressure readings:

1. **Systolic pressure.** The pressure in the arteries during ventricular systole is known as the systolic pressure. This is the larger of the two readings, averaging between 100 and 120 mmHg.

2. **Diastolic pressure.** The pressure in the arteries during ventricular diastole is the diastolic pressure. This is the smaller of the two readings, averaging between 60 and 80 mmHg.

Arterial blood pressure is measured by placing the cuff of the sphygmomanometer around the upper arm. When the cuff is inflated, it compresses the brachial artery and cuts off blood flow. When the pressure is released to the level of the systolic arterial pressure, blood flow through the brachial artery resumes but becomes turbulent. This results in sounds known as **sounds of Korotkoff,** which may be auscultated with a stethoscope.

Procedure Measuring Blood Pressure

Practice is necessary to develop the skills to accurately measure arterial blood pressure. Following are the steps that you may use to practice using the sphygmomanometer and stethoscope together. All readings should be taken with your lab partner seated and relaxed.

1 Obtain a stethoscope and sphygmomanometer of the appropriate size (about 80% of the circumference of the arm).

2 Clean the earpieces and diaphragm as in Exercise 1.

3 Wrap the cuff around your partner's arm. It should not be noticeably tight, but it should stay in place when you are not holding it. It should be about 1½ inches proximal to the antecubital fossa.

4 Place the diaphragm of your stethoscope over the brachial artery. You should *not* hear anything at this point.

5 Support your partner's arm by cradling it in your arm, or have your partner rest his or her arm on the lab table.

6 Locate the screw of the sphygmomanometer near the bulb, and close it by turning it clockwise. Inflate the cuff by squeezing the bulb several times. Pay attention to the level of pressure you are applying by watching the pressure gauge. You should not inflate it beyond about 30 mmHg above your partner's normal systolic pressure (for most people, this is no higher than 180 mmHg). Your lab partner will not likely be happy with you if you inflate it above about 200 mmHg, as this can be uncomfortable.

7 Slowly open the screw by turning it counter-clockwise. Watch the pressure gauge, and listen to the brachial artery with your stethoscope.

8 Eventually you will see the needle on the pressure gauge begin to bounce; at about the same time, you will begin to hear the pulse in the brachial artery. Record the pressure at which this first happens as the *systolic pressure*.

9 Continue to listen and watch the gauge until you can no longer hear the pulse. At this point, the needle on the gauge will stop rhythmically bouncing. Record the pressure at which this happens as the *diastolic pressure*. The numbers should be recorded as a fraction—e.g., 110/70, where 110 is the systolic pressure and 70 is the diastolic pressure.

Practice reading 1: _____

Practice reading 2: _____

 Procedure Measuring the Effects of the Autonomic Nervous System on the Blood Pressure and Heart Rate

The autonomic nervous system (ANS) exerts a great deal of control over blood pressure through its influence on cardiac output, peripheral resistance, and blood volume. This exercise limits us to measuring only its effects on cardiac output and peripheral resistance (unless you want to stay in lab for the next 2 days and measure urine output—but I'm guessing you don't want to do that). When you have completed this activity, answer Check Your Understanding questions 5 and 6 (p. 432).

Note: Students with health problems or known cardiovascular disorders should not engage in this activity.

1 Have your lab partner remain seated and relaxed for 3 minutes. After 3 minutes, measure your partner's blood pressure and heart rate.

2 Have your partner place one of his or her hands into a bucket of ice water.

3 Repeat the blood pressure and pulse measurements with your partner's hand still in the ice water. (*Note:* Be kind to your partner and do this quickly!)

4 Have your partner remove his or her hand from the ice water.

5 Wait 5 minutes, then repeat the blood pressure and pulse measurements. Record your results in Table 18.6.

TABLE **18.6**	Blood Pressure and Pulse Readings: Ice Water Experiment	
Test Situation	**Blood Pressure**	**Pulse Rate**
At rest		
After immersing in ice water		
5 minutes after removing from ice water		

6 Interpret your results:

   a. Which situation(s) represented activation of the sympathetic nervous system?

   b. What situation(s) represented activation of the parasympathetic nervous system?

Procedure Measuring the Effects of Caffeine on the Blood Pressure and Heart Rate

You are likely already familiar with caffeine's stimulant effects on the central nervous system, but caffeine also has stimulant effects on the cardiovascular system. This exercise will allow you to examine and measure these effects.

Note: Students with known cardiovascular problems, phenylketonuria, or sensitivity to caffeine-containing products should not engage in this activity.

---

1 Choose two classmates from your lab table who have not consumed caffeine for a minimum of 2 hours prior to lab to be your test subjects.

2 Have each subject sit quietly for 5 minutes. After 5 minutes, measure each subject's blood pressure and heart rate. Record the results in Table 18.7.

3 Give subject 1 a diet caffeine-free soda, and give subject 2 a diet caffeine-containing soda. Have each subject drink his or her entire can of soda as quickly as possible, ideally in 2 minutes or less.

4 Have each subject remain seated for 15 minutes, then measure each subject's blood pressure and pulse rate. Record the results in Table 18.7.

TABLE 18.7 | Blood Pressure and Pulse Readings: Caffeine Experiment

Blood Pressure and Heart Rate Measurements	Subject 1: Caffeine-Free Soda	Subject 2: Caffeinated Soda
Blood pressure before soda		
Blood pressure after soda		
Heart rate before soda		
Heart rate after soda		

5 Report these data to your lab instructor, who will pool the data for the entire class. Calculate the average blood pressure and heart rate of all subjects before and after consuming the soda, and record these data in Table 18.8.

TABLE 18.8 | Average Blood Pressure and Pulse Readings: Caffeine Experiment

Blood Pressure and Heart Rate Measurements	Average of All Subject 1s: Caffeine-Free Soda	Average of All Subject 2s: Caffeinated Soda
Average blood pressure before soda		
Average blood pressure after soda		
Average heart rate before soda		
Average heart rate after soda		

6 Interpret your results:

a. How did the average blood pressure and heart rate change before and after consuming the soda?

b. What can you conclude about the effects of caffeine on blood pressure and heart rate from this experiment?

Exercise 4

The Ankle-Brachial Index

MATERIALS NEEDED

- Doppler ultrasound device
- Sphygmomanometer
- Stethoscope
- Alcohol and cotton ball
- Ultrasound gel

Peripheral vascular disease, also called *peripheral artery disease*, is any disease of the arteries outside of the brain and coronary circuit—in particular the arteries of the legs. It can be caused by diabetes, atherosclerosis, and numerous other conditions. One commonly performed test to assess the severity of peripheral vascular disease is the **ankle-brachial index** (**ABI**). The ABI compares the blood pressure in the legs (the "ankle" portion) to the blood pressure in the arms (the "brachial" portion). Generally, the blood pressure in the ankles declines with worsening disease in the arteries of the legs.

The ABI is obtained by dividing the systolic pressure in the ankle by the systolic pressure in the arm. Typically, the ABI is a decimal number because ankle pressure is slightly lower than the brachial pressure. This is because the legs are farther from the heart than the arms, which causes a slight decline in blood pressure. Interpretation of the ABI is as follows:

0.9–1.0 = normal

0.5–0.9 = peripheral vascular disease

0.2–0.5 = intermittent claudication (temporary blockages to blood flow)

<0.2   = severe disease causing death of tissue

Note that in younger or more athletic persons, the ABI is often greater than 1. This is simply a result of more muscle mass in the lower limb, which is difficult to completely compress with a sphygmomanometer cuff. This falsely elevates the ankle pressure and makes it appear that it is higher than the brachial pressure.

 Procedure Ankle-Brachial Index

The brachial pressure in this test is performed in the standard way, with a sphygmomanometer and a stethoscope. You cannot easily auscultate either of the pedal pulses, however, so a Doppler ultrasound device is used to hear the blood flow instead. A Doppler ultrasound device uses sound waves transmitted through a liquid medium (ultrasound gel) to produce audible sounds of the blood flowing through the vessel. Following is the procedure for measuring the ankle pressure.

1 Take the brachial pressure in the standard way with a sphygmomanometer and a stethoscope. Record *only* the systolic pressure below.

2 Wrap the sphygmomanometer cuff around the ankle.

3 Palpate for either the dorsalis pedis or the posterior tibial pulses.

4 Place a small amount of ultrasound gel over the pulse. Place the Doppler probe in the gel so it is lightly touching the skin. (*Note:* If you press too hard with the probe, it is possible to cut off blood flow. If you do this, you will hear no sound.)

5 Once you can hear sounds of the blood flowing, hold the probe in place as your lab partner inflates the sphygmomanometer cuff. Continue to inflate the cuff until you no longer hear any sound coming from the Doppler.

6 Have your lab partner slowly deflate the cuff while you continue to hold the probe in place. Record the pressure at which you first hear the blood flowing again. Note that this is a systolic pressure only; you will not obtain a diastolic pressure for the ankle. Remember that we record the diastolic pressure as the number when we can no longer hear blood flowing. But with the Doppler, you will continue to hear blood flow; therefore, there will be no diastolic reading.

7 Calculate the ABI by dividing the systolic pressure of the ankle by the systolic pressure of the arm. You most likely will get a decimal number, and this is your ABI.

Systolic pressure of the ankle: _____

Systolic pressure of the arm (brachial):   _____

Ankle pressure/Brachial pressure: _____

NAME _____

SECTION _____ DATE _____

Check Your Recall

1 Fill in the blanks: The first heart sound is known as _____ and is caused by closure of the

_____. The second heart sound is known as _____ and is caused by

closure of the _____.

2 The process of listening to the heart sounds is known as
a. palpation.
b. bruits.
c. auscultation.
d. heart murmur.

3 Matching: Match the following terms with the correct definition.

a. Bruit   _____   A. A noise heard in the heart due to defective valves

b. Murmur   _____   B. Extra heart sounds that may signify pathology

c. Tachycardia _____   C. A decreased heart rate below 60 beats per minute

d. Rhythm   _____   D. A noise heard in an artery as a result of turbulent blood flow

e. S3 and S4   _____   E. The pattern and regularity with which the heart beats

f. Bradycardia _____   F. An elevated heart rate above 100 beats per minute

4 Which of the following statements is false?
a. The diaphragm of a stethoscope is used to hear higher-pitched sounds.
b. The bell of a stethoscope is used to hear lower-pitched sounds.
c. The aortic area is the area where the aortic valve is best heard.
d. Pulse palpation is the process of listening to an artery with a stethoscope.

5 What is the baroreceptor reflex?

6 Which of the following is *not* a factor that determines blood pressure?

a. Cardiac output
b. Ankle-brachial index
c. Peripheral resistance
d. Blood volume

7 Fill in the blanks: The _____ is the pressure in the arteries during ventricular systole and

averages about _____. The _____ is the pressure in the arteries during

ventricular diastole and averages about _____.

8 The sounds heard in an artery when measuring the blood pressure are called

a. bruits.
b. sounds of Korotkoff.
c. Doppler waves.
d. S3 and S4

9 The actions of the sympathetic nervous system _____ blood pressure, and the actions of the parasympathetic
nervous system _____ blood pressure.

a. increase; decrease
b. increase; increase
c. decrease; increase
d. decrease; decrease

10 The ankle-brachial index compares the

a. diastolic pressure in the arm to the diastolic pressure in the ankle.
b. systolic pressure in the arm to the diastolic pressure in the ankle.
c. diastolic pressure in the arm to the diastolic pressure in the ankle.
d. systolic pressure in the arm to the systolic pressure in the ankle.

Check Your Understanding | **Critical Thinking and Application Questions**

1 What might it mean if a pulse were normal (grade 2/4) on the right limb, and absent (grade 0/4) on the left limb?

2 Palpation of both carotid arteries at the same time initiates a reflex known as the *baroreceptor reflex*. If this reflex is initiated in a healthy person with normal blood pressure, the person may temporarily lose consciousness. Explain why this happens.

3 The baroreceptor reflex can be used to our benefit in a procedure known as a *carotid sinus massage*, which is performed when the blood pressure or heart rate is pathologically high. Why would a "massage" of the carotid sinus help to treat these conditions?

4 What might be signified if the capillary refill time measures 10 seconds in the digits of one hand and 2 seconds in the digits of the other hand? What might it mean if the capillary refill time measures 10 seconds in all digits?

5 Predict the effects of a drug that activates the sympathetic nervous system on blood pressure and heart rate. How would this drug affect the three factors that determine blood pressure?

6 Predict the effects of a drug that activates the parasympathetic nervous system on blood pressure and heart rate. How would this drug affect the three factors that determine blood pressure?

Blood

OBJECTIVES

Once you have completed this unit,
you should be able to:

- Identify the formed elements of blood.

- Perform blood typing of the ABO and Rh blood groups using simulated blood.

- Explain the basis for blood typing and matching for blood donation.

- Determine appropriate blood donors for a given recipient.

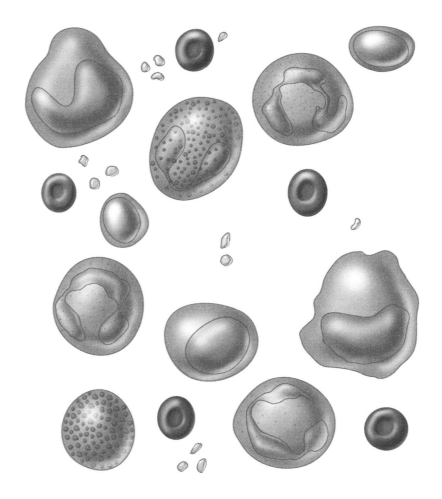

Pre-Lab Exercise 1
Key Terms

Table 19.1 lists the key terms with which you should be familiar before coming to lab.

TABLE **19.1**	Key Terms

Term	Definition
Formed Elements	
Erythrocyte	_____
	_____
Leukocyte	_____
	_____
Granulocyte	_____
	_____
Neutrophil	_____
	_____
Eosinophil	_____
	_____
Basophil	_____
	_____
Agranulocyte	_____
	_____
Lymphocyte	_____
	_____
Monocyte	_____
	_____
Platelets	_____
	_____

Blood Typing

Antigen _____

_____

Antiserum _____

_____

Antibody _____

_____

Blood Donation

Universal donor _____

_____

Universal recipient _____

_____

Pre-Lab Exercise 2
Formed Elements

In this unit we will identify the formed elements of blood on a peripheral blood smear. Each formed element has unique morphological characteristics and functions. Use your text and Exercise 1 in this unit to fill in Table 19.2 with these functions and characteristics.

TABLE 19.2	Properties of Formed Elements			
Formed Element	Nucleus Shape	Cytoplasm and/or Granule Color	Function	Prevalence
Erythrocyte				
Neutrophil				
Eosinophil				
Basophil				
Lymphocyte				
Monocyte				
Platelet				

Exercises

Safety concerns often preclude the use of real blood in the laboratory, but we can still demonstrate important principles of blood by viewing prepared microscope slides of blood cells and by using simulated blood to learn about blood typing. There is no real concern over bloodborne diseases with simulated blood, but do keep in mind that the simulated blood contains chemicals that may be hazardous. As such, use appropriate safety protocols when handling all materials in this lab.

In this unit you will identify the formed elements of blood on microscope slides. You will also play a murder mystery game in which you use simulated blood to apply blood typing techniques, after which you use the same cast of characters to examine blood donation.

 Exercise 1

Formed Elements (Cells) of Blood

MATERIALS NEEDED

- Blood slide
- Light microscope
- Colored pencils

Formed elements, which make up the cellular portion, or "living matrix," of blood, account for about 45% of the volume of whole blood. Each class of formed element has a distinctive appearance, as you learned in the pre-lab exercises. Formed elements can be divided into three classes, each of which is illustrated in **Figure 19.1**.

1. **Erythrocytes.** Erythrocytes, also known as **red blood cells**, are the most numerous blood cells. These cells carry oxygen around the body on an iron-containing molecule called **hemoglobin**. Erythrocytes are easily distinguished from the other formed elements by their red color and the fact that mature erythrocytes lack nuclei.

2. **Leukocytes.** Leukocytes, also known as **white blood cells,** play a role in the immune system. The two subclasses of leukocytes are based upon the presence or absence of granules in their cytoplasm.

   a. **Granulocytes.** As implied by their name, granulocytes are cells containing cytoplasmic granules that are visible when stained. The three types of granulocytes stain differently when treated with the dyes hematoxylin and eosin, and are named for the type of stain with which they interact.

      i. **Neutrophils** do not interact strongly with either type of dye, and they stain a light violet color. They are the most numerous type of leukocyte and have multi-lobed nuclei. Their nuclei often vary in appearance and for this reason they are often called *polymorphonucleocytes*.

      ii. **Eosinophils** interact strongly with the red dye eosin, and their granules stain bright red. They are far less numerous than neutrophils and tend to have bilobed nuclei.

      iii. **Basophils** take up the dark purple stain hematoxylin (it is a basic dye, hence their name *baso*phil) and their granules appear dark blue-purple. They tend to have bilobed nuclei but their nuclei are often obscured by their dark granules. They are the least numerous of the leukocytes and will likely be the most difficult to find on your slide.

   b. **Agranulocytes.** The cells known as agranulocytes lack cytoplasmic granules. The two types of agranulocytes are the following:

      i. **Lymphocytes** tend to be smaller than granulocytes and have large, spherical nuclei. They are the second most numerous type of leukocyte.

      ii. **Monocytes** are the largest of the leukocytes and have "U"-shaped or horseshoe-shaped nuclei. They are the third most numerous type of leukocyte.

3. **Platelets.** Note in **Figure 19.1** that platelets aren't actually cells at all but are instead just small cellular fragments. As such, they lack nuclei and are much smaller than the other formed elements. Platelets are involved in blood clotting.

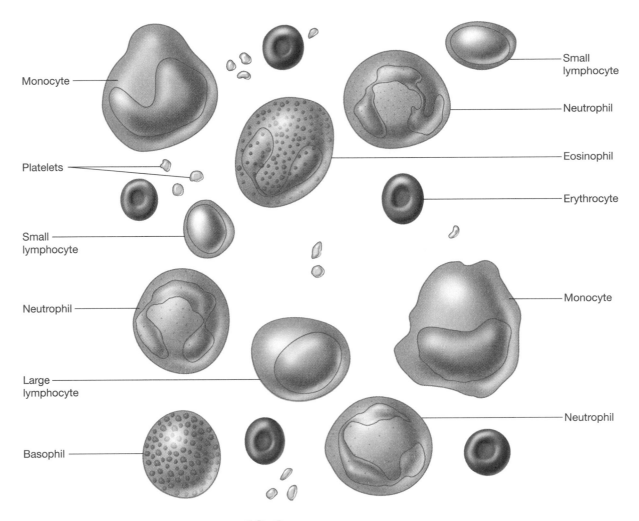

FIGURE **19.1** Formed elements of the blood

Procedure

In this exercise, you will examine a blood slide called a **peripheral blood smear**. Examine the peripheral blood smear on high power, and scroll through to find each of the formed elements. Note that you may have to find a second slide to locate certain cells, because some types are more difficult to find (in particular the eosinophils and basophils). In the spaces below, use colored pencils to draw and describe each formed element that you locate. When you have finished, answer Check Your Understanding questions 1 and 2 (p. 451).

1. *Erythrocyte*:

2. *Neutrophil*:

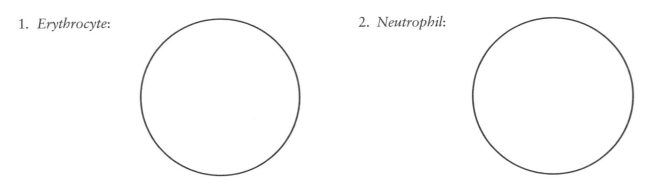

3. Eosinophil:

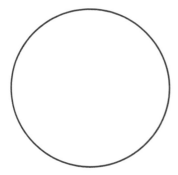

4. Basophil:

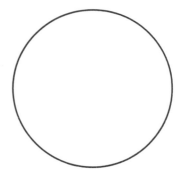

5. Monocyte:

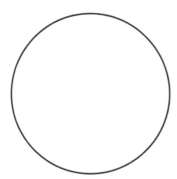

6. Lymphocyte:

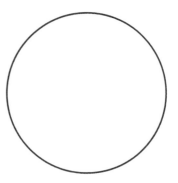

7. Platelets:

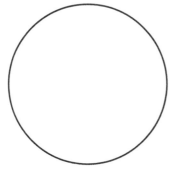

Exercise 2

ABO and Rh Blood Groups

MATERIALS NEEDED

- Well plates
- Simulated blood types A+, B−, AB+, and O−
- Antisera: anti-A, anti-B, anti-Rh

Blood typing is done by checking the blood for the presence or absence of specific glycoproteins found on the cell surface called **antigens**. Two clinically relevant antigens are: the **A antigen** and the **B antigen**. The blood type is named based upon which of the antigens is present.

- **Type A** blood has A antigens on the cell surface.
- **Type B** blood has B antigens on the cell surface.

- **Type AB** blood has both A and B antigens on the cell surface.
- **Type O** blood has neither A nor B antigens on the cell surface.

An additional clinically relevant antigen is the Rh antigen.

- Blood that has the Rh antigen is denoted as **Rh positive** (e.g., A+).
- Blood that lacks the Rh antigen is denoted as **Rh negative** (e.g., A−).

The antigens present on the surface of an erythrocyte can be determined by combining it with a solution called an **antiserum**. An antiserum is a solution that contains proteins from the immune system called **antibodies** that bind to specific antigens. When antibodies bind to antigens on erythrocytes, they cause **agglutination,** or clumping of the erythrocytes in the sample. In this exercise we are using simulated blood, so you won't see agglutination unless your instructor sets up a demonstration or permits you to type your blood. The antisera that are used to determine the blood type of a sample are named according to the antibody they bind:

- **Anti-A antiserum** contains anti-A antibodies that bind to erythrocytes with A antigens.
- **Anti-B antiserum** contains anti-B antibodies that bind to erythrocytes with B antigens.
- **Anti-Rh antiserum** contains anti-Rh antibodies that bind to erythrocytes with Rh antigens.

Antigen-Antibody Reactions

Note: Gloves and safety glasses are required.

This exercise allows you to examine the antigen–antibody reactions of known blood types. Each table should take one spot plate and one set of dropper bottles. The bottles are labeled A+, AB−, B+, and O− to represent each of those blood types, and anti-A, anti-B, and anti-Rh to represent the different antisera.

Procedure

Use **Figure 19.2** as a guide to placement of samples in the wells.

1 Label wells on the spot plate as wells 1–12.

2 Drop two drops of type A+ blood in well 1, well 2, and well 3.

3 Drop two drops of type B− blood in well 4, well 5, and well 6.

4 Drop two drops of type AB+ blood in well 7, well 8, and well 9.

5 Drop two drops of type O− blood in well 10, well 11, and well 12.

6 Add two drops of the anti-A antiserum to wells 1, 4, 7, and 10.

7 Add two drops of the anti-B antiserum to wells 2, 5, 8, and 11.

8 Add two drops of the anti-Rh antiserum to wells 3, 6, 9, and 12.

9 Observe the samples for changes in color symbolizing the agglutination or clumping that would normally occur between antisera and specific blood types. Record your results in Table 19.3.

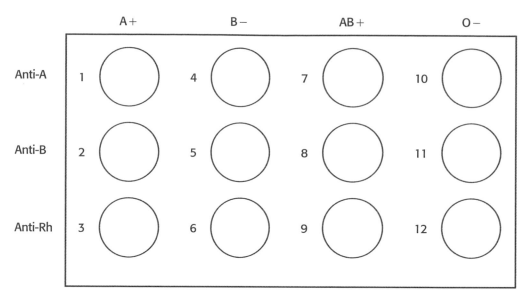

FIGURE **19.2** Well plate diagram

TABLE **19.3**	Blood Typing Results			
Blood Type	Reacted with anti-A? (yes/no)	Reacted with anti-B? (yes/no)	Reacted with anti-Rh? (yes/no)	Antigens Present on Cell Surface
A+				
B−				
AB+				
O−				

Exercise 3

Murder Mystery Game

MATERIALS NEEDED

▸ Well plate
▸ Antisera: anti-A, anti-B, anti-Rh
▸ Murder Mystery game

In this game you will be applying the blood typing techniques that you learned in Exercise 2 while solving a series of murder mysteries. Each of the following cases presents a victim, a murderer, three suspects, three possible murder rooms, and three possible murder weapons. Your job is to play the role of detective and determine the identity of the murderer, which weapon he or she used, and in which room the crime was committed.

Procedure

For each case a unique set of bottles is marked with a number that corresponds to the specific cases (i.e., the bottles are marked 1 for Case 1, 2 for Case 2, and 3 for Case 3). The murderer, rooms, and murder weapons are different for each case, but the cast of characters remains the same.

1 To begin the game, assemble into groups of two or three students. Obtain a well plate and choose one set of samples to test (e.g., the rooms from Case 1, the suspects from Case 2, or the weapons from Case 3).

2 Test the samples by placing drops of the sample in three separate wells. Add two drops of anti-A antiserum to the first well, add two drops of anti-B antiserum to the second well, and add two drops of anti-Rh antiserum to the third well. After you have tested each of the samples, return them to their proper places in the front of the lab.

3 Read and record the blood type by watching for a reaction with the antisera. Remember that this is simulated blood, just like in Exercise 2. A positive reaction is denoted by a color change.

4 To determine the

▶ *Murderer*: Match the blood type of one of the *suspects* to that of the murderer.

▶ *Weapon and room*: Match the blood type of the *victim* to the blood types found in the rooms and on the weapon.

Case 1: Ms. Magenta

We enter the scene to find the dearly departed Ms. Magenta. Forensic analysis determines that there are two types of blood on the body. One blood type is Ms. Magenta's, and the other blood type is a trace amount left behind from the murderer.

Ms. Magenta's blood type: _____   **Murderer's blood type:** _____

We have three suspects:

1. *Mrs. Blanc* was being blackmailed by Ms. Magenta and Col. Lemon. They had discovered that Mrs. Blanc had murdered her late husband. Mrs. Blanc knew this would ruin her reputation at the country club.

2. *Col. Lemon* wanted to keep the blackmail money for himself, and wanted Ms. Magenta out of the way.

3. *Mr. Olive* had been secretly in love with Ms. Magenta for years, and when he told her of his feelings, she rejected him harshly.

Mrs. Blanc's blood type: _____   **Col. Lemon's blood type:** _____

Mr. Olive's blood type: _____

We have three possible murder rooms:

Ballroom blood type: _____   **Library blood type:** _____

Den blood type: _____

We have three possible murder weapons:

Candlestick blood type: _____   **Noose blood type:** _____

Knife blood type: _____

Case 1 Conclusion:

Ms. Magenta was killed by _____, in the _____,

with the _____.

Case 2: Col. Lemon

Our next victim is poor Col. Lemon. On his body we find his blood, and also trace amounts of the blood of another person, presumably the murderer.

Colonel Lemon's blood type: _____   **Murderer's blood type:** _____

We have three potential suspects:

1. *Mrs. Blanc.* Now, with Ms. Magenta out of the way, Mrs. Blanc could easily rid herself of her problem by disposing of the only other person who knows her secret—Col. Lemon.

2. *Professor Purple* believed the Colonel had stolen his groundbreaking research into collagen lip injections.

3. *Mr. Olive* couldn't stand the Colonel because of his close relationship with Ms. Magenta.

Mrs. Blanc's blood type: _____   **Professor Purple's blood type:** _____

Mr. Olive's blood type: _____

We have blood in three different rooms:

Hall blood type: _____   **Kitchen blood type:** _____

Billiards blood type: _____

Forensics found blood on three different weapons:

Copper pipe blood type: _____   **Hammer blood type:** _____

Revolver blood type: _____

Case 2 Conclusion:

Col. Lemon was killed by _____, in the _____,

with the _____.

Case 3: Mr. Olive

Our next (and hopefully last) victim is Mr. Olive. Analysis demonstrates two blood types—one belonging to Mr. Olive, and trace amounts of another belonging to the murderer.

Mr. Olive's blood type: _____   **Murderer's blood type:** _____

We have three potential suspects:

1. *Ms. Feather* had always secretly loved Mr. Olive, but he spurned her advances in favor of Ms. Magenta.

2. *Mrs. Blanc* was worried that Mr. Olive knew her secret and wanted him out of the way.

3. *Professor Purple* discovered that Mr. Olive—not Col. Lemon—had actually stolen the collagen lip implant research. Whoops!

Ms. Feather's blood type: _____   Mrs. Blanc's blood type: _____

Professor Purple's blood type: _____

Blood was found in three rooms:

Lounge blood type: _____   Dining room blood type: _____

Greenhouse blood type: _____

We have three potential murder weapons:

Noose blood type: _____   Hammer blood type: _____

Revolver blood type: _____

Case 3 Conclusion:

Mr. Olive was killed by _____, in the _____,

with the _____.

 # Exercise 4

Blood Donation

Blood transfusion, the infusion of a recipient with a donor's blood cells, is a commonly performed medical procedure. Before a recipient is given a blood transfusion, the medical team must first learn the patient's blood type and then find a suitable or "matching" donor. This is necessary because of the A, B, and Rh antigens on the surface of the donor erythrocytes and the presence of preformed antibodies in the recipient's blood. If a donor's blood has antigens that the recipient's immune system recognizes as foreign, the recipient's antibodies will agglutinate the foreign erythrocytes and the immune system will destroy them, a process known as **hemolysis.** This is called a **transfusion reaction,** and it is a medical emergency that can lead to kidney failure and death.

To ensure that a transfusion reaction does not occur, we must make sure that the donor blood does not have antigens that the recipient's immune system will recognize as foreign. Our immune systems produce antibodies to any antigen that is *not* present on the surface of our own cells.

- People with type A blood have A antigens and so produce anti-B antibodies.
- People with type B blood have B antigens and so produce anti-A antibodies.
- People with type O blood have neither A nor B antigens and so produce anti-A and anti-B antibodies.
- People with type AB blood have both A and B antigens and so produce neither anti-A nor anti-B antibodies.

If you're wondering about the Rh factor, wait just a moment—we're getting there. Let's do an example with the ABO blood groups first:

Patient 1 has type B blood, which means that he has anti-A antibodies. What will happen if we give him blood from a donor with:

- **Type A blood?** There are A antigens on these erythrocytes, and his anti-A antibodies would agglutinate them. **X**
- **Type B blood?** Patient 1's anti-A antibodies would have no effect on the B antigens on these erythrocytes, so this blood is safe. ✔

▶ **Type O blood?** There are no antigens on these donor erythrocytes, so Patient 1's anti-A antibodies would have no effect on them, and this blood is safe. ✔

▶ **Type AB blood?** There are both A and B antigens on these erythrocytes, and Patient 1's anti-A antibodies would agglutinate the erythrocytes. **X**

Now that's easy, isn't it?

Next let's address the Rh factor. The blood of an Rh negative person does *not* contain pre-formed antibodies to the Rh antigen. If an Rh negative person is exposed to the Rh antigen, however, he or she *does* produce anti-Rh antibodies. In an emergency setting, it is generally not possible to determine if an Rh negative person has been exposed to the Rh antigen, so healthcare professionals err on the side of caution and assume that the person has anti-Rh antibodies. For the sake of simplicity, we will assume the same thing in this exercise. So, for our purposes:

▶ People with Rh positive blood *do not* produce anti-Rh antibodies.

▶ People with Rh negative blood *do* produce anti-Rh antibodies.

Let's do one more example, taking into account the Rh factor this time:

Patient 2 has A− blood, which means that she has anti-B and anti-Rh antibodies. What will happen if we give her blood from a donor with:

▶ **Type A+ blood?** There are A and Rh antigens on these erythrocytes, and her anti-Rh antibodies would agglutinate them. **X**

▶ **Type B− blood?** There are B antigens on these erythrocytes, and her anti-B antibodies would agglutinate them. **X**

▶ **Type O− blood?** There are no antigens on these erythrocytes, so Patient 2's anti-B and anti-Rh antibodies would have no effect on them, and this blood is safe. ✔

▶ **Type AB+ blood?** There are A, B, and Rh antigens on these erythrocytes and Patient 2's anti-B and anti-Rh antibodies would agglutinate them. **X**

Procedure Blood Type Matching Practice

Use the information above and your text to fill in Table 19.4. After you have filled in the table, answer Check Your Understanding question 3 (p. 452).

TABLE **19.4**	Blood Donation			
Blood Type	**Antigens Present**	**Antibodies Present**	**Can Donate Safely to Which Blood Types?**	**Can Receive Safely from Which Blood Types?**
A+				
A−				
B+				
B−				
AB+				
AB−				
O+				
O−				

You should notice something from Table 19.4: AB+ blood can receive from any blood type, and O− blood can donate to any blood type. This is because AB+ blood has all antigens but no antibodies, which is why it is often called the **universal recipient**. But Type O− has no antigens for a recipient's antibodies to bind, so it can be donated to any blood type. For this reason, type O− is often called the **universal donor**.

Procedure Type Matching for Transfusions

Gasp! It turns out that Ms. Magenta, Colonel Lemon, and Mr. Olive all survived their injuries! But they have lost blood and are in need of blood transfusions. All of the suspects have had a sudden change of heart and have offered to help the three victims by donating blood. Your job is to determine who among the suspects could safely donate blood to whom. You do not need to retest each person's blood type, as you may use your results from Exercise 3. After you have finished this activity, answer Check Your Understanding questions 4 and 5 (p. 452).

HINTS & TIPS

Remember—when trying to work out who can donate blood to whom, you are concerned with the recipient's antibodies and the donor's antigens. So first work out which antibodies the recipient has, then make sure the recipient's antibodies won't bind any antigens on the donor's erythrocytes.

Recipient 1: Ms. Magenta

Ms. Magenta's blood type   _____

Donors:

Ms. Feather's blood type: _____

Mrs. Blanc's blood type: _____

Professor Purple's blood type: _____

Who could safely donate blood to Ms. Magenta?

Who could not safely donate blood to Ms. Magenta?

Recipient 2: Colonel Lemon

Colonel Lemon's blood type   _____

Donors:

Ms. Feather's blood type: _____

Mrs. Blanc's blood type: _____

Professor Purple's blood type: _____

Who could safely donate blood to Colonel Lemon?

Who could not safely donate blood to Colonel Lemon?

Recipient 3: Mr. Olive

Mr. Olive's blood type _____

Donors:

Ms. Feather's blood type: _____

Mrs. Blanc's blood type: _____

Professor Purple's blood type: _____

Who could safely donate blood to Mr. Olive?

Who could not safely donate blood to Mr. Olive?

EXPLORING ANATOMY & PHYSIOLOGY IN THE LABORATORY

NAME _____

SECTION _____ DATE _____

Check Your Recall

1 Mark the following statements as true (T) or false (F). If the statement is false, correct it to make it a true statement.

_____ Red blood cells are also known as leukocytes.

_____ White blood cells with granules in their cytoplasm are known as agranulocytes.

_____ The granulocytes include the neutrophils, eosinophils, and basophils.

_____ Erythrocytes carry oxygen through the body on the protein hemoglobin.

_____ Both lymphocytes and monocytes are granulocytes.

_____ Platelets are fully formed cells involved in blood clotting.

2 Label the formed elements on the peripheral blood smear in **Figure 19.3**.

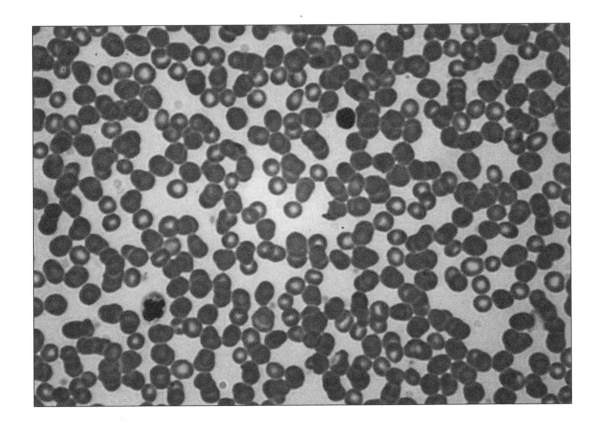

FIGURE **19.3** Peripheral blood smear

3 Which of the following is not an antigen that may be found on the surface of an erythrocyte?
   a. A antigen
   b. B antigen
   c. O antigen
   d. Rh antigen

4 State which antigens are present on the surface of erythrocytes of the following blood types:
   a. B−

   b. O+

   c. AB−

   d. A+

5 Antibodies cause _____ of erythrocytes.
   a. agglutination
   b. aggregization
   c. neutralization
   d. They have no effect on erythrocytes.

6 A person with type A blood has
   a. anti-A antibodies.
   b. anti-B antibodies.
   c. anti-O antibodies.
   d. no antibodies.

7 A person with type B− blood has which of the following antibodies? (Assume the person has been exposed to Rh antigens. Circle all that apply.)
   a. anti-A antibodies
   b. anti-B antibodies
   c. anti-Rh antibodies
   d. no antibodies

8 To which blood types could the following people donate?
   a. Person 1: Type A−

   b. Person 2: Type O+

   c. Person 3: Type AB−

   d. Person 4: Type B+

9 Which blood type is considered the universal donor? Why can this blood type be given to all other blood types?

10 Which blood type is considered the universal recipient? Why can people with this blood type receive blood from all other blood types?

Check Your Understanding | **Critical Thinking and Application Questions**

1 When someone is admitted to the hospital, one of the first procedures that health professionals perform is a blood draw. One lab value that is checked is a white blood cell (WBC) count. If the WBC count is elevated (normal range = 4,000–11,000/mm^3), this could indicate the presence of inflammation. Further analysis (called a differential) is performed to determine the relative prevalence of different types of white blood cells and the potential cause of the inflammation. What might be causing the inflammation if . . . (use your textbook for reference):

a. neutrophils were elevated? Explain your reasoning.

b. lymphocytes were elevated? Explain your reasoning.

c. eosinophils were elevated? Explain your reasoning.

2 Another lab value that is monitored routinely is the total red blood cell (RBC) count. A normal RBC count is typically 4.2–5.9 million/mm^3. Having a low RBC count is called *anemia*. Predict the possible effects of having anemia, considering the functions of erythrocytes.

3 The disease *erythroblastosis fetalis* develops in a fetus or a newborn infant with Rh-positive blood and an Rh-negative mother. Symptoms result when maternal anti-Rh antibodies cross the placenta and interact with the fetus' erythrocytes. Why are the children of Rh-positive mothers not at risk for this disease? Why are Rh − fetuses not at risk for this disease?

4 When Colonel Lemon arrived at the hospital, the staff determined that his blood had been mistyped and he was in fact blood type AB − . Which of our three suspects (Mrs. Blanc, Ms. Feather, and/or Professor Purple) could safely donate blood to Colonel Lemon now?

5 What would have happened to Colonel Lemon if he had received blood from an incompatible donor?

Lymphatics and Immunity

OBJECTIVES

Once you have completed this unit, you should be able to:

- Identify structures of the lymphatic system.

- Describe the role of the immune system in processing lymph as it is returned to the heart.

- Trace the pathway of lymph as it is returned to the cardiovascular system.

- Identify histological features of lymphoid organs on microscope slides.

- Trace the response of the immune system to an invading pathogen.

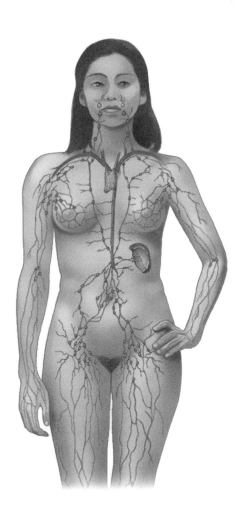

Pre-Lab Exercise 1
Key Terms

Table 20.1 lists the key terms with which you should be familiar before coming to lab.

TABLE **20.1**	Key Terms

Term	Definition
Lymphatic Structures	
Lymphatic capillary	_____
Lymph	_____
Lymph collecting vessel	_____
Lymph trunk	_____
Lymph duct	_____
Spleen	_____
Thymus	_____
Tonsil	_____
Mucosal-associated lymphatic tissue	_____
Lymph node	_____

Components of the Immune System

Nonspecific response _____

_____

Inflammation _____

_____

Granulocytes _____

_____

Macrophages _____

_____

Specific response _____

_____

Antigen _____

_____

Humoral immunity _____

_____

B lymphocyte _____

_____

Antibody _____

_____

Cell-mediated immunity _____

_____

CD4 (helper) T lymphocyte _____

_____

CD8 (cytotoxic) T lymphocyte _____

_____

✍ Pre-Lab Exercise 2

Anatomy of the Lymphatic System

Label and color the structures of the lymphatic system depicted in Figure 20.1 with the terms from Exercise 1. Use your text and Exercise 1 in this unit for reference.

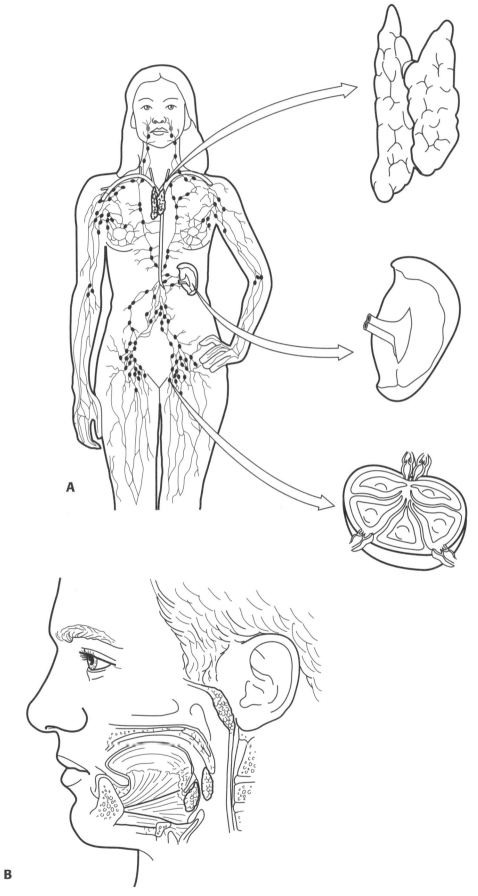

FIGURE 20.1 Lymphoid organs: (A) overview of the lymphatic organs and vessels; (B) the tonsils

EXPLORING ANATOMY & PHYSIOLOGY IN THE LABORATORY

Exercises

The **lymphatic system** serves numerous homeostatic functions in the body. For one, it is an important part of the immune system and combats harmful agents in the internal and external environments. It also works with the cardiovascular system to maintain fluid homeostasis in the extracellular fluid, and with the gastrointestinal system to absorb fats.

The exercises in this unit introduce you to this important system. In the first exercise, you will study the lymphatic system's anatomy and trace the flow of lymph throughout the body. In the second exercise, you will examine three types of lymph tissue on microscope slides. Finally, you will examine the role of the lymphatic structures in the immune system response.

 # Exercise 1

Lymphatic System Anatomy

MATERIALS NEEDED

- Human torso models
- Head and neck models
- Intestinal villus model
- Laminated outline of the human body
- Water-soluble marking pens

As you have learned, the lymphatic system consists of a diverse group of organs that have three primary functions:

1. **Transporting excess interstitial fluid back to the heart.** Hydrostatic pressure is stronger than colloid osmotic pressure in blood capillaries, which forces fluid out of the blood capillaries and into the interstitial space. Approximately 1.5 ml/min of fluid is lost from the circulation in this manner. This may not sound like a lot, but if this fluid were not returned to the blood vessels, we could lose our entire plasma volume in about a day! Fortunately, the lymphatic system picks up this lost fluid, carries it through lymphatic vessels, and returns it to the cardiovascular system. The fluid is picked up first by small blind-ended **lymph capillaries** that surround blood capillary beds. These lymph capillaries are distinct from blood capillaries and contain highly permeable walls that allow large substances and large volumes of fluid to enter and exit. Once inside the lymph capillaries, the fluid is called **lymph** and is delivered to larger **lymph-collecting vessels**. Lymph-collecting vessels drain the fluid into larger **lymph trunks**, and these drain into **lymph ducts** that return the lymph to the cardiovascular system (**Figure 20.2A**). Note in **Figure 20.2A** that there are two lymph ducts—the **right lymphatic duct**, which drains the right arm and the right side of the head, neck, and thorax, and the **thoracic duct**, which drains lymph from the remainder of the body. The right lymphatic and thoracic ducts drain lymph into the blood at the junctions of the right and left subclavian and internal jugular veins, respectively.

2. **Activating the immune system.** Several of the lymphatic organs activate the immune system. These include: the **thymus**, the organ in which T lymphocytes mature; the **spleen**, which filters the blood and houses phagocytes; and the **tonsils**, aggregates of unencapsulated lymphoid tissue found in the oropharynx and nasopharynx (**Figure 20.2B**). As you can see in **Figure 20.2B**, there are three main sets of tonsils: the **pharyngeal tonsils**, which are located in the posterior nasopharynx; the **palatine tonsils**, located in the posterior oropharynx; and the **lingual tonsil**, located at the base of the tongue. Clusters of lymphatic tissue similar in structure to the tonsils are scattered throughout the gastrointestinal tract, where it is called **mucosal-associated lymphatic tissue**, or **MALT**. The lymphoid organs called **lymph nodes** are similar to tonsils and MALT, but they are surrounded by a connective tissue capsule (note that lymph nodes are often called "lymph glands," but this is a misnomer as they do not secrete any products). Lymph nodes are found along lymphatic vessels, where they filter lymph and remove pathogens, toxins, and cells (such as cancer or virally infected cells).

3. **Absorbing dietary fats.** Fats are not absorbed from the small intestine directly into the blood capillaries because they are too large to enter these small vessels. Instead, fats enter a lymphatic capillary called a **lacteal**. The lacteal delivers the fats to the lymph in a large lymphatic vessel called the **cisterna chyli**, which then drains into the thoracic duct.

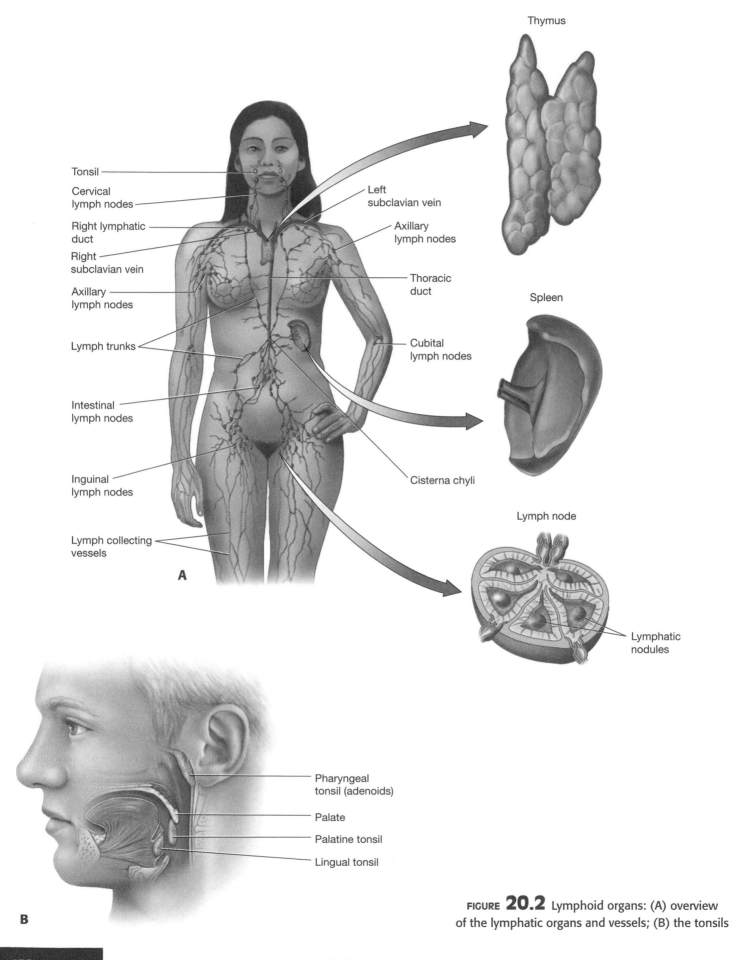

Tonsil

Cervical
lymph nodes

Right lymphatic
duct

Right
subclavian vein

Axillary
lymph nodes

Lymph trunks

Intestinal
lymph nodes

Inguinal
lymph nodes

Lymph collecting
vessels

A

Thymus

Left
subclavian vein

Axillary
lymph nodes

Thoracic
duct

Cubital
lymph nodes

Cisterna chyli

Spleen

Lymph node

Lymphatic
nodules

Pharyngeal
tonsil (adenoids)

Palate

Palatine tonsil

Lingual tonsil

B

FIGURE **20.2** Lymphoid organs: (A) overview
of the lymphatic organs and vessels; (B) the tonsils

EXPLORING ANATOMY & PHYSIOLOGY IN THE LABORATORY

In the following activities you will identify structures of the lymphatic system, and then trace the flow of lymph through the vessels on its way back to the cardiovascular system. Your instructor may also wish you to dissect a preserved small mammal to identify some of the structures that are difficult to see on models, such as the thymus. If so, follow the procedure outlined in Unit 1 (p. 10) to open the animal and identify the required structures.

Procedure Model Inventory

Identify the following structures of the lymphatic system on models and diagrams, using your textbook and this unit for reference. As you examine the anatomical models and diagrams, record on the model inventory in Table 20.2 the name of the model and the structures you were able to identify. When you have completed this activity, answer Check Your Understanding questions 1 through 3 (p. 471).

1. Lymph vessels
   a. Thoracic duct
   b. Right lymphatic duct
   c. Lymph trunks
   d. Lymph collecting vessels
   e. Lacteal
   f. Cisterna chyli
2. Lymph nodes
   a. Cervical lymph nodes
   b. Axillary lymph nodes
   c. Inguinal lymph nodes
   d. Intestinal lymph nodes

3. Spleen
4. Thymus (this is best viewed on a fetal pig)
5. Mucosal-associated lymphoid tissue
6. Tonsils
   a. Palatine tonsil
   b. Pharyngeal tonsil
   c. Lingual tonsil

TABLE 20.2	Model Inventory for the Lymphatic System
Model	Structures Identified

Procedure Tracing the Flow of Lymph Through the Body

In this exercise you will trace the pathway of lymph flow from the starting point to the point at which the lymph is delivered to the cardiovascular system. You will trace the flow through the major lymph-collecting vessels, trunks, and ducts, and highlight clusters of lymph nodes through which the lymph passes as it travels.

1 Write the sequence of the flow.

2 Then use differently colored water-soluble markers to draw the pathway on a laminated outline of the human body. If no outline is available, use colored pencils and **Figure 20.3**.

Trace the flow from the following locations:

Start: Right foot

Start: Right arm

Start: Intestines (with fat)

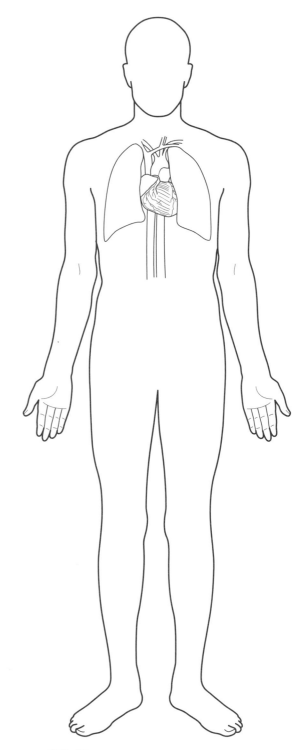

FIGURE **20.3** Outline of human body, anterior view

Exercise 2

Lymphatic Organ Histology

MATERIALS NEEDED

- Spleen section slide
- Lymph node slide
- Peyer's patch slide
- Light microscope
- Colored pencils

In this exercise, you will examine the microscopic anatomy of three different lymphoid organs: the spleen, a lymph node, and a special type of MALT found in the terminal portion of the small intestine called a **Peyer's patch**. Following are some hints regarding what to look for on each slide.

1. **Spleen**: Note in **Figure 20.4A** that the spleen consists of two types of tissue:

   a. **red pulp**, which is involved in the destruction of old and worn-out red blood cells, and

   b. **white pulp**, which contains phagocytes and lymphocytes that play a role in the immune system.

   Be careful here in identifying the red and white pulp. Notice that the red pulp is reddish in color but the white pulp doesn't show up on the slide as white; instead, it appears as purplish "islands" within the red pulp. The purple color is caused by the presence of white blood cells, which have prominent nuclei that stain purple with commonly used stains.

2. **Lymph nodes**: Lymph nodes consist of an outer **cortex**, which contains spherical clusters of cells (primarily B lymphocytes) called **lymphatic nodules** or **germinal centers** (**Figure 20.4B**). Deep to the lymphatic nodules we find an area of the cortex that houses primarily T lymphocytes. The innermost region of the node, called the **medulla**, houses macrophages, which are highly active phagocytes. The entire lymph node is surrounded by a capsule made of dense irregular collagenous connective tissue.

3. **Peyer's patch**: Peyer's patches somewhat resemble lymph nodes, although their lymphatic nodules are less well defined and their capsules are incomplete (**Figure 20.4C**). To have the best chance of finding them, examine your section on low power and look for the epithelial lining of the small intestine. Scroll downward (deep to the epithelium) into the connective tissue, where you will note large, oval or teardrop-shaped, purplish clusters. These are Peyer's patches.

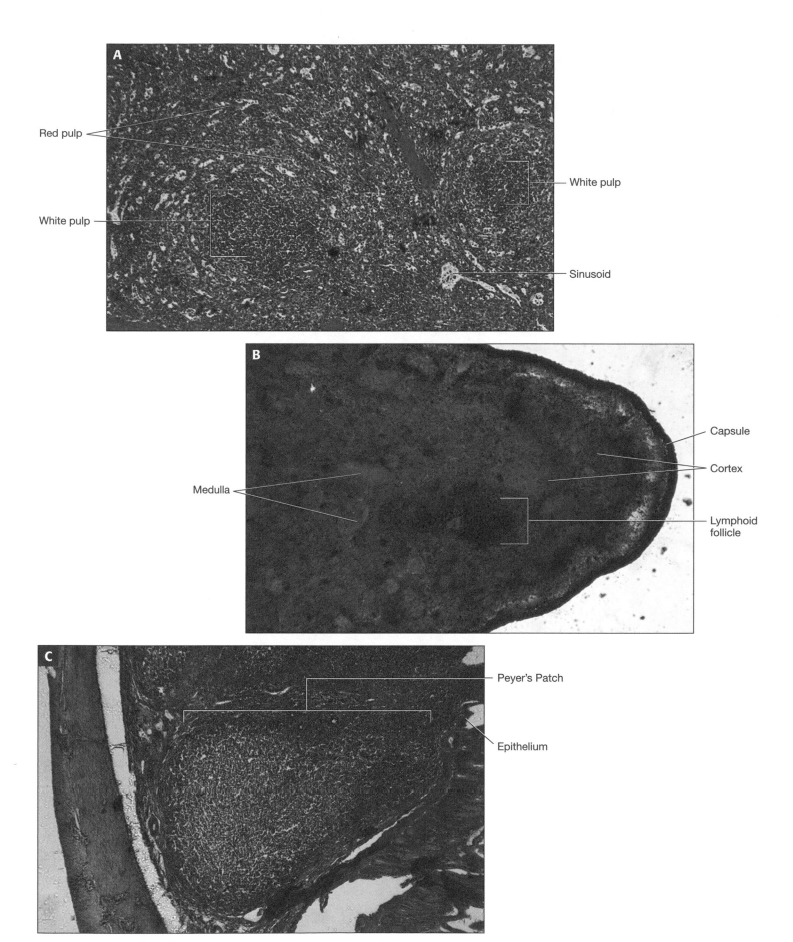

Red pulp

White pulp

White pulp

Sinusoid

Capsule

Cortex

Medulla

Lymphoid follicle

Peyer's Patch

Epithelium

FIGURE **20.4** Lymphatic organs, photomicrographs: (A) spleen; (B) lymph node; (C) Peyer's patch in the ileum

EXPLORING ANATOMY & PHYSIOLOGY IN THE LABORATORY

Procedure Microscopy

Obtain prepared slides of the spleen, a lymph node, and the small intestine (the *ileum*). Use your colored pencils to draw what you see in the field of view and label your drawing with the terms below.

Spleen

1. Red pulp

2. White pulp

Lymph Node

1. Capsule

2. Cortex

3. Medulla

4. Lymphatic nodule

Peyer's Patch

1. Mucosa (simple columnar epithelial tissue)

2. Peyer's patch

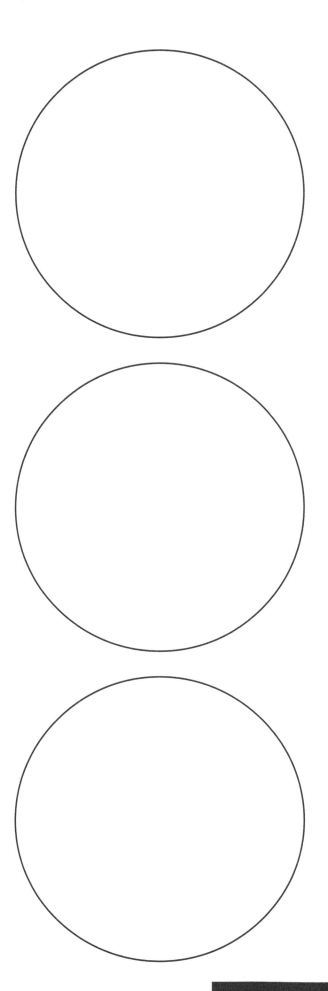

 **Exercise 3**

The Immune Response

The immune system may be divided into two major types of responses:

1. the **nonspecific response,** which reacts the same way to any invading organism or cellular injury, and

2. the **specific response,** which responds to foreign cells and molecules with a unique set of cells and molecules for each individual threat.

The nonspecific responses are sometimes called **innate immunity** because we are born with these responses and they do not require exposure to a foreign cell or molecule to be produced. Specific responses, in contrast, do require exposure to a foreign agent to be produced, and for this reason they are often referred to as **acquired immunity.**

The nonspecific responses may be subdivided into the following:

1. **Surface barriers.** Surface barriers include the skin and mucous membranes. Both surfaces provide continuous, avascular surfaces that secrete substances that either deter the growth of microorganisms or trap foreign particles.

2. **Inflammation and fever.** When cells are damaged by microorganisms, foreign cells, toxins, or trauma, injured cells (primarily macrophages) release chemicals called **inflammatory mediators.** The inflammatory mediators attract other immune cells to the area, especially neutrophils and other macrophages. They also cause local *vasodilation*, which causes the area to become red and warm, *increased capillary permeability*, which causes the area to become swollen, and *increased sensitivity to pain*. Certain inflammatory mediators called **pyrogens** act on the hypothalamus to raise the body temperature "set point," which results in the elevated temperature of **fever.**

3. **Granulocytes and macrophages.** The body's main phagocytic cells, the granulocytes (**neutrophils, basophils,** and **eosinophils**) and the **macrophages** (mature monocytes), will nonspecifically ingest and destroy any foreign or damaged cell. Macrophages are generally the first cells to respond to tissue damage and release inflammatory mediators. These mediators attract other phagocytes, particularly neutrophils, to the damaged area in large numbers. Macrophages can ingest and destroy foreign and damaged cells repeatedly; however, neutrophils can ingest a cell only once, after which they die as well. When tissue damage is extensive, many neutrophils in one area will die and accumulate, leaving a whitish liquid called **pus.**

4. **Antimicrobial proteins.** Several proteins in the blood will nonspecifically attack and lyse invading organisms when activated. The largest such system is the **complement system,** which is a collection of more than 20 proteins, most of which are made by the liver. Complement proteins lyse foreign cells, enable phagocytes to better function, and enhance the inflammatory response.

The specific immune response involves primarily B and T lymphocytes, which respond specifically to unique **antigens,** or chemical markers. Antigens are present in many molecules, such as toxins and allergens, and on the surfaces of all cells, including our own cells. There are two types of specific immunity:

1. **Humoral immunity.** Humoral immunity is mediated by B lymphocytes. The humoral response is initiated when a macrophage or T lymphocyte presents a B lymphocyte with an antigen for which it has a specific receptor. This causes the B lymphocyte to differentiate into two cell types: **plasma cells,** which secrete molecules called **antibodies** that bind to antigens, and **memory cells,** which will "remember" the antigen and permit a more rapid antibody response upon subsequent exposures. Antibodies are proteins that bind and **agglutinate** antigens and facilitate their destruction by phagocytes. Antibodies also activate the nonspecific response, particularly complement, and neutralize toxins such as venoms.

2. **Cell-mediated immunity.** As its name implies, cell-mediated immunity is mediated by cells rather than by antibodies, specifically T lymphocytes. Of the several different kinds of T lymphocytes, we will consider two here—the **CD4** or **helper T cells,** and the **CD8** or **cytotoxic T cells.** The cell-mediated response begins when a macrophage ingests an antigen and presents it to a helper T cell. The helper T cell then secretes chemicals called **cytokines,** which activate cells of the nonspecific response such as macrophages, encourage inflammation, activate B lymphocytes, and activate cytotoxic T cells. The cytotoxic T cells locate cells infected with viruses or intracellular bacteria, cancer cells, and

transplanted cells, and lyse and destroy them. Both helper and cytotoxic T cells may become **memory T cells**, which help to prevent re-infection with the same agent.

Note that all types of cell damage and/or invading organisms are potentially met with multiple facets of the non-specific *and* specific responses. For example, cancer cells are prevented from spreading by surface barriers, are phagocytized by granulocytes and macrophages in the blood and in the lymph nodes, are lysed by antimicrobial proteins that are released when healthy cells are damaged, cause local and systemic inflammation and fever, are met with an antibody response from B lymphocytes, and are located and lysed by helper T and cytotoxic T cells. This same cascade of responses is seen with viral infections, bacterial infections, envenomations, severe burns, major trauma, organ and tissue transplantations, fungal infections, parasitic infections, and more.

The immune response is . . . well, *complicated*, to say the least, and you could nearly fill this entire book tracing through everything that happens when you get a cold. The concept can be simplified somewhat because most types of invading organisms or cellular damage will trigger one type of response slightly more than the others. For example, bacterial infections tend to lead to elevated levels of neutrophils that phagocytize the bacterial cells. In addition, viral infections and cancer cells tend to trigger a predominantly cell-mediated response and lead to elevated levels of cytotoxic T lymphocytes that lyse affected cells.

Procedure Tracing Exercises

In this exercise, you will be tracing two invaders: bacteria in the extracellular fluid that enter the lymphatic vessels, and virus-infected cells in the blood. To keep things simple, you will trace the *main* pathway and the *main* immune response to the invading organism. Do note, however, that the actual response is much more complex, and your instructor may wish you to include more information. I offer hints and the number of steps that you should aim to include for each organism, but you will likely want to consult your book for more help. After you have completed the activity, answer Check Your Understanding questions 4 and 5 (p. 472).

Start: bacteria in the extracellular fluid $\longrightarrow$ bacteria damage local macrophages and damaged macrophages release

_____ $\longrightarrow$ _____

$\longrightarrow$ neutrophils _____ $\longrightarrow$ bacteria enter lymph capillaries

$\longrightarrow$ _____ $\longrightarrow$ _____

$\longrightarrow$ bacteria encounter macrophages in lymph node $\longrightarrow$ _____

$\longrightarrow$ macrophages present antigens to _____

$\longrightarrow$ _____ proliferate and release _____

which _____ **End**

Start: virus-infected cells in the blood $\longrightarrow$ viruses damage tissue and damaged macrophages release

_____ $\longrightarrow$ _____

$\longrightarrow$ _____

$\longrightarrow$ inflammation and fever begin $\longrightarrow$ virus-infected cells enter spleen $\longrightarrow$ virus-infected cells encounter macrophages in

spleen $\longrightarrow$ _____ $\longrightarrow$

macrophages present antigens to _____ $\longrightarrow$ _____ prolifer-

ate and secrete _____ which activate _____

$\longrightarrow$ _____ **End**

NAME _____

SECTION _____ DATE _____

Check Your Recall

1 Label **Figure 20.5** with the terms below.

Figure 20.5A:

- Thymus
- Spleen
- Lymph nodes: cervical, axillary, intestinal, and inguinal
- Right lymphatic duct
- Thoracic duct
- Cisterna chyli

Figure 20.5B:

- Lingual tonsil
- Palatine tonsil
- Pharyngeal tonsil

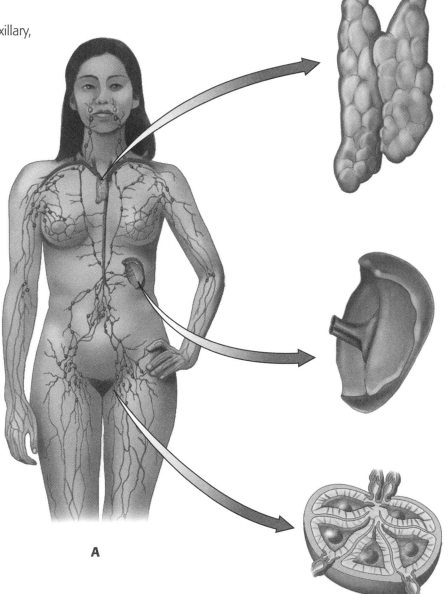

A

FIGURE 20.5 Lymphoid organs: (A) overview of the lymphatic organs and vessels

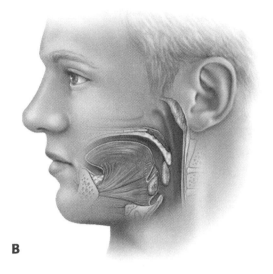

B

FIGURE **20.5** Lymphoid organs (cont.): (B) the tonsils

2 Which of the following is *not* a function of the lymphatic system?

a. Maintaining blood pressure
b. Absorbing dietary fats
c. Activating the immune system
d. Transporting excess interstitial fluid back to the heart

3 Lymph nodes filter _____, and the spleen filters _____.

a. lymph; lymph
b. blood; blood
c. lymph; blood
d. blood; lymph

4 The vessels into which fats are absorbed are called

a. blood capillaries.
b. lymph ducts.
c. lacteals.
d. fats are not absorbed.

5 Describe the structure and location of the tonsils, lymph nodes, and MALT.

6 The cortex of a lymph node contains spherical clusters called _____, which contain primarily _____.

  a. lymphatic nodules; T lymphocytes
  b. lymphatic nodules; B lymphocytes
  c. medullary centers; T lymphocytes
  d. medullary centers; B lymphocytes

7 Explain how each of the following sets of terms differ.

  a. Specific immune response and nonspecific immune response:

  b. Cell-mediated immunity and humoral immunity:

  c. Granulocytes and lymphocytes:

  d. CD4 (helper) T lymphocytes and CD8 (cytotoxic) T lymphocytes:

8 Which of the following is *not* a type of nonspecific immune response?

a. Inflammation and fever
b. Antimicrobial proteins
c. Cell-mediated immunity
d. Surface barriers
e. Neutrophils and macrophages

9 Fill in the blanks: When tissues are damaged, macrophages release inflammatory mediators that cause

_____, _____, and _____.

Certain inflammatory mediators called _____ increase the hypothalamus' "set point"

for temperature and cause fever.

10 True or false: The immune system's response to an infection involves either the nonspecific response or the specific response, but not both.

Check Your Understanding | **Critical Thinking and Application Questions**

1 Predict some potential consequences of removing the spleen.

2 In a condition called DiGeorge syndrome, infants are born with either an absent thymus or a thymus that isn't functional. Predict the consequences of this disease.

3 Explain why blockage or removal of the lymphatic vessels can result in significant edema (accumulation of fluid in a limb or body part).

4 A common symptom of pharyngitis is swelling of the anterior cervical lymph nodes. Why may lymph nodes swell in the presence of an infection? What else may cause swollen lymph nodes? (*Hint:* What part of the immune response causes swelling?)

5 Explain why conditions that cause destruction of B lymphocytes and/or T lymphocytes cause malfunctioning of both the specific and nonspecific immune responses.

Respiratory System Anatomy

OBJECTIVES

Once you have completed this unit, you should be able to:

- Identify structures of the respiratory system.

- Identify histological structures of the respiratory system on microscope slides.

- Describe the anatomical changes associated with inflation of the lungs.

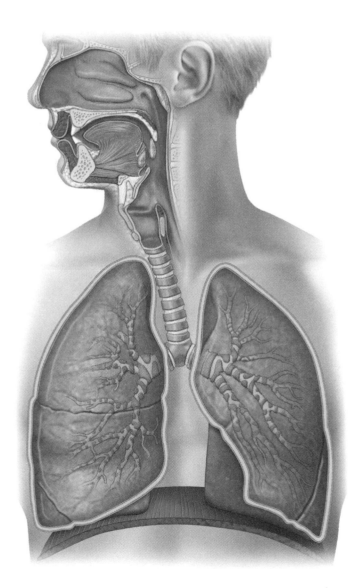

Pre-Lab Exercises | Complete the following exercises prior to coming to lab, using your textbook and lab manual for reference.

Pre-Lab Exercise 1
Key Terms

Table 21.1 lists the key terms with which you should be familiar before coming to lab.

TABLE **21.1**	Key Terms

Term   **Definition**

General Structures of the Respiratory System

Respiratory tract _____

_____

Parietal pleura _____

_____

Visceral pleura _____

_____

Pleural cavity _____

_____

Lungs and lobes _____

_____

Structures of the Respiratory Tract

Nasal cavity _____

_____

Nasopharynx _____

_____

Oropharynx _____

_____

Laryngopharynx _____

_____

       EXPLORING ANATOMY & PHYSIOLOGY IN THE LABORATORY

Larynx _____

_____

Trachea _____

_____

Primary bronchi _____

_____

Secondary bronchi _____

_____

Terminal bronchioles _____

_____

Respiratory bronchioles _____

_____

Alveolar ducts _____

_____

Alveoli _____

_____

Structures of the Larynx

Thyroid cartilage _____

_____

Cricoid cartilage _____

_____

True vocal cords _____

_____

False vocal cords _____

_____

Respiratory System Anatomy

Label and color the diagrams of the structures of the respiratory system in **Figure 21.1** with the terms from Exercise 1. Use your text and Exercise 1 in this unit for reference.

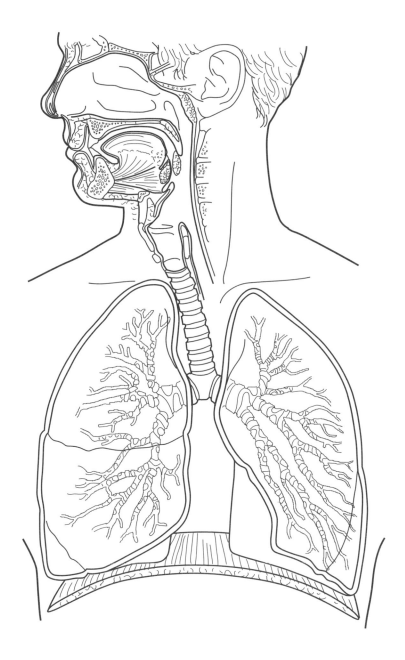

A

FIGURE 21.1 Structures of the respiratory system: (A) the lungs and respiratory tract

EXPLORING ANATOMY & PHYSIOLOGY IN THE LABORATORY

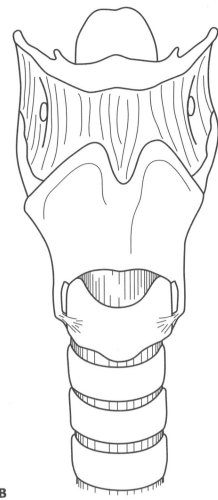

B

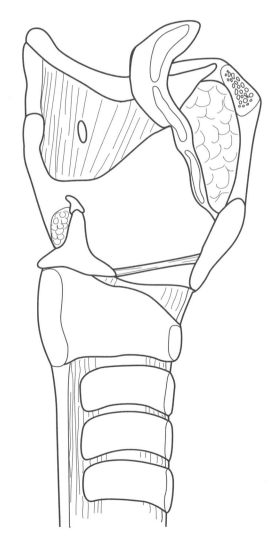

FIGURE **21.1** Structures of the respiratory system (cont.): (B) the larynx, anterior view; (C) the larynx, midsagittal section

C

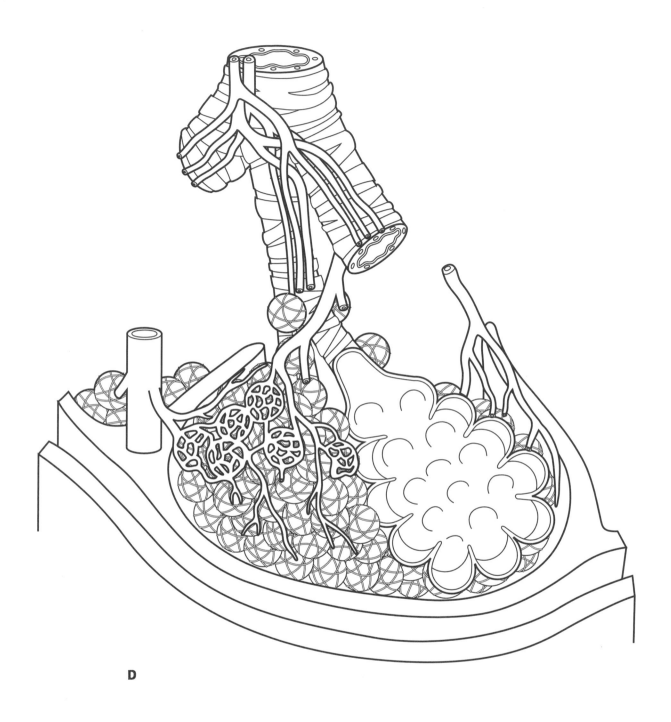

D

FIGURE **21.1** Structures of the respiratory system (cont.): (D) bronchiole and alveolar sac

EXPLORING ANATOMY & PHYSIOLOGY IN THE LABORATORY

Exercises

Cells require oxygen in the reactions that synthesize ATP, and these reactions produce carbon dioxide as a waste product. The respiratory system and the cardiovascular system work together to supply the cells with the oxygen that they need and to rid them of carbon dioxide.

The exercises in this unit will familiarize you with the organs and histology of the respiratory system, including the paired **lungs** and the collection of airway passages known as the **respiratory tract**. In the final exercise you will observe the changes in the texture, shape, and size of the lungs as they inflate and deflate with the process known as **ventilation**.

 # Exercise 1

Respiratory System Anatomy

MATERIALS NEEDED

- Lung models
- Larynx models
- Alveolar sac model
- Head and neck model

The lungs are composed of elastic connective tissue and tiny air sacs called **alveoli**, where gas exchange takes place. Each lung is divided into smaller structures called **lobes**. The right lung has three lobes (upper, middle, and lower), and the left lung has two lobes (upper and lower) (**Figure 21.2**). The lobes are separated from each another by deep indentations called **fissures**. The **horizontal fissure** separates the right upper and right middle lobes; the **right oblique fissure** separates the right middle and right lower lobes; and the **left oblique fissure** separates the left upper and left lower lobes.

Each lung is surrounded by a double-layered serous membrane similar to the pericardial membranes called the **pleural membranes**. The two layers of the pleural membranes are the following:

1. **Parietal pleura.** The outer parietal pleura lines the interior of the thoracic cavity and the superior surface of the diaphragm.

2. **Visceral pleura.** When the parietal pleura reaches the structures of the mediastinum, it folds inward to become the visceral pleura. The visceral pleura adheres tightly to the surface of the lung.

There is a very thin potential space between the parietal and visceral pleurae called the **pleural cavity**. The space is only a "potential" space because it is filled with a thin layer of serous fluid that reduces friction as the lungs change in shape and size during ventilation.

Air is delivered to the lungs' alveoli through the passageways of the respiratory tract. The respiratory tract may be divided into two regions according to both structure and function. Structurally, it is divided into:

1. The **upper respiratory tract**, which consists of the passages superior to the thoracic cavity, and

2. The **lower respiratory tract**, which consists of the passages within the thoracic cavity.

Functionally, it is divided into:

1. The **conducting zone**, where passages of the conducting zone carry, or "conduct," air to the lower passages where gas exchange takes place. The majority of the passages of the respiratory tract belong to the conducting zone.

2. The **respiratory zone**, which consists of the passages in which gas exchange takes place, is located in the terminal respiratory tract.

The conducting zone begins with the **nasal cavity** and the **paranasal sinuses**, which filter, warm, and humidify the inhaled air. Air from the nasal cavity next enters the **pharynx** (pronounced "fair-inks"), also known as the throat, which has the following three divisions:

1. **Nasopharynx.** The nasopharynx is the region posterior to the nasal cavity. The muscles of the soft palate move superiorly to close off the nasopharynx during swallowing to prevent food from entering the passage. Sometimes this

mechanism fails (such as when a person is laughing and swallowing simultaneously), and the unfortunate result is that food or liquid comes out of the nose. The nasopharynx is lined with pseudostratified ciliated columnar epithelium with copious mucus-secreting **goblet cells.**

2. **Oropharynx.** The oropharynx is the region posterior to the oral cavity. Both food and air pass through the oropharynx, and it is therefore lined with stratified squamous epithelium. This tissue provides more resistance to mechanical and thermal stresses.

3. **Laryngopharynx.** The laryngopharynx is the intermediate region between the larynx and the esophagus. Like the oropharynx, both food and air pass through the laryngopharynx and it is lined with stratified squamous epithelium.

Air passes from the pharynx to the **larynx** (pronounced "lair-inks"), also called the *voice box,* which is a short passage framed by nine cartilages (**Figure 21.3**). The "lid" of the larynx is a piece of elastic cartilage called the **epiglottis.** During swallowing, muscles of the pharynx and larynx move the larynx superiorly and the epiglottis seals off the larynx from food and liquids. The largest cartilage

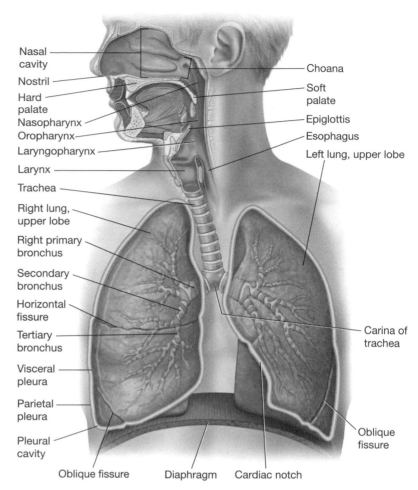

FIGURE **21.2** The lungs and respiratory tract

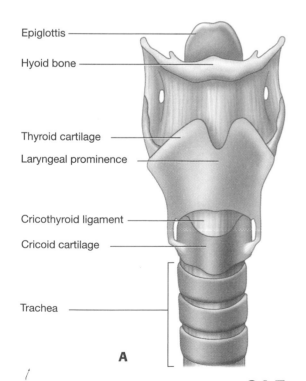

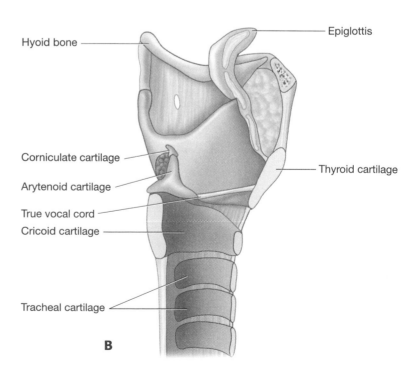

FIGURE **21.3** Larynx: (A) anterior view; (B) midsagittal section

EXPLORING ANATOMY & PHYSIOLOGY IN THE LABORATORY

of the larynx is the shield-like **thyroid cartilage**, which forms the larynx's anterior and lateral walls. Inferior to the thyroid cartilage is a smaller cartilage called the **cricoid cartilage**. Between the two is a soft piece of connective tissue called the **cricothyroid ligament**.

As its common name "voice box" implies, the larynx is the structure where sound is produced. It contains two sets of elastic ligaments known as the **vocal folds** or **vocal cords**. The superior set of vocal folds, the **false vocal cords** (also called the *vestibular folds*), play no role in sound production. They do, however, serve an important sphincter function and can constrict to close off the larynx. The inferior set of vocal folds, called the **true vocal cords**, vibrate as air passes over them to produce sound.

Inspired air passes from the larynx into a tube supported by C-shaped rings of hyaline cartilage called the **trachea**. The trachea bifurcates in the mediastinum into two **primary bronchi**, which begin the large and branching **bronchial tree**. The right primary bronchus is short, fairly straight, and wide, and the left primary bronchus is long, more horizontal, and narrow.

Each primary bronchus divides into smaller **secondary bronchi**, each of which serves one lobe of the lung. The two left secondary bronchi serve the two lobes of the left lung, and the three right secondary bronchi serve the three lobes of the right lung. The bronchi continue to branch and become tertiary bronchi, quaternary bronchi, and so on until the air reaches tiny air passages smaller than 1 millimeter in diameter called **bronchioles** (Figure 21.4).

Bronchioles smaller than 0.5 mm in diameter are called the **terminal bronchioles**, and these mark the end of the conducting zone. The respiratory zone begins with small branches called **respiratory bronchioles**, which have thin-walled sacs called **alveoli** in their walls. As the respiratory bronchioles progressively branch, the number of alveoli in the walls

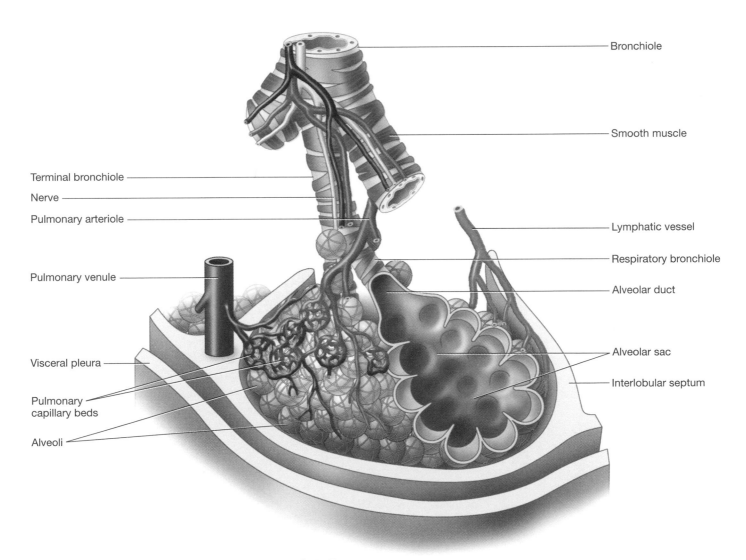

FIGURE 21.4 Bronchiole and alveolar sac

increases until the wall is made up exclusively of alveoli, at which point it is termed an **alveolar duct**. The terminal portions of the respiratory zone, called **alveolar sacs**, are grapelike clusters of alveoli surrounded by **pulmonary capillaries.** This junction is where pulmonary gas exchange takes place: Oxygen from the alveoli diffuses into the blood, and carbon dioxide in the blood diffuses into the alveoli to be exhaled.

 ## Procedure Model Inventory

Identify the following structures of the respiratory system on models and diagrams, using your textbook and this unit for reference. As you examine the anatomical models and diagrams, record on the model inventory in Table 21.2 the name of the model and the structures you were able to identify. After you have completed the activity, answer Check Your Understanding questions 1 and 2 (p. 495).

1. Lungs
   a. Parietal pleura
   b. Visceral pleura
   c. Pleural cavity
2. Right lung
   a. Upper, middle, lower lobes
   b. Horizontal fissure
   c. Oblique fissure
3. Left lung
   a. Upper and lower lobes
   b. Oblique fissure
   c. Cardiac notch
4. Nasal cavity
   a. Nares
   b. Vestibule
   c. Nasal septum
   d. Nasal conchae
      (1) Superior nasal conchae
      (2) Middle nasal conchae
      (3) Inferior nasal conchae
5. Paranasal sinuses
   a. Sphenoid sinus
   b. Ethmoid sinus
   c. Frontal sinus
   d. Maxillary sinus
6. Hard palate
7. Soft palate and uvula
8. Pharynx
   a. Nasopharynx
   b. Oropharynx
   c. Laryngopharynx

9. Larynx
   a. Epiglottis
   b. Thyroid cartilage
   c. Cricoid cartilage
   d. Cricothyroid ligament
   e. False vocal cords
   f. True vocal cords
10. Trachea
    a. Hyaline cartilage rings
    b. Carina
11. Bronchi
    a. Right and left primary bronchi
    b. Secondary bronchi
    c. Tertiary bronchi
12. Bronchioles
    a. Terminal bronchioles
    b. Respiratory bronchioles
    c. Alveolar duct
13. Alveoli and alveolar sacs
14. Muscles
    a. Diaphragm
    b. External intercostals
    c. Internal intercostals
15. Vascular structures
    a. Pulmonary arteries
    b. Pulmonary arterioles
    c. Pulmonary capillaries
    d. Pulmonary venules
    e. Pulmonary veins

TABLE **21.2**	Model Inventory for the Respiratory System
Model/Diagram	**Structures Identified**

Procedure Time to Trace

Part 1

You are a molecule of oxygen floating happily through the atmosphere when all of a sudden you are inhaled by Ms. Magenta.

1 Trace your pathway through Ms. Magenta's respiratory tract, beginning in her nasal cavity to the point where you enter the pulmonary capillaries.

2 Trace your pathway from Ms. Magenta's pulmonary capillaries through her lungs and heart and to your final destination in a hepatocyte (liver cell). You likely will have to review Unit 17 (p. 385) for some help with this.

3 Explain what happens to you in Ms. Magenta's hepatocyte. What do you do, and what do you become? (You can be somewhat general here in your answer unless your lab instructor says otherwise. You may have to refer to your textbook and read about the basics of oxidative catabolism and the electron transfer system.)

Part 2

While in her hepatocyte, you notice that a molecule of carbon dioxide has just been produced.

1 How was the molecule of carbon dioxide produced in Ms. Magenta's hepatocyte? (Again, you can be somewhat general in your answer unless your lab instructor says otherwise. You may have to refer to your textbook and read about the basics of glycolysis and the citric acid cycle.)

2 Trace the carbon dioxide's pathway from Ms. Magenta's hepatocyte through her heart and to her pulmonary capillaries. You will likely need to review Unit 17 (p. 385) for some help with this.

3 Trace the pathway of the carbon dioxide from the pulmonary capillaries through the respiratory tract to the point where it exits from Ms. Magenta's body through her nares.

 Exercise 2

Histology of the Respiratory Tract

MATERIALS NEEDED

- Trachea slide
- Lung tissue slide
- Bronchiole slide
- Light microscope
- Colored pencils

As you learned in Exercise 1, the respiratory tract is a series of branching tubes that conduct air to and from the respiratory zone for gas exchange. The "tubes" of the respiratory tract are similar in structure histologically to other hollow organs of the body and consist of the following three tissue layers (**Figure 21.5A**):

1. **Mucosa.** Mucous membranes, or **mucosae,** line all passageways that open to the outside of the body, including the passages of the respiratory tract. Mucosae consist of epithelial tissue overlying the **basement membrane.** As we have seen elsewhere in the body, there is a clear relationship of structure and function in the respiratory tract, as the epithelial tissue of the mucosae is adapted for each region's function. Some examples of the epithelia found in different regions include:

   a. the nasopharynx, larynx (inferior to the vocal cords), trachea, and bronchi: pseudostratified ciliated columnar epithelium with mucus-secreting cells called **goblet cells** (sometimes called **respiratory epithelium**)

   b. the oropharynx, laryngopharynx, and larynx (superior to the vocal cords): stratified squamous epithelium

   c. the bronchioles: simple cuboidal epithelium

   d. alveoli: simple squamous epithelium. Note in **Figure 21.5B** that the alveoli consist of a mucosa only and have no other tissue layers. This minimizes the distances that gases must diffuse across the alveoli and into the pulmonary capillaries. The structure though which gases must diffuse, called the **respiratory membrane,** consists of the squamous epithelial cells, the endothelial cells of the pulmonary capillaries, and their shared basal lamina.

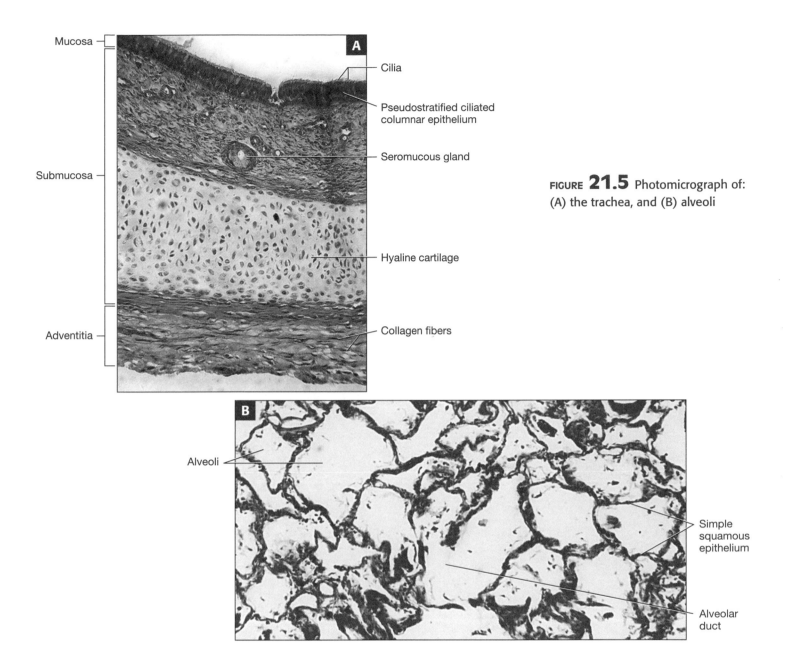

FIGURE **21.5** Photomicrograph of: (A) the trachea, and (B) alveoli

2. **Submucosa**. Deep to the mucosa we find the submucosa, a layer of loose connective tissue. The submucosa contains specialized **seromucous glands**, which secrete watery mucus. The submucosa of larger passages contains hyaline cartilage for support, it also contains smooth muscle that controls the diameter of the airways.

3. **Adventitia**. The outermost layer of the respiratory tract, the adventitia, consists of dense irregular collagenous connective tissue with elastic fibers for support.

As you examine different regions of the respiratory tract, you will note a few trends:

▶ The epithelial tissue changes from taller (pseudostratified) in the upper passages to shorter (cuboidal) in the lower passages and finally to flat (squamous) in the alveoli.

▶ The amount of hyaline cartilage gradually decreases (it is absent in bronchioles), as does the number of goblet cells.

▶ The amount of smooth muscle and elastic fibers increases.

These trends can help you to determine the portion of the respiratory tract from which the section on your slide was taken. In the following exercise you will be able to see these trends as you examine photomicrographs of the trachea, a bronchiole, and the alveoli.

 **Procedure Microscopy**

View prepared slides of a section of the trachea, a bronchiole, and alveoli. Note that the bronchiole may be on the same slide as the alveoli. Begin your examination of the slides on low power, and advance to medium and high power to observe details. Use colored pencils to draw what you see under the microscope and label your drawing with the terms indicated. When you have completed the activity, answer Check Your Understanding questions 3 and 4 (pp. 495–496).

Trachea

Label the following on your drawing:

1. Pseudostratified columnar epithelium
2. Cilia
3. Goblet cells
4. Submucosa
5. Seromucous glands
6. Hyaline cartilage

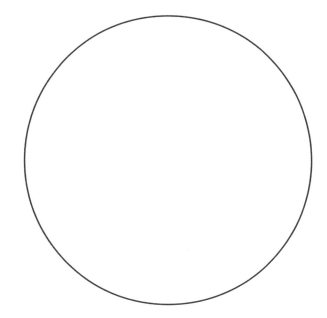

Bronchioles

Label the following on your drawing:

1. Simple columnar or cuboidal epithelium
2. Smooth muscle
3. Pulmonary arteriole

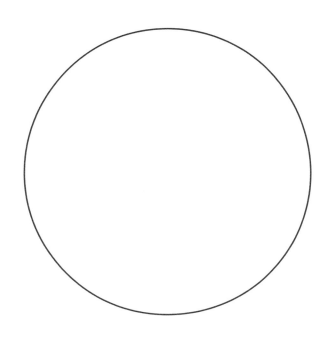

EXPLORING ANATOMY & PHYSIOLOGY IN THE LABORATORY

Alveoli

Label the following on your drawing:

1. Simple squamous epithelium

2. Alveolar duct

3. Alveolus

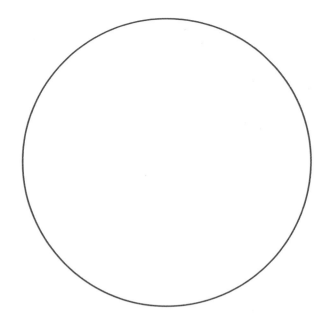

 Exercise 3

Lung Inflation

MATERIALS NEEDED

- Preserved small mammal
- Dissection equipment
- Dissection trays
- Straw
- Air hose
- Fresh lungs

In movies or television programs you may have seen a hero rescuing a choking victim by poking a hole in the victim's neck (usually with a pocket knife) and inserting a straw (or pen) into the hole to restore the airway and save the victim's life. This is actually a somewhat crude version of a legitimate procedure called a **cricothyroidotomy**—but don't try this at home!

A cricothyroidotomy is performed when the upper respiratory tract is blocked and air is prevented from moving into the lungs. The first step in the procedure is to place an incision (hopefully with a scalpel, not a pocket knife!) in the cricothyroid ligament, which is the soft spot between the thyroid and cricoid cartilages. A tube is then inserted into the opening and the patient is ventilated artificially. This restores the patient's airway by bypassing the upper respiratory tract.

In the following procedure you will be able to see the effects of a cricothyroidotomy firsthand by performing the procedure on a fetal pig (or other preserved small mammal) and inflating its lungs. You will also inflate fresh lungs from a sheep, a pig, or a cow. As you perform both procedures, note the difference in the textures and appearances of the preserved lungs and fresh lungs. After you have completed both procedures, answer Check Your Understanding question 5 (p. 496).

 Procedure **Cricothyroidotomy and Inflating Preserved Lungs**

Note: Safety glasses and gloves are required.

---

1 Obtain a fetal pig or other preserved small mammal and dissection equipment.

2 Carefully dissect the animal's neck and remove tissue so the larynx is clearly visible.

3 Locate the cricothyroid ligament and make a small incision with a scalpel into its anterior surface.

4 Insert a small straw into the hole you have just made.

5 Attach the straw to a small hose, and attach the hose to an air outlet.

6 Turn on the air slowly and watch the lungs inflate. Crimp the hose to cut off the airflow, and watch the lungs deflate.

 Procedure Inflating Fresh Lungs

Note: Safety glasses and gloves are required.

---

1 Obtain a fresh specimen and a large air hose.

2 Examine the specimen for structures covered in Exercise 1, in particular the epiglottis and vocal folds, the other laryngeal cartilages, the trachea and its hyaline cartilage rings, and the pleural membranes.

3 Squeeze the deflated lungs between your fingertips and note their texture. Compare their texture with that of the lungs of the preserved mammal:

4 Insert the air hose into the larynx and feed it down into the trachea. Take care not to get the hose stuck in one of the primary bronchi.

5 Attach the hose to the air outlet and turn it on slowly. You may have to squeeze the trachea and the hose to prevent air leakage.

6 Observe the lungs as they inflate. You may inflate the lungs quite fully. Don't worry—they're very unlikely to pop.

7 Squeeze the inflated lungs between your fingers and note their texture:

8 Crimp the air hose and watch the lungs deflate. Again, feel the lungs and note changes in texture:

NAME _____

SECTION _____ DATE _____

1 Label Figure 21.6A and B with the terms below. Note that 21.6b is on the next page.

- Nasal cavity
- Nasopharynx
- Oropharynx
- Laryngopharynx
- Larynx
- Cricoid cartilage
- Epiglottis
- Hyoid bone
- Right primary bronchus
- Left primary bronchus
- Secondary bronchus
- Visceral pleura
- Parietal pleura

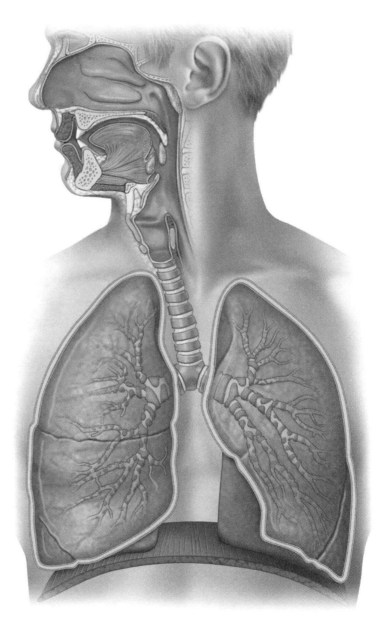

A

FIGURE **21.6** Respiratory structures: (A) the lungs and respiratory tract

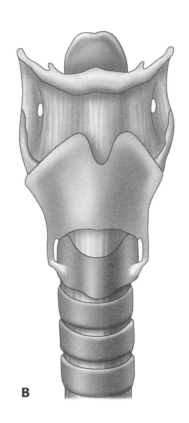

FIGURE **21.6** Respiratory structures
(cont.): (B) the larynx, anterior view

B

2 Label **Figure 21.7** with the terms below.
  • Terminal bronchiole
  • Respiratory bronchiole
  • Alveolar duct
  • Alveolar sac
  • Alveolus
  • Pulmonary capillaries
  • Pulmonary arteriole
  • Pulmonary venule

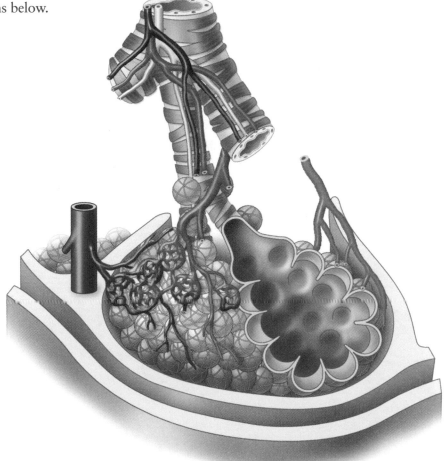

FIGURE **21.7** Bronchiole
and alveolar sac

3 Fill in the blanks: The _____ is the lungs' outer membrane that adheres to the inner wall of the thoracic cavity. At the lungs' root, it folds inward on itself to become the inner membrane called the _____, which adheres to the lungs' surface. Between the two layers of membrane is the _____.

4 How do the respiratory zone and the conducting zone differ?

5 Number the following structures of the respiratory tract in the proper order. The structure that comes into contact with oxygenated air first should be number 1, and the structures where gas exchange takes place should be number 11.

_____ Oropharynx

_____ Trachea

_____ Nasal cavity

_____ Bronchiole

_____ Alveolar sac

_____ Larynx

_____ Nasopharynx

_____ Bronchi

_____ Terminal bronchiole

_____ Laryngopharynx

_____ Respiratory bronchiole

6 The piece of elastic cartilage that seals off the larynx during swallowing is called the
   a. uvula.
   b. false vocal cord.
   c. true vocal cord.
   d. epiglottis.

7 Passages in the respiratory tract smaller than 1 mm in diameter are called

    a. bronchi.
    b. bronchioles.
    c. alveolar ducts.
    d. paranasal sinuses.

8 Which of the following is *not* one of the tissue layers of the respiratory tract?

    a. Muscularis
    b. Mucosa
    c. Adventitia
    d. Submucosa

9 What three structures comprise the respiratory membrane?

    a. _____

    b. _____

    c. _____

10 Which of the following is *not* a trend that we find within the respiratory tract?

    a. The epithelium gets progressively shorter in the lower passages.
    b. The amount of hyaline cartilage gradually increases as we move to smaller passages.
    c. The amount of smooth muscle increases in the smaller passages.
    d. The amount of elastic fibers increases in the smaller passages.

Check Your Understanding | Critical Thinking and Application Questions

1 Why do you think the hyaline cartilage rings of the trachea are "C"-shaped rather than "O"-shaped? (*Hint:* Think about the structure behind the trachea.)

2 Conditions such as pneumonia and lung cancer can result in what is known as a *pleural effusion*, in which the pleural cavity becomes filled with a large amount of fluid. What effects do you think a pleural effusion would have on ventilation? Explain.

3 Explain how the epithelium in each region of the respiratory tract is adapted so its form follows its function.

4 Why do you think that goblet cells are more numerous in the upper respiratory passages than they are in the lower respiratory passages?

5 The most common site for obstruction in a choking victim is the right primary bronchus. Explain why the obstruction tends to lodge in the right primary bronchus rather than the left primary bronchus. (*Hint:* Consider the structure of the right versus the left primary bronchus.)

EXPLORING ANATOMY & PHYSIOLOGY IN THE LABORATORY

Respiratory System Physiology

OBJECTIVES

Once you have completed this unit, you should be able to:

- Describe the pressure-volume relationships in the lungs.

- Describe and measure respiratory volumes and capacities.

- Explain the effects of carbon dioxide concentration on blood pH.

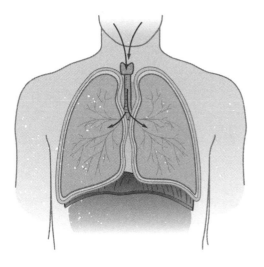

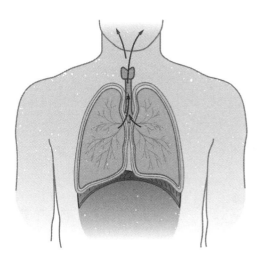

Pre-Lab Exercises | Complete the following exercises prior to coming to lab, using your textbook and lab manual for reference.

Pre-Lab Exercise 1

Key Terms

Table 22.1 lists the key terms with which you should be familiar before coming to lab. Please note that the respiratory volumes and capacities are covered in Pre-Lab Exercises 2 and 3.

TABLE 22.1	Key Terms

Term	Definition
Basic Respiratory Processes	
Pulmonary ventilation	_____
External respiration	_____
Gas transport	_____
Internal respiration	_____
Pulmonary Ventilation Terms	
Inspiration	_____
Expiration	_____
Boyle's law	_____
Intrapulmonary pressure	_____
Atmospheric pressure	_____

Spirometer _____

_____

Respiratory volume _____

_____

Respiratory capacity _____

_____

Restrictive disease _____

_____

Obstructive disease _____

_____

Gas Transport and External Respiration Terms

_____

Bicarbonate _____

_____

Carbonic acid-bicarbonate buffer system _____

_____

Carbonic anhydrase _____

_____

Hyperventilation _____

_____

Respiratory alkalosis _____

_____

Hypoventilation _____

_____

Respiratory acidosis _____

_____

Pre-Lab Exercise 2

Defining Pulmonary Volumes and Capacities

Define and give the normal value for males and females for each of the following respiratory volumes and capacities listed in Table 22.2.

TABLE 22.2	Respiratory Volumes and Capacities	
Volume/Capacity	Definition	Normal Value (in ml)
Tidal volume		
Inspiratory reserve volume		
Expiratory reserve volume		
Residual volume		
Inspiratory capacity		
Functional residual capacity		
Vital capacity		
Total lung capacity		

Pre-Lab Exercise 3

Labeling Pulmonary Volumes and Capacities

Label and color the respiratory volumes and capacities depicted in **Figure 22.1** using the terms from Exercise 2. Use your text and Exercise 2 in this unit for reference.

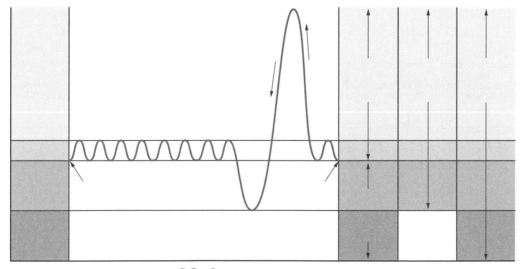

FIGURE **22.1** Respiratory volumes and capacities

Exercises

In the previous unit we covered respiratory system anatomy, and now we turn our attention to respiratory system physiology. The respiratory system carries out four basic physiologic processes:

1. **pulmonary ventilation**, the physical movement of air into and out of the lungs,

2. **external respiration**, the movement of gases across the respiratory membrane,

3. **gas transport**, the movement of gases through the blood, and

4. **internal respiration**, the exchange of gases between the blood and the tissues.

In this unit we examine the first three processes by studying a model of the lungs, measuring respiratory volumes, and measuring the effect of exercise on carbon dioxide exchange.

 # Exercise 1

Pressure–Volume Relationships in the Lungs

MATERIALS NEEDED

▶ Bell-jar model of the lungs

Pulmonary ventilation, the physical movement of air in and out of the lungs, consists of two phases:

1. **inspiration**, during which air is brought into the lungs, and

2. **expiration**, during which air is expelled from the lungs.

The movement of air during inspiration and expiration is driven by changes in the lungs' volume and pressure (**Figure 22.2**). The relationship of gas pressure and volume is expressed in what is known as **Boyle's law**, expressed mathematically as:

$$P_1 V_1 = P_2 V_2$$

Stated simply, this means that pressure and volume are inversely proportional: As the volume of a container increases, the pressure decreases, and as the volume of a container decreases, the pressure increases.

The changes in volume during the phases of ventilation are driven by the inspiratory muscles (the **diaphragm** and the **external intercostals**). When the inspiratory muscles contract, they increase both the height and the diameter of the thoracic cavity, which increases its volume. Recall that the lungs are attached to the thoracic cavity directly by the pleural membranes. Therefore, the lungs increase in volume when the thoracic cavity increases in volume. As the lungs' volume increases, their pressure, called the **intrapulmonary pressure**, decreases. When intrapulmonary pressure is lower than the **atmospheric pressure**, inspiration occurs and air rushes into the lungs. Because the intrapulmonary pressure must be *lower* than the atmospheric pressure for inspiration to occur, it sometimes is referred to as **negative pressure**.

Expiration is achieved primarily by the elastic recoil of the lungs. As the inspiratory muscles relax, the lungs' elastic tissue causes them to recoil to their original, smaller size. This decreases the volume of the lungs and increases the intrapulmonary pressure. Once the intrapulmonary pressure is higher than the atmospheric pressure, air exits the lungs and expiration occurs. In the event of forced expiration, several muscles, including the internal intercostals, will decrease the height and diameter of the thoracic cavity.

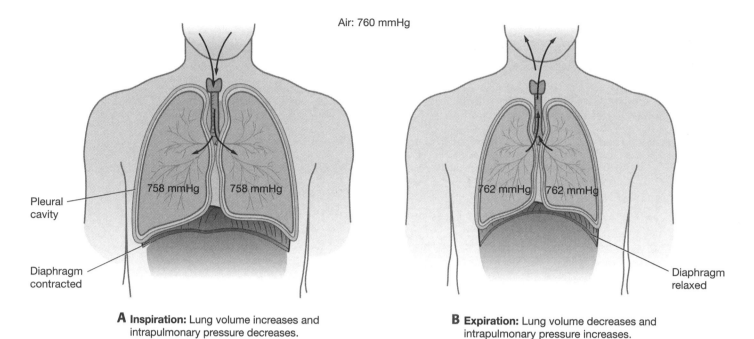

Air: 760 mmHg

758 mmHg   758 mmHg

Pleural cavity

Diaphragm contracted

762 mmHg   762 mmHg

Diaphragm relaxed

A Inspiration: Lung volume increases and intrapulmonary pressure decreases.

B Expiration: Lung volume decreases and intrapulmonary pressure increases.

FIGURE **22.2** Pressure-volume relationships in the lungs during inspiration and expiration

Procedure Model Ventilation with the Bell-Jar Model

In this procedure we will use a bell-jar model of the lungs to view the effects of pressure and volume on ventilation. The bell jar model has two balloons, each representing one lung, and a flexible membrane on the bottom, which represents the diaphragm. After you have completed this activity, answer Check Your Understanding questions 1 and 2 (p. 511).

1 Apply upward pressure to the diaphragm. This represents how the diaphragm looks when it is relaxed. What has happened to the pressure of the system (has it increased or decreased)? What happened to the volume of the lungs?

2 Now slowly release the diaphragm. This represents the diaphragm flattening out as it contracts. What is happening to the pressure as you release the diaphragm? What happened to the volume of the lungs?

3 If your bell-jar model has a rubber stopper in the top, you can use it to demonstrate the effects of a **pneumothorax** on lung tissue. A pneumothorax generally is caused by a tear in the pleural membranes that allows air to enter the pleural cavity. With the diaphragm flat and the lungs (balloons) inflated, loosen the rubber stopper. What happens to the lungs? Why?

Measuring Pulmonary Volumes and Capacities

MATERIALS NEEDED

- Wet or computerized spirometer
- Disposable mouthpiece

Respiratory volumes are the volumes of air that are exchanged with each breath. These volumes of air are measured with an instrument called a **spirometer** and include the following (**Figure 22.3**):

1. **Tidal volume (TV).** The tidal volume, or TV, is the amount of air exchanged with each breath during normal, quiet breathing. It measures about 500 ml in a healthy adult.

2. **Expiratory reserve volume (ERV).** The expiratory reserve volume, or the ERV, is the volume of air that may be expired after a tidal expiration. It averages between 700 and 1200 ml of air.

3. **Inspiratory reserve volume (IRV).** The inspiratory reserve volume, or the IRV, is the amount of air that may be inspired after a tidal inspiration. It averages between 1900 and 3100 ml of air.

Note that there is also a fourth volume, called the **residual volume (RV)**, that cannot be measured with a spirometer. It is defined as the amount of air that remains in the lungs after maximal expiration, and is generally equal to about 1100–1200 ml of air. This amount of air accounts for the difference between the IRV and the ERV.

Two or more respiratory volumes may be combined to give **respiratory capacities.** As you can see in **Figure 22.3**, there are four respiratory capacities:

1. **Inspiratory capacity (IC).** The inspiratory capacity, or IC, is equal to the TV plus the IRV and is the amount of air that a person can maximally inspire after a tidal expiration. It averages between 2400 and 3600 ml of air.

2. **Functional residual capacity (FRC).** The functional residual capacity, or the FRC, is the amount of air that is normally left in the lungs after a tidal expiration. It is the sum of the ERV and the RV and averages between 1800 and 2400 ml of air. It is not measurable with general spirometry.

3. **Vital capacity (VC).** The vital capacity, or VC, represents the total amount of exchangeable air that moves in and out of the lungs. It is equal to the sum of the TV, IRV, and the ERV and averages between 3100 and 4800 ml of air.

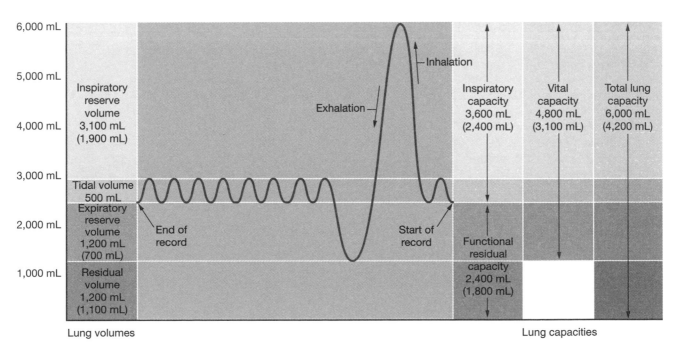

FIGURE **22.3** Respiratory volumes and capacities

4. **Total lung capacity (TLC).** The total lung capacity, or TLC, represents the total amount of exchangeable and non-exchangeable air in the lungs. It is the total of all four respiratory volumes and is not measureable with general spirometry. It averages between 4200 and 6000 ml of air.

Spirometry is a useful tool with which to assess pulmonary function. The respiratory volumes and capacities are especially helpful in differentiating the two primary types of respiratory disorders—**restrictive diseases** and **obstructive diseases.** Restrictive diseases, such as pulmonary fibrosis, are characterized by a loss of elasticity (decreased compliance) of the lung tissue. As a result, patients' ability to inspire is affected adversely. This decreases the IRV, IC, VC, and TLC.

Obstructive diseases, such as chronic obstructive pulmonary disease (COPD) and asthma, are characterized by increased airway resistance caused by narrowing of the bronchioles, increased mucus secretion, and/or an obstructing body such as a tumor. It may seem counterintuitive, but obstructive diseases make expiration difficult. This is because the increased intrapulmonary pressure during expiration naturally tends to shrink the diameter of the bronchioles. When the bronchioles are already narrowed, as in an obstructive disease, the increased intrapulmonary pressure can actually collapse the bronchioles and trap oxygen-poor air in the distal respiratory passages. Therefore, patients with obstructive diseases often exhale slowly and through pursed lips to minimize the pressure changes and maximize the amount of air exhaled. This decreases the ERV and VC and increases the RV and FRC.

 Procedure **Measure Respiratory Volumes with a Wet Spirometer**

Two types of spirometers are commonly found in anatomy and physiology labs:

1. a bell, or "wet" spirometer, and

2. a computerized spirometer.

Many types of wet spirometers allow you to assess only expiratory volumes, whereas computerized spirometers typically allow you to assess both inspiratory and expiratory volumes. Both types of devices allow measurement of one or more respiratory capacities.

The following procedure is intended for a wet spirometer. If your lab has a computerized spirometer, however, you can either follow the procedure below or follow the prompts by the computer program. When you have completed the activity, answer Check Your Understanding question number 3 (p. 511).

Note: You should not be the subject of this experiment if you have cardiovascular or pulmonary disease or are prone to dizziness or fainting.

---

1 Obtain a disposable mouthpiece and attach it to the end of the tube.

2 Before you begin, practice exhaling through the tube several times. Note that you are supposed to only *exhale* into the tube, as it likely has no filter.

3 *Measure the tidal volume:* Sit in a chair with your back straight and your eyes closed. Inhale a normal tidal inspiration, and exhale this breath through the tube. Getting a true representation of the tidal volume is often difficult, because people have a tendency to force the expiration. To get the most accurate tidal volume, take several measurements and average the numbers.

Measurement 1: _____

Measurement 2: _____

Measurement 3: _____

Average Tidal Volume: _____

4 *Measure the expiratory reserve volume:* Before taking this measurement, inhale and exhale a series of tidal volumes. Then inspire a normal tidal inspiration, breathe out a normal tidal expiration, put the mouthpiece to your mouth, and exhale as forcibly as possible. Don't cheat and take in a large breath first! As before, perform several measurements and average the numbers.

Measurement 1: _____

Measurement 2: _____

Measurement 3: _____

Average Expiratory Reserve Volume: _____

5 *Measure the vital capacity*: As before, inhale and exhale a series of tidal volumes. Then bend over and exhale maximally. Once you have exhaled as much air as you can, raise yourself upright and inhale as much air as you possibly can (until you feel like you are nearly about to "pop"). Quickly place the mouthpiece to your mouth and exhale as forcibly and as long as possible. Take several measurements (you may want to give yourself a minute to rest between measurements), and record the data below.

Measurement 1: _____

Measurement 2: _____

Measurement 3: _____

Average Vital Capacity: _____

6 *Calculate the inspiratory reserve volume*: Even though the bell spirometer cannot measure inspiratory volumes, you can calculate this volume now that you have the vital capacity (VC), tidal volume (TV), and expiratory reserve volume (ERV). Recall that VC = TV + IRV + ERV. Rearrange the equation: IRV = VC − (TV + ERV).

Average IRV: _____

7 How do your values compare with the average values? What factors, if any, do you think may have affected your results?

8 Pool the results for your class and divide the results into four categories: female cigarette smokers, female nonsmokers, male cigarette smokers, and male nonsmokers. Calculate the average TV, ERV, and VC for each group, and record these data in Table 22.3.

TABLE 22.3 Respiratory Volumes and Capacities for Smokers and Nonsmokers				
Group	Tidal Volume	Expiratory Reserve Volume	Inspiratory Reserve Volume	Vital Capacity
Female smokers				
Female nonsmokers				
Male smokers				
Male nonsmokers				

9 Interpret your results:

   a. How did the average values differ for males and females?

   b. How did the average values differ for smokers and nonsmokers?

   c. What disease pattern would you expect to see with the smokers? Did your results follow this expectation?

 # Exercise 3

pH and Ventilation

MATERIALS NEEDED

- One 200 ml beaker
- Deionized water
- pH meter
- Drinking straw

Pulmonary ventilation is required not only to bring oxygen into the body but also to *release carbon dioxide*. Carbon dioxide is a nonpolar gas that does not fully dissolve in the water of plasma. For this reason, carbon dioxide is transported in another form that is water soluble: the weak base **bicarbonate** (HCO_3^-). Carbon dioxide reacts with water to form **carbonic acid** (H_2CO_3), which dissociates to form a hydrogen ion and bicarbonate. This forms a system known as the **carbonic acid-bicarbonate buffer system**. This reversible reaction is catalyzed by an enzyme called **carbonic anhydrase** and proceeds as follows:

$$H_2O + CO_2 \rightleftharpoons H_2CO_3 \rightleftharpoons HCO_3^- + H^+$$

Because carbon dioxide reacts to form an acid, the concentration of carbon dioxide in the blood is a major determinant of blood pH. The concentration of carbon dioxide in the blood is determined directly by the rate and depth of ventilation. When the rate and depth of ventilation increase during **hyperventilation**, more carbon dioxide is expelled from the body and the carbon dioxide concentration of the blood decreases. This in turn decreases the amount of carbonic acid formed, which raises the pH of the blood. Conversely, when the rate and depth of ventilation decrease during **hypoventilation**, carbon dioxide is retained and its concentration in the blood increases. This increases the amount of carbonic acid formed and decreases the blood pH.

When the pH of the blood increases or decreases due to carbon dioxide or other factors, the body attempts to return the blood pH to normal by adjusting the rate and depth of ventilation. If the blood pH decreases, the brainstem triggers an increase in respiratory rate to remove carbon dioxide from the blood and raise the blood pH. This is something you witness every time you engage in even minor exercise—when you exercise, your metabolic rate and carbon dioxide production increase, and your brainstem triggers an increase in respiratory rate to remove the excess carbon dioxide from the blood. If the pH of the blood increases, the brainstem will slow the rate of ventilation to retain carbon dioxide and lower the blood pH.

Procedure Measure the Effect of Carbon Dioxide on the pH of a Solution

We can't measure the effects of ventilatory rate on blood pH directly in the lab because it requires an analysis of arterial blood. However, we can observe and measure the effects of carbon dioxide on the pH of water and correlate this with the changes you would see in the blood, which is what you will do in the following experiment. When you have completed the activity, answer Check Your Understanding questions 4 and 5 (p. 512).

1 Measure your partner's respiratory rate while he or she is seated and resting. Measuring the respiratory rate can be tricky because as soon as you tell your partner to breathe normally, he or she automatically becomes conscious of breathing and alters the rate. A trick you may want to try is to place two fingers on your partner's radial artery and take his or her pulse. This will cause your partner to focus more on his or her pulse rate, and he or she will breathe more naturally. Record this value in Table 22.4.

2 Obtain a 200 ml beaker about half-full of deionized water. Measure the pH of the water with a pH meter. Record this value below:

pH of deionized water: _____

3 Place a drinking straw into the water and have your partner exhale through the straw into the water (the water should bubble) for 20 seconds. Repeat the pH measurement of the water with the pH meter and record the value in Table 22.4.

4 Empty your beaker and refill it with new water. Have your partner do vigorous exercise for at least 5 minutes (running up and down stairs usually does the job). Immediately upon finishing the exercise, have your partner blow through the straw into the beaker of water for 20 seconds. Measure the pH of the water with the pH meter and record the value in Table 22.4.

5 After you record the pH, measure your partner's respiratory rate (you may want to use the pulse trick again) and record the value in Table 22.4.

6 Empty your beaker again and refill it with new water. Allow your partner to have a (well-deserved) rest for 5 minutes. Again, have your partner blow through the straw into the beaker of water for 20 seconds, and measure the pH of the water with the pH meter. Record the value in Table 22.4.

7 After you record the pH, measure your partner's respiratory rate and record the value in Table 22.4.

TABLE 22.4	pH of Water and Ventilatory Rate	
Setting	Respiratory Rate	pH
At rest		
Immediately after exercise		
After 5 minutes of rest		

8 Interpret your results:
   a. Start with the basics: What happens to the metabolism during exercise? What is generated as a byproduct of glucose metabolism?

b. What does this byproduct do to the pH of the blood? Why?

c. What sort of ventilatory response does this change in pH cause (hyperventilation or hypoventilation)? Why?

d. Now apply this to your results:

(1) What happened to the respiratory rate during and after exercise? Why? (*Hint:* See item c. above.)

(2) What happened to the pH of the water after exercise? Why?

(3) What happened to the respiratory rate and the pH of the water after resting? Why?

UNIT
22
REVIEW

NAME _____

SECTION _____ DATE _____

Check Your Recall

1 Which of the following is *not* a basic physiologic process carried out by the respiratory system?

a. Internal respiration
b. Cellular respiration
c. Pulmonary ventilation
d. Gas transport
e. External respiration

2 Fill in the blanks: According to Boyle's law, as the volume of a container increases, the pressure _____.

Conversely, as the volume of a container decreases, the pressure _____.

3 Air moves into the lungs when

a. intrapulmonary pressure is less than atmospheric pressure.
b. intrapulmonary pressure is more than atmospheric pressure.
c. blood pressure is more than intrapulmonary pressure.
d. blood pressure is less than intrapulmonary pressure.

4 Passive expiration is achieved primarily by the

a. contraction of the diaphragm and external intercostals.
b. decrease in intrapulmonary pressure.
c. increase in atmospheric pressure.
d. elastic recoil of the lungs.

5 Matching: Match the following pulmonary volumes and capacities with their correct definition or description.

a. Total lung capacity   _____   A.   Volume of air in the lungs after a tidal expiration

b. Residual volume   _____   B.   Volume of air forcibly expired after a tidal expiration

c. Functional residual capacity   ____   C.   Volume of air forcibly inspired after a tidal inspiration

d. Vital capacity   _____   D.   Equal to the TV + ERV + IRV + RV

e. Inspiratory reserve volume   ____   E.   Equal to the TV + IRV

f. Tidal volume   _____   F.   Volume of air remaining in the lungs after forceful expiration

g. Inspiratory capacity   _____   G.   Equal to the TV + IRV + ERV

h. Expiratory reserve volume   _____   H.   Volume of air exchanged with normal, quiet breathing

6 How do restrictive and obstructive diseases differ?

7 Carbon dioxide is transported through the blood primarily as
   a. dissolved gas.
   b. bicarbonate.
   c. carbonic anhydrase.
   d. carbaminohemoglobin.

8 Fill in the blanks: Hyperventilation _____ the amount of carbon dioxide exhaled and therefore

_____ the blood pH. Hypoventilation _____ the amount of carbon dioxide

exhaled and therefore _____ the blood pH.

9 The enzyme that converts carbon dioxide and water into carbonic acid is called
   a. bicarbonate dehydrogenase.
   b. adenosine deaminase.
   c. carbonic anhydrase.
   d. hemoglobin carboxylase.

10 Why does the ventilatory rate increase during exercise?

Check Your Understanding | **Critical Thinking and Application Questions**

1 The condition *emphysema* results in loss of elastic recoil of the lung tissue. Would this make inspiration or expiration difficult? Explain.

2 Why might you have difficulty breathing at higher altitudes, where the atmospheric pressure is low?

3 A patient presents with the following respiratory volumes and capacities: TV = 500 ml, ERV = 300 ml, IRV = 1900 ml. What is this patient's VC? Are these values normal? If not, are they more consistent with an obstructive or restrictive disease pattern? Explain.

4 Why should a person breathe into a paper bag when he or she is hyperventilating abnormally?

5 A patient presents in a state of *ketoacidosis* (a type of metabolic acidosis in which the pH of the blood drops), after going on a no-carb, all-fat diet. Will the patient be hyperventilating or hypoventilating? Why?

EXPLORING ANATOMY & PHYSIOLOGY IN THE LABORATORY

Urinary System Anatomy

OBJECTIVES

Once you have completed this unit,
you should be able to:

- Identify gross structures of the urinary system.

- Identify structures of the nephron.

- Identify histological structures of the urinary system.

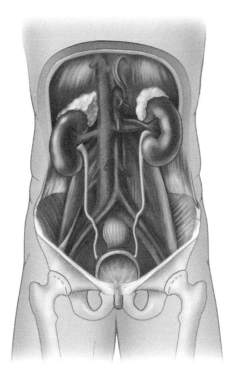
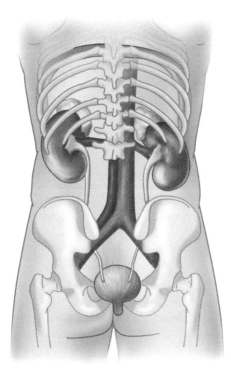

Pre-Lab Exercises

Complete the following exercises prior to coming to lab, using your textbook and lab manual for reference.

Pre-Lab Exercise 1

Key Terms

Table 23.1 lists the key terms with which you should be familiar before coming to lab.

TABLE 23.1 Key Terms

Term	Definition
Gross Structures of the Kidney	
Renal capsule	______________________
Renal cortex	______________________
Renal medulla	______________________
Medullary pyramid	______________________
Renal column	______________________
Renal pelvis	______________________
Major and minor calyces	______________________
Other Structures of the Urinary System	
Ureter	______________________
Urinary bladder	______________________
Trigone	______________________

Urethra _____

_____

Internal and external urethral sphincters _____

_____

Transitional epithelium _____

_____

Blood Vessels and Tubules of the Nephron

Nephron _____

_____

Afferent arteriole _____

_____

Glomerulus _____

_____

Efferent arteriole _____

_____

Peritubular capillaries _____

_____

Proximal tubule _____

_____

Nephron loop (loop of Henle) _____

_____

Distal tubule _____

_____

Juxtaglomerular apparatus _____

_____

Collecting duct _____

_____

Pre-Lab Exercise 2

Structures of the Urinary System

Label and color the structures of the urinary system depicted in **Figures 23.1** and **23.2** with the terms from Exercise 1. Use your text and Exercise 1 in this unit for reference.

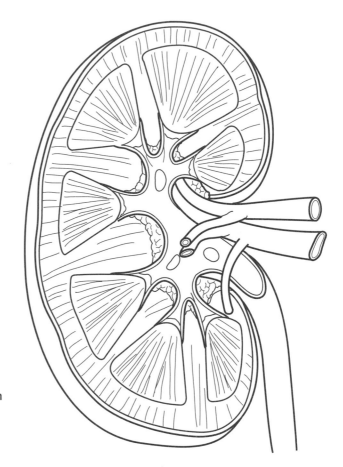

FIGURE 23.1 Right kidney, frontal section

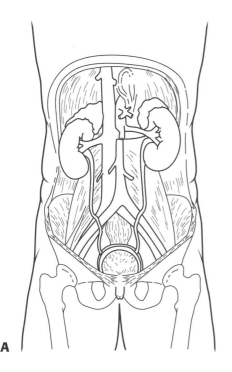 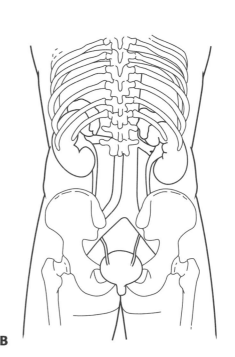

A   B

FIGURE 23.2 Organs of the urinary system: (A) anterior view; (B) posterior view

EXPLORING ANATOMY & PHYSIOLOGY IN THE LABORATORY

Pre-Lab Exercise 3

Structures of the Nephron

Label and color the structures of the nephron depicted in **Figure 23.3** with the terms from Exercise 1. Use your text and Exercise 1 in this unit for reference.

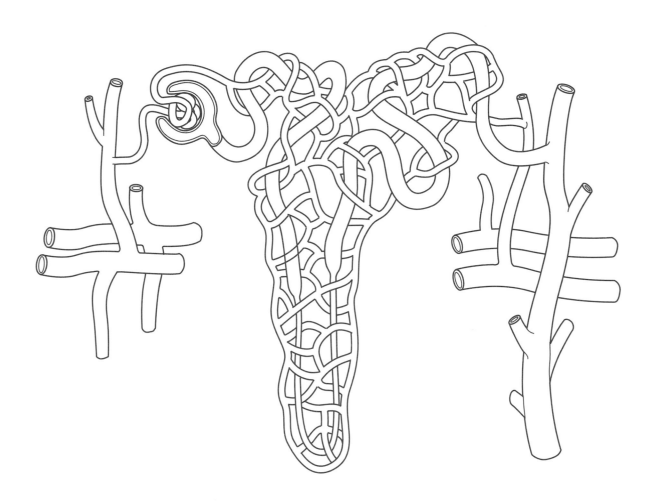

FIGURE **23.3** The nephron

Exercises

The urinary system performs many functions critical to homeostasis, including removing waste products from the blood, regulating the body's fluid, electrolyte, and acid-base balance, and producing the hormone erythropoietin, which regulates blood cell formation. In addition to these roles, the urinary system helps the liver detoxify certain compounds and makes glucose during times of starvation. These critical functions are carried out primarily by the main organs of the urinary system—the kidneys. The other organs in the urinary system include the ureters, the urinary bladder, and the urethra. In this unit you will become acquainted with the anatomy and the histology of the urinary system. In Unit 24 you will use this knowledge to study urinary system physiology.

Exercise 1

Urinary System Anatomy

MATERIALS NEEDED

- Urinary system models
- Kidney models
- Nephron models
- Preserved kidney
- Dissection equipment
- Dissecting tray

Urine formation takes place in the paired **kidneys**, which are situated against the posterior body wall. The kidneys are said to be **retroperitoneal** because they are located posterior to the peritoneal membranes that surround many of the digestive organs. The kidneys are encased within three layers of connective tissue, the thickest of which is a middle layer of adipose tissue. The innermost layer of connective tissue encases each kidney like plastic wrap and is called the **renal capsule**.

Each kidney has three distinct regions internally when viewed in a frontal section (**Figure 23.4**):

1. **Renal cortex.** The most superficial region is known as the renal cortex. It is dark brown because it consists of many blood vessels that serve the tiny blood-filtering structures of the kidney, the **nephrons**.

2. **Renal medulla.** The kidney's middle region is known as the medulla and consists of triangular structures known as **renal** or **medullary pyramids**. The medullary pyramids are separated from one another by inward extensions of the renal cortex called **renal columns**. Like the renal cortex, the renal columns contain many blood vessels. Each pyramid contains looping tubules of the nephron, as well as structures that drain fluid from the nephron. These tubes give the pyramids a striped (or *striated*) appearance.

3. **Renal pelvis.** The tubes that drain the fluid from the nephron drain into larger tubes called **minor calyces**, which in turn drain into even larger **major calyces**. The major calyces drain into the kidney's innermost region called the renal pelvis which serves as a basin for collecting urine.

The blood flow through the kidney follows a unique pattern that allows it to carry out its function

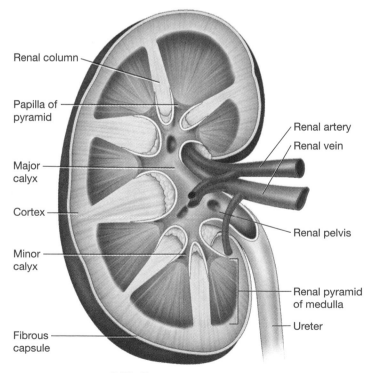

FIGURE 23.4 Right kidney, frontal section

Renal column
Papilla of pyramid
Major calyx
Cortex
Minor calyx
Fibrous capsule
Renal artery
Renal vein
Renal pelvis
Renal pyramid of medulla
Ureter

EXPLORING ANATOMY & PHYSIOLOGY IN THE LABORATORY

of maintaining the homeostasis of the blood (**Figure 23.5**). The large **renal arteries** deliver about 1200 milliliters of blood per minute to the kidney to be filtered. The renal arteries branch into progressively smaller arteries as they pass through the medulla to the cortex, including the **segmental arteries** and **lobar arteries** in the renal pelvis, the **interlobar arteries** between the medullary pyramids, the **arcuate arteries** that curve around the top of the pyramids, and finally the small **interlobular arteries** in the renal cortex. The interlobular arteries branch into tiny **afferent arterioles**, each of which supplies a ball of capillaries known as the **glomerulus**. This is where the blood is filtered. Note that the glomerulus is not the primary site for gas and nutrient exchange for the tissues of the kidneys.

You learned in Unit 17 that a capillary bed generally drains into a venule, but note in **Figure 23.5** that the capillaries of the glomerulus drain into a second *arteriole* called the **efferent arteriole**. The efferent arteriole then branches to form a second capillary bed known as the **peritubular capillaries**. These capillaries surround the tubules of the nephron, where they provide them with the oxygen and nutrients and also take substances that have been reabsorbed by the tubules back into the blood. This second capillary bed then drains out through the small **interlobular veins**, which drain into **arcuate veins**, then into **interlobar veins**, and finally into the large **renal vein**. This pattern of blood flow allows the kidneys to both filter blood and reclaim most of the fluid and solutes that were filtered.

Microscopically, each kidney is composed of more than a million tiny units called **nephrons** (**Figure 23.6**). Blood first enters the high-pressure circuit of the glomerulus, where fluid and small solutes are forced out of the capillaries and into a space called the **capsular space**. The capsular space is found within a structure known as the **glomerular capsule** or **Bowman's capsule**. Note in **Figure 23.6A** that the glomerular capsule has two layers: (a) the outer **parietal layer**, which is

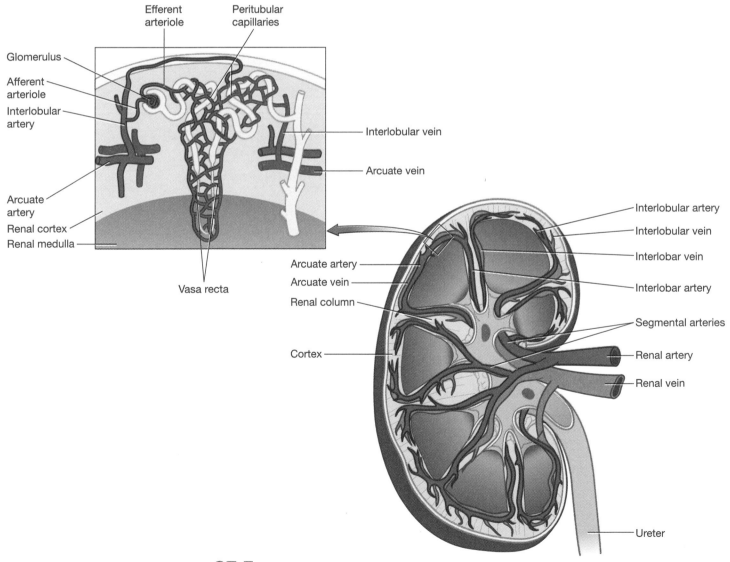

FIGURE **23.5** Pathway of blood flow and microanatomy of the kidney

simple squamous epithelium, and (b) the inner **visceral layer**, which consists of cells called **podocytes** that surround the capillaries of the glomerulus. The glomerulus and its surrounding capsule together are known as the **renal corpuscle**.

The podocytes of the visceral layer have extensions called **foot processes**, which interlock to form narrow **filtration slits**. As you can see in **Figure 23.6B**, the glomerular endothelial cells, the podocytes, and their shared basal lamina together are called the **filtration membrane**. The filtration membrane prevents large substances in the blood, such as blood cells and most proteins, from exiting the glomerular capillaries.

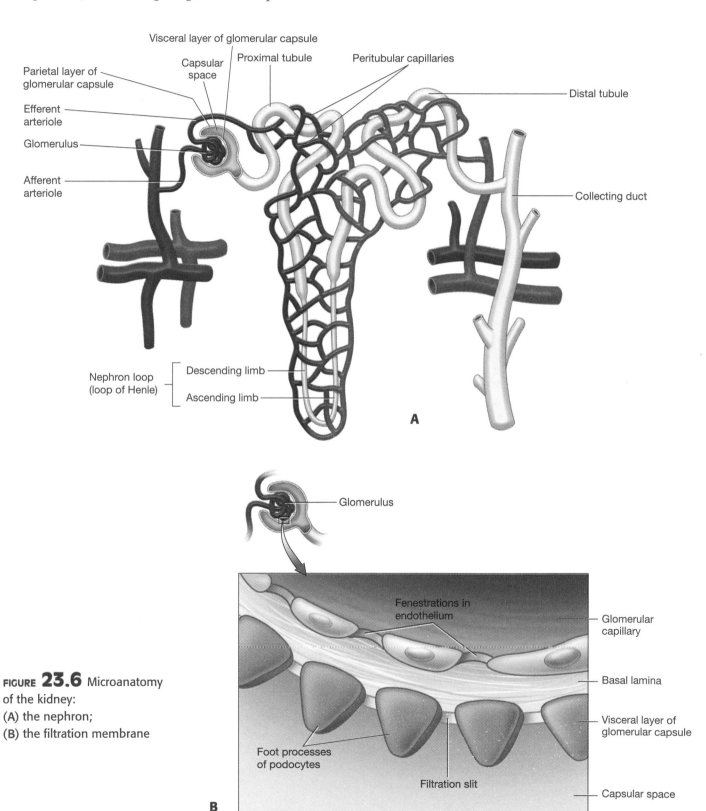

FIGURE 23.6 Microanatomy of the kidney:
(A) the nephron;
(B) the filtration membrane

EXPLORING ANATOMY & PHYSIOLOGY IN THE LABORATORY

The fluid in the capsular space, known as **filtrate**, next enters a series of small tubes or *tubules* that can be likened to the "plumbing" of the kidney. These tubes are called collectively the **renal tubule**, which consists of three parts:

1. the **proximal tubule** (sometimes called the *proximal convoluted tubule*),

2. the ascending and descending limbs of the **nephron loop** (also called the **loop of Henle**), and

3. the **distal tubule** (sometimes called the *distal convoluted tubule*).

Several distal tubules drain into one **collecting duct**. Note that the majority of the renal tubule is confined to the renal cortex; only the nephron loops of certain nephrons dip down into the renal medulla.

At the junction between the ascending limb of the nephron loop and the distal tubule we find a group of tall, closely packed cells known as the **macula densa**. The macula densa comes into contact with a portion of the afferent arteriole that contains specialized cells called **juxtaglomerular (JG) cells**. The JG cells and macula densa together are called the **juxtaglomerular apparatus (JGA)**. The JGA plays a role in controlling the flow of filtrate through the nephron and the blood pressure within the glomerulus.

Filtrate in the renal tubule and collecting duct is modified, and most of the water and solutes are reclaimed. From the collecting ducts, filtrate drains into larger tubules called **papillary ducts**, where more water is reclaimed. After the fluid leaves the papillary ducts, it is known as **urine** and drains into the minor calyces, major calyces, and finally the renal pelvis. From the renal pelvis, urine enters the next organs of the urinary system—the tubes called the **ureters** (Figure 23.7). Ureters are lined by a type of epithelium called **transitional epithelium**, and their walls contain smooth muscle that massages the urine inferiorly via peristalsis.

The ureters drain urine into the posteroinferior wall of the organ known as the **urinary bladder**. Like the ureters, the urinary bladder is lined with transitional epithelium and contains smooth muscle (sometimes called the **detrusor muscle**) in its wall. The majority of the urinary bladder contains folds called **rugae**, which allow it to expand when it is filled with urine. The smooth inferior portion of the urinary bladder wall, a triangular-shaped area known as the **trigone**, contains the opening called the **internal urethral orifice**. This opens into the final organ of the urinary system—the **urethra**. The urethra contains two rings of smooth muscle—the involuntary **internal urethral sphincter** and the voluntary **external urethral sphincter**. When both of these sphincters relax, urine is expelled from the body via a process called **micturition**.

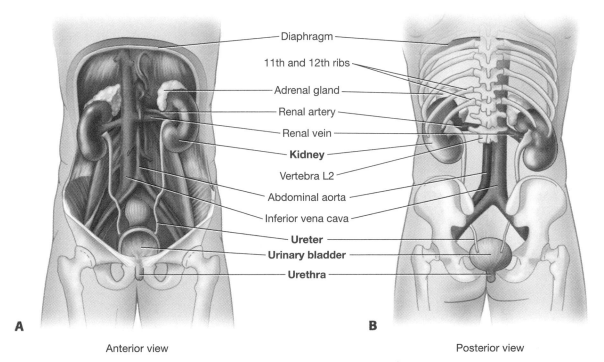

Diaphragm
11th and 12th ribs
Adrenal gland
Renal artery
Renal vein
Kidney
Vertebra L2
Abdominal aorta
Inferior vena cava
Ureter
Urinary bladder
Urethra

A

Anterior view

B

Posterior view

FIGURE 23.7 Organs of the urinary system: (A) Anterior view; (B) Posterior view

Procedure Model Inventory

Identify the following structures of the urinary system on models and diagrams, using your textbook and this unit for reference. As you examine the anatomical models and diagrams, record on the model inventory in Table 23.2 the name of the model and the structures you were able to identify. When you have completed the activity, answer Check Your Understanding questions 1 and 2 (p. 533).

Kidney Anatomy

1. Surrounding connective tissue:
   a. Renal fascia
   b. Adipose capsule
   c. Renal capsule
2. Hilum
3. Regions:
   a. Renal cortex
   b. Renal medulla
      (1) Renal columns
      (2) Renal pyramids
      (3) Renal papilla
   c. Minor calyces
   d. Major calyces
   e. Renal pelvis
4. Blood supply:
   a. Renal artery
      (1) Segmental artery
      (2) Lobar artery
      (3) Interlobar artery
      (4) Arcuate artery
      (5) Interlobular artery
   b. Renal vein
      (1) Interlobular vein
      (2) Arcuate vein
      (3) Interlobar vein

Structures of the Nephron

1. Renal corpuscle
   a. Glomerulus
      (1) Afferent arteriole
      (2) Efferent arteriole
      (3) Peritubular capillaries
      (4) Vasa recta
   b. Glomerular (Bowman's) capsule
      (1) Parietal layer
      (2) Visceral layer
      (3) Podocytes
      (4) Capsular space
2. Renal tubule
   a. Proximal tubule
   b. Nephron loop (loop of Henle)
      (1) Descending limb
      (2) Ascending limb
   c. Distal tubule
3. Juxtaglomerular apparatus
   a. JG cells
   b. Macula densa
4. Collecting duct
5. Papillary duct

Other Urinary Structures

1. Ureter
2. Urinary bladder
   a. Detrusor muscle
   b. Ureteral orifices
   c. Internal urethral orifice
   d. Trigone
3. Urethra
   a. Internal urethral sphincter
   b. External urethral sphincter
   c. External urethral orifice

TABLE 23.2	Model Inventory for Urinary Anatomy
Model/Diagram	**Structures Identified**

Procedure Kidney Dissection

In this exercise you will identify several of the structures that you just identified on models and diagrams by dissecting a preserved kidney.

Note: Safety glasses and gloves are required.

---

1 Obtain a fresh or preserved kidney specimen and dissection supplies.

2 If the thick surrounding connective tissue coverings are intact, note their thickness and amount of adipose tissue.

3 Use scissors to cut through the connective tissue coverings, and remove the kidney.

4 List surface structures that you are able to identify:

5 Distinguishing between the ureter, the renal artery, and the renal vein is often difficult. Following are some hints to aid you:

  a. The renal artery typically has the thickest and most muscular wall, and it branches into several segmental arteries prior to entering the kidney.

  b. The renal vein is thinner, flimsier, and often larger in diameter than the renal artery.

  c. The ureter has a thick, muscular wall, too, but it does not branch once it leaves the kidney. Also, its diameter is usually smaller than either the renal artery or the renal vein.

Keeping these points in mind, determine the location of the renal artery, the renal vein, and the ureter on your specimen. Use **Figure 23.8** for reference. Sketch the arrangement of the three structures below:

6 Use a scalpel to make a frontal section of the kidney. List the internal structures you are able to identify (**Figure 23.8**):

EXPLORING ANATOMY & PHYSIOLOGY IN THE LABORATORY

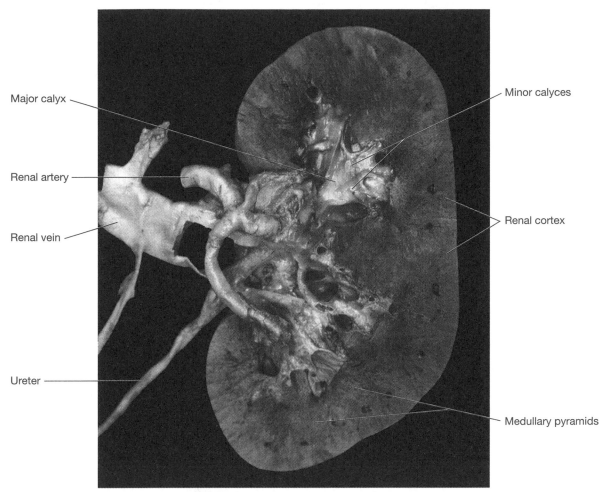

Major calyx

Renal artery

Renal vein

Ureter

Minor calyces

Renal cortex

Medullary pyramids

FIGURE **23.8** A frontal section of the left kidney

 Exercise 2

Urinary Organ Histology

MATERIALS NEEDED

- Urinary bladder slide
- Renal cortex slide
- Renal medulla slide
- Light microscope
- Colored pencils

As we discussed in Exercise 1, the ureters and urinary bladder are lined with a special type of epithelium called **transitional epithelium** (**Figure 23.9**). Recall from Unit 5 that transitional epithelium is stratified epithelium with cells that differ in appearance on the apical and basal sides. The cells at the apical edge are dome-shaped (or sometimes squamous) in appearance, and those nearer the basal edge are typically cuboidal. The apical cells can change shape to accommodate stretching of the urinary bladder.

In this exercise you will also examine the tissues in two different regions of the kidney—the renal cortex and the renal medulla. Both regions are composed primarily of blood vessels and tubules made of simple epithelia. The simple epithelia of the renal tubules provide an example of form following function: A main function of these tubules is to reabsorb the water and solutes in the filtrate. Were the tubules made of stratified epithelium, reabsorption would not take place efficiently and excessive water and solutes would be lost to the urine. Note in **Figure 23.10A** the abundance of small, ball-shaped glomeruli in the renal cortex. The glomeruli are surrounded by a space (the capsular space) that is lined by the ring of simple squamous epithelial cells that make up the parietal layer of the glomerular capsule. In between the glomeruli, note the presence of nephron tubules (the proximal and distal tubules and sections of nephron loops).

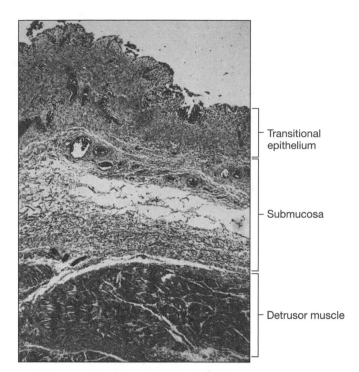

FIGURE 23.9 Transitional epithelium and other tissues of the urinary bladder wall

- Transitional epithelium
- Submucosa
- Detrusor muscle

Note in **Figure 23.10B** that the renal medulla is composed mostly of collecting ducts and the nephron loops. You will not see any glomeruli in the medulla, as they are confined to the cortex. This makes the renal cortex and renal medulla easily distinguishable.

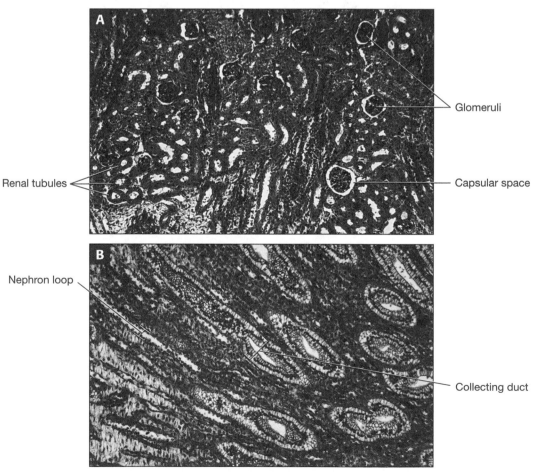

Glomeruli

Capsular space

Renal tubules

Nephron loop

Collecting duct

FIGURE 23.10 Tissues of the kidney: (A) Renal cortex; (B) Renal medulla

EXPLORING ANATOMY & PHYSIOLOGY IN THE LABORATORY

Procedure Microscopy

Now you will view prepared slides of a section of the urinary bladder, the renal cortex, and the renal medulla. Begin your examination of the slides on low power, and advance to medium and high power to observe details. Use colored pencils to draw what you see under the microscope, and label your drawing with the terms indicated. When you have completed the activity, answer Check Your Understanding questions 3 and 4 (p. 534).

Transitional Epithelium (Urinary Bladder)

Label the following on your drawing:

1. Transitional epithelium

2. Submucosa

3. Muscularis (detrusor muscle)

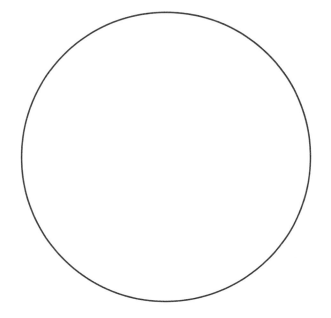

Renal Cortex

Label the following on your drawing:

1. Glomerulus

2. Glomerular capsule with simple squamous epithelium

3. Capsular space

4. Kidney tubules

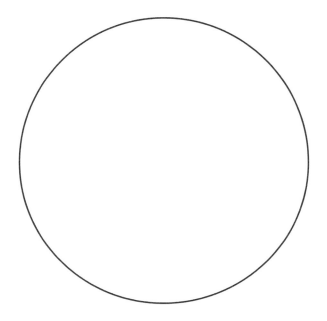

Renal Medulla

Label the following on your drawing:

1. Nephron loop

2. Collecting duct

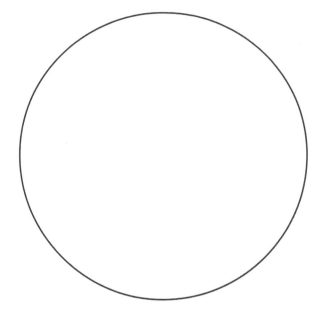

EXPLORING ANATOMY & PHYSIOLOGY IN THE LABORATORY

NAME _____

SECTION _____ DATE _____

 Check Your Recall

1 Label the following parts of the kidney on **Figure 23.10.**

- Renal capsule
- Renal cortex
- Renal medulla
- Renal pyramid
- Renal column
- Minor calyx
- Major calyx
- Renal pelvis
- Renal artery
- Renal vein
- Ureter

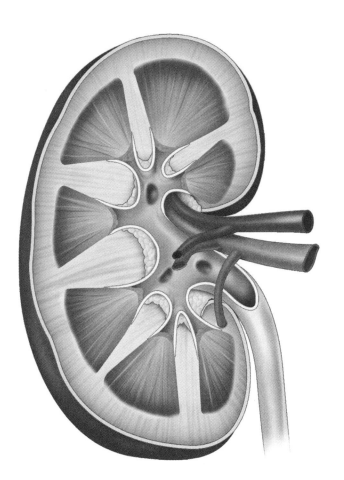

FIGURE **23.10** Right kidney, frontal section

2 Label the following parts of the nephron on **Figure 23.11**.

- Glomerulus
- Glomerular capsule – parietal layer
- Glomerular capsule – visceral layer
- Capsular space
- Afferent arteriole
- Efferent arteriole

- Peritubular capillaries
- Proximal tubule
- Nephron loop
- Distal tubule
- Collecting duct

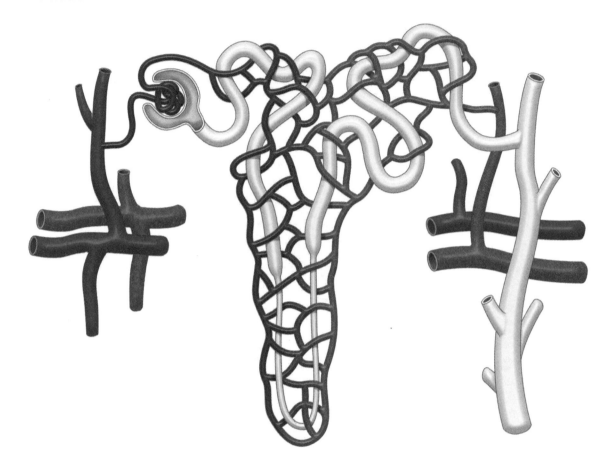

FIGURE **23.11** The nephron

3 Which of the following is *not* one of the urinary system's functions?

a. Regulating fluid, electrolyte, and acid-base balance
b. Regulating blood cell formation
c. Regulating production of insulin and glucagon
d. Removing waste products from the blood

4 The blood flow through the kidney is unique because

a. its first capillary beds drain into arterioles.
b. its second capillary beds drain into arterioles.
c. it is supplied by three renal arteries.
d. it contains no capillary beds.

EXPLORING ANATOMY & PHYSIOLOGY IN THE LABORATORY

5 Number the following from the point that the filtrate is first formed (with a number 1) to the point that it drains into the renal pelvis (with a number 7).

Minor calyx _____

Proximal tubule _____

Collecting duct _____

Capsular space _____

Nephron loop _____

Papillary duct _____

Distal tubule _____

6 Fill in the blanks: At the junction between the ascending limb of the nephron loop and the distal tubule we find a

group of densely-packed cells called the _____. These cells come into contact with a portion

of the afferent arteriole that contains cells called the _____. Together, the two are called the

_____.

7 Urine drains from the kidneys via the
a. urinary bladder.
b. urethra.
c. ureters.
d. papillary calyces.

8 Urine is expelled from the body by a process called
a. micturition.
b. parturition.
c. defecation.
d. procrastination.

9 The urinary bladder and ureters are lined by
a. simple squamous epithelium.
b. transitional epithelium.
c. pseudostratified columnar epithelium.
d. stratified cuboidal epithelium.

10 How can one easily discern the renal cortex from the renal medulla on a microscope slide?

EXPLORING ANATOMY & PHYSIOLOGY IN THE LABORATORY

Check Your Understanding | **Critical Thinking and Application Questions**

1 Predict the effects that renal failure might have on the body's homeostasis.

2 How is the pattern of blood flow unique in the kidney? How does this pattern of blood flow allow the kidney's form to follow its function?

3 Why do you think the epithelium of the urinary bladder is stratified? What might happen if the epithelial tissue were simple rather than stratified? (*Hint*: Remember that form follows function.)

4 How does the form of the epithelial tissue making up the renal tubules follow its function?

EXPLORING ANATOMY & PHYSIOLOGY IN THE LABORATORY

Urinary System Physiology

OBJECTIVES

Once you have completed this unit, you should be able to:

- Model the physiology of the kidney and test for chemicals that are in the filtrate.

- Perform and interpret urinalysis on simulated urine specimens.

- Research the possible causes of abnormalities detected on urinalysis.

- Trace an erythrocyte, a glucose molecule, and a urea molecule through the gross and microscopic anatomy of the kidney.

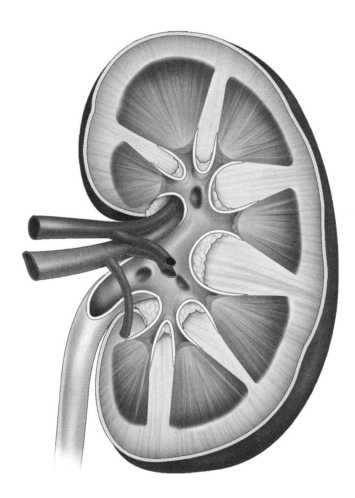

Pre-Lab Exercises

Complete the following exercises prior to coming to lab, using your textbook and lab manual for reference.

Pre-Lab Exercise 1
Key Terms

Table 24.1 lists the key terms with which you should be familiar before coming to lab.

TABLE 24.1	Key Terms

Term	Definition
Glomerular filtration	_____
Glomerular filtration rate (GFR)	_____
Tubular reabsorption	_____
Filtration membrane	_____
Filtrate	_____
Tubular secretion	_____
Urine	_____
Urinalysis	_____
Specific gravity	_____
Urobilinogen	_____
Ketones	_____

Pre-Lab Exercise 2
Nephron Structure and Function

Review the structure of the nephron by labeling and color-coding the diagram in **Figure 24.1**. Then use your textbook to determine which substances are absorbed or secreted from each region of the nephron and enter this information in the blank boxes to the side of the tubules. Place substances that are secreted into the urine into the boxes whose arrows are pointing toward the tubules. Place substances that are re-absorbed into the blood-stream into boxes whose arrows are pointing away from the tubules.

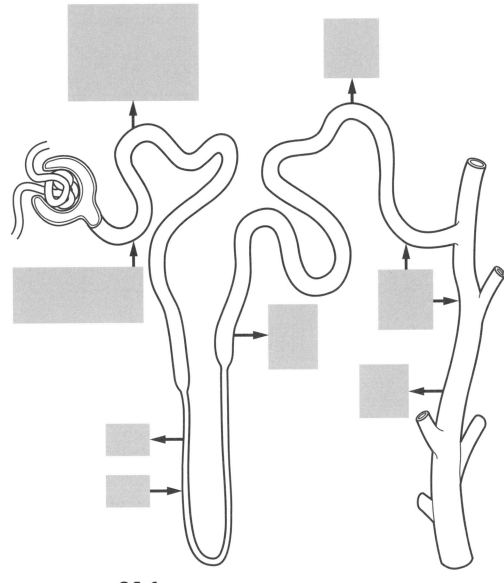

FIGURE 24.1 The nephron

Pre-Lab Exercise 3
Glomerular Filtration and Tubular Reabsorption

Table 24.2 lists chemicals and cells found in blood. Do the following for each chemical, using your textbook for reference:

1. Determine whether each listed substance is filtered at the glomerulus (i.e., if it is able to leave the blood and enter the nephron to become part of the filtrate).

2. If the substance is found in the filtrate, determine if the substance is reabsorbed into the blood or if it is found in the urine.

3. If the substance is reabsorbed into the blood, determine where this substance is reabsorbed in the nephron.

TABLE 24.2	Substances Found in the Blood and Their Filtration and Reabsorption by the Nephron		
Substance	Filtered in the Glomerulus? (yes/no)	Reabsorbed? (yes/no)	Location where Reabsorption Takes Place
Erythrocyte			
Leukocyte			
Water			
Glucose			
Proteins			
Amino acids			
Urea			
Creatinine			
Electrolytes (sodium, potassium, chloride)			
Uric acid			
Ketones			

Exercises

The ball-shaped tufts of capillaries called glomeruli filter the blood that is delivered to them. The fluid that results from this filtration process enters the proximal tubule and is called **filtrate**. Filtrate is similar in composition to blood, but it lacks most of the proteins and cells that we find in blood. This selectivity is possible because of the **filtration membrane**, which consists of the glomerular endothelial cells, the basal lamina, and specialized cells called **podocytes**. The filtration membrane acts in a similar manner to the filter in your coffee maker: It prevents large substances such as proteins and cells from leaving the blood while allowing small substances such as water, glucose, amino acids, electrolytes, and waste products to leave the blood and enter the filtrate.

As you can see, the filtrate contains a number of substances that our bodies need to reclaim, which occurs by a process known as **tubular reabsorption**. About 99% of the water in the filtrate and most of the solutes are reabsorbed through the epithelium of the nephron tubules and returned to the blood. The importance of this function cannot be overstated. If the water were not reabsorbed in the renal tubule and the collecting duct, we would lose our entire plasma volume in less than 30 minutes! In addition to tubular reabsorption, the nephron tubules remove substances from the blood that were not filtered at the glomerulus and put them into the filtrate, a process known as **tubular secretion**.

The rate of filtrate production averages about 120 ml/min and is called the **glomerular filtration rate** (**GFR**). The GFR is an important determinant of kidney function. If the GFR is too low, the kidneys will not be able to carry out their homeostatic functions adequately. Conversely, if the GFR is too high, the nephron tubules will not have time to reabsorb all of the substances in the filtrate.

In the following exercises you will examine the processes of filtration and reabsorption by constructing a model kidney and testing samples of simulated urine. In the final exercise you will trace the pathway of substances through the general circulation and the microanatomy of the kidney to contrast the filtration and reabsorption of three substances: erythrocytes, urea, and glucose.

Exercise 1

The Model Kidney

MATERIALS NEEDED

- 4-inch piece of dialysis tubing
- 2 pieces of string
- Animal blood or simulated blood
- One 200 ml beaker
- Deionized water
- URS-10 and bottle with key

In this exercise you will examine the process of glomerular filtration by constructing a model kidney and testing for substances that appear in the filtrate. A kidney is easily modeled with either animal blood or simulated blood and simple dialysis tubing, which has a permeability similar to that of the filtration membrane. Glomerular filtration is mimicked by immersing the blood-filled tubing in water. Substances to which the "filtration membrane" is permeable will enter the surrounding water (the "filtrate"), whereas those to which the membrane is not permeable will remain in the tubing. Recall from Unit 23 (p. 520) that the filtration membrane consists of the glomerular endothelial cells, the basal lamina, and specialized cells called **podocytes** (Figure 24.2).

The results of this test are analyzed using **urinalysis reagent test strips** (URS-10; Figure 24.3). Each strip consists of 10 small, colored pads that change color in the presence of certain chemicals. The strip is interpreted by watching the pads for color changes and comparing the color changes to a color-coded key on the side of the bottle. The color that is closest to the color on the strip is recorded as your result. Please note that for this exercise you will read only four of the 10 boxes: blood, leukocytes, protein, and glucose.

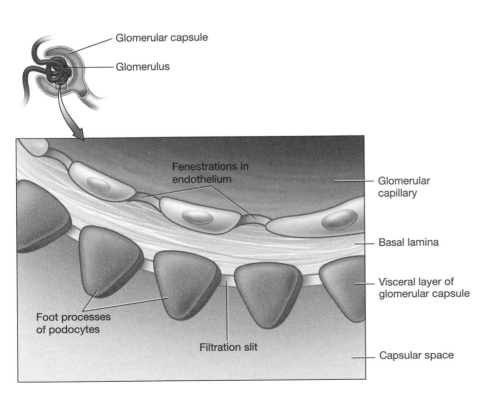

FIGURE **24.2** The filtration membrane

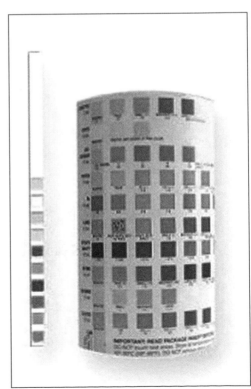

FIGURE **24.3** URS-10 vial and strip

Procedure Making a Model Kidney and Testing Glomerular Filtration

Now let's construct our model kidneys. The model kidney must sit in water for 25 minutes while you wait for your results. While you wait, answer Check Your Understanding questions 1 through 3 (p. 547).

Note: Safety glasses and gloves are required.

---

1 Cut a 4-inch piece of dialysis tubing.

2 Securely tie off one end of the tubing with string. Open the other end of the tubing by wetting the end of the tube and rubbing it between your fingers.

3 Fill the dialysis tubing about half-full with either animal blood or simulated blood, and securely tie off the open end of the tube with string.

4 Place the tied-off tube in a 200 ml beaker containing about 150 ml of deionized water.

5 Leave the tubing in the water for approximately 25 minutes.

6 After 25 minutes have passed, remove the tubing from the water.

7 Dip a URS-10 strip in the water and remove it quickly. Turn the bottle on its side and compare the colors of the pads for glucose, blood, leukocytes, and protein. You will notice that on the side of the bottle, time frames are listed for each substance that is tested. Wait this amount of time to watch for a reaction; otherwise, you could obtain a false negative result. If you wait too long to read the results, though, the colors will tend to darken and may blend with adjacent colors.

8 Record your results below.

    a. Glucose:

    b. Erythrocytes (blood):

    c. Leukocytes:

    d. Protein:

9 Which substances were filtered out? Which substances stayed in the tubing?

 # Exercise 2

Urinalysis

MATERIALS NEEDED

- Samples of simulated urine
- Graduated cylinder or test tube
- URS-10 and bottle with key

For hundreds of years, healthcare providers have recognized the utility of urinalysis as a diagnostic tool. Historically, the urine was evaluated for color, translucency, odor, and taste (yes, taste!). Today, while these characteristics are still examined (except, thankfully, taste), we also utilize urinalysis test strips, as we did in Exercise 1, to test for the presence of various chemicals in the urine. Both normal and abnormal readings give a healthcare provider a wealth of information about a patient's renal function and overall health.

   In this exercise you will analyze different urine samples using a URS-10, as in Exercise 1. For each urine sample, one or more results will be read as abnormal. In the second part of the exercise, you will research potential causes of the abnormalities that you detected.

Procedure Urinalysis Part One

Test a minimum of five samples of simulated urine using your urinalysis test strips. Use a new strip for each sample, and take care to clean your graduated cylinder thoroughly between each sample that you test. Please note that this is *simulated* urine rather than real urine, although it contains the same chemicals as real urine and should be handled with equal caution.

Note: Safety glasses and gloves are required.

---

1 Randomly choose one sample of simulated urine, and pour approximately 3 ml of the sample into a test tube or graduated cylinder.

2 Submerge one URS-10 in the urine, and quickly remove it.

3 Compare the resulting colors and patterns on the test strip to those on the key on the bottle. Be sure to wait the appropriate amount of time to read the results.

4 Record your results in Table 24.3, and note any anomalous results.

5 Repeat this procedure for the remaining samples.

TABLE 24.3	Urinalysis Results				
Reading	Sample 1	Sample 2	Sample 3	Sample 4	Sample 5
pH					
Leukocytes					
Nitrite					
Urobilinogen					
Protein					
Blood					
Specific Gravity					
Ketones					
Bilirubin					
Glucose					

Procedure Urinalysis Part Two

Each sample that you tested should have had at least one abnormality. Use your textbook and/or the Internet as a resource to determine which disease state(s) could lead to the abnormality you detected in each sample. Record your results in Table 24.4. After you have completed the activity, answer Check Your Understanding questions 4 and 5 (p. 548).

TABLE 24.4	Urinalysis Abnormalities and Disease States	
Sample	Primary Abnormality	Potential Causes
1		
2		
3		
4		
5		

 Exercise 3

Time to Trace!

It's time to trace again! In this exercise you will trace the pathway of different molecules through the general circulation, the vasculature of the kidney, and the microanatomy of the kidney. You will examine three different substances:

1. An erythrocyte,

2. A molecule of glucose, and

3. A molecule of urea.

Following are some hints to use as you trace:

▶ Refer to Unit 17 (Blood Vessel Anatomy) to review the pathway of blood flow through the heart and the body.

▶ Don't forget the basic rules of blood flow: If you start in a capillary bed (which you will for each of these), you first must go through a *vein*, and then the heart and pulmonary circuit, before you may enter the arterial system.

▶ Refer to Table 24.2 in Pre-Lab Exercise 3 before you trace each substance. This will help you determine if the substance gets filtered at the glomerulus or if it remains in the blood, if a filtered substance gets reabsorbed, and where the reabsorption occurs.

▶ Refer to Figure 24.4 for help with the pathway of blood flow through the kidney and the kidney's microanatomy.

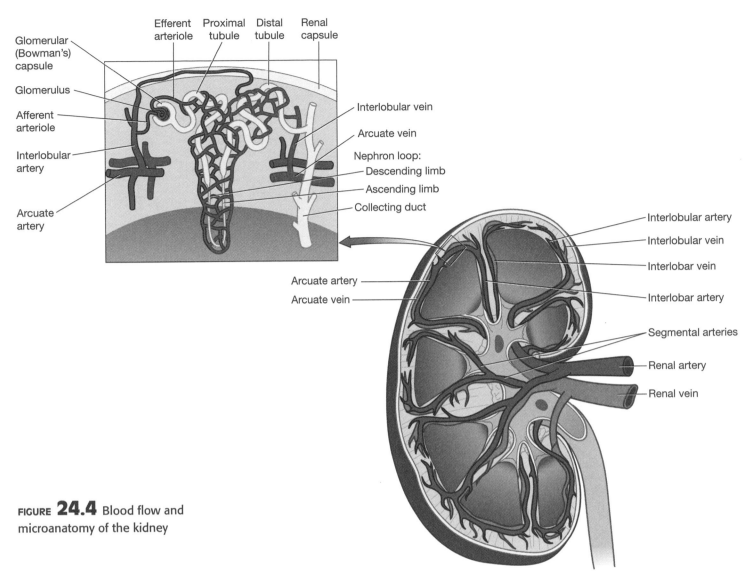

Glomerular (Bowman's) capsule
Glomerulus
Afferent arteriole
Interlobular artery
Arcuate artery

Efferent arteriole
Proximal tubule
Distal tubule
Renal capsule

Interlobular vein
Arcuate vein
Nephron loop:
Descending limb
Ascending limb
Collecting duct

Interlobular artery
Interlobular vein
Interlobar vein
Interlobar artery
Segmental arteries
Renal artery
Renal vein

Arcuate artery
Arcuate vein

FIGURE **24.4** Blood flow and microanatomy of the kidney

Part 1: Erythrocyte

Trace an erythrocyte from its origin in the bone marrow of the left humeral head to the renal *vein*. Keep in mind that the erythrocyte will enter a capillary in the bone marrow before it is drained by a local vein.

Part 2: Glucose

Trace a molecule of glucose from its point of absorption into the capillaries of the small intestine to the renal vein. Don't forget the hepatic portal system!

Part 3: Urea

Trace a molecule of urea from its origin in the liver to its final destination outside the body.

EXPLORING ANATOMY & PHYSIOLOGY IN THE LABORATORY

NAME _____

SECTION _____ DATE _____

Check Your Recall

1 What is filtrate, and how is it formed?

2 Fill in the blanks: The rate of filtrate formation is known as the _____ and averages about

_____.

3 What is the purpose of tubular reabsorption?
   a. To remove substances in the blood and place them into the filtrate
   b. To reclaim 99% of the water that was filtered
   c. To reclaim electrolytes, glucose, and amino acids
   d. Both b and c are correct.
   e. Both a and b are correct.

4 How do tubular reabsorption and tubular secretion differ?

5 Which of the following is *not* a component of the filtration membrane?
   a. Glomerular endothelial cells
   b. Podocytes
   c. Smooth muscle
   d. Basal lamina

6 How does the filtration membrane prevent certain substances from entering the filtrate?

7 Which of the following is *true*?

   a. The fluid and solutes in the filtrate have been removed from the blood and are located in the renal tubules.
   b. The filtration membrane allows blood cells and other large substances to pass but prevents small substances from moving through it.
   c. Tubular secretion involves the reclamation of substances in the filtrate and their return to the blood.
   d. Substances such as glucose, proteins, and erythrocytes are secreted into the filtrate.

8 Which of the following substances would you *not* expect to find in the filtrate?

   a. Erythrocytes
   b. Glucose
   c. Electrolytes
   d. Amino acids

9 Urinalysis can tell you

   a. if there is an infection in the kidney or urinary tract.
   b. if blood sugar is elevated.
   c. if a person is dehydrated.
   d. if a person's kidneys function normally.
   e. All of the above.

10 Which of the following substances would you expect to find in normal urine?

   a. Glucose
   b. Proteins
   c. Urea
   d. Ketones
   e. Nitrites

EXPLORING ANATOMY & PHYSIOLOGY IN THE LABORATORY

Check Your Understanding | **Critical Thinking and Application Questions**

1 Why didn't all of the substances exit your model kidney to enter the "filtrate?"

2 Should glucose, erythrocytes, leukocytes, and protein appear in the urine? Why or why not? What happens to each of these substances in the kidney?

3 What other substances would you expect to find in the "filtrate" surrounding your model kidney? Explain.

4 The presence of blood and protein in the urine can indicate inflammation of the glomerulus. Why would glomerular inflammation result in these substances passing into the filtrate and the urine? (*Hint*: What happens to capillaries during an inflammatory response?)

5 Poorly controlled diabetes mellitus, which results in a high blood glucose, often causes glucose to be found in the urine. Explain this finding. (You may need to use your textbook for help.)

EXPLORING ANATOMY & PHYSIOLOGY IN THE LABORATORY

Digestive System

OBJECTIVES

Once you have completed this unit, you should be able to:

- Identify structures of the digestive system.

- Identify histological structures of the digestive system.

- Demonstrate and describe the action of emulsifying agents on lipids.

- Demonstrate and describe the functions of digestive enzymes.

- Trace the pathway of physical and chemical digestion of carbohydrates, proteins, and lipids.

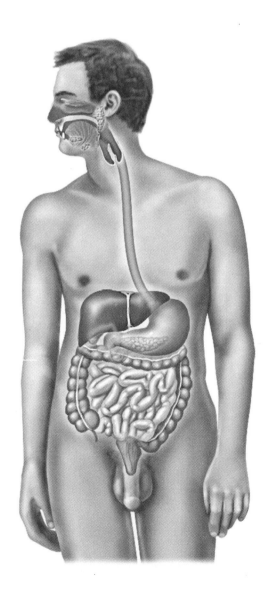

Pre-Lab Exercise 1
Key Terms

Table 25.1 lists the key terms with which you should be familiar before coming to lab. Please note that this list in not all-inclusive, as the terminology for the digestive system is extensive. Digestive enzymes are covered separately in Pre-Lab Exercise 3.

TABLE 25.1	Key Terms

Term   **Definition**

Digestive System Structures

Alimentary canal _____

_____

Accessory organ _____

_____

Peritoneum: visceral, parietal _____

_____

Peritoneal cavity _____

_____

Cardioesophageal sphincter _____

_____

Rugae   _____

_____

Pyloric sphincter _____

_____

Duodenum   _____

_____

Jejunum _____

_____

Ileum _____

_____

Colon _____

_____

Salivary glands _____

_____

Pancreas _____

_____

Liver _____

_____

Gallbladder _____

_____

Digestive Histology

Mucosa _____

_____

Submucosa _____

_____

Muscularis externa _____

_____

Serosa _____

_____

Gastric gland _____

_____

Goblet cell _____

_____

Mucosal-associated lymphoid tissue (MALT) _____

_____

Acinar cells _____

_____

Pancreatic islet _____

_____

Liver lobule _____

_____

Portal triad _____

_____

Digestive Physiology

Digestive enzyme _____

_____

Chemical digestion _____

_____

Emulsification _____

_____

Bile _____

_____

Pre-Lab Exercise 2

Anatomy of the Digestive System

Label and color the structures of the digestive system depicted in **Figures 25.1–25.4** with the terms from Exercise 1. Use your text and Exercise 1 in this unit for reference.

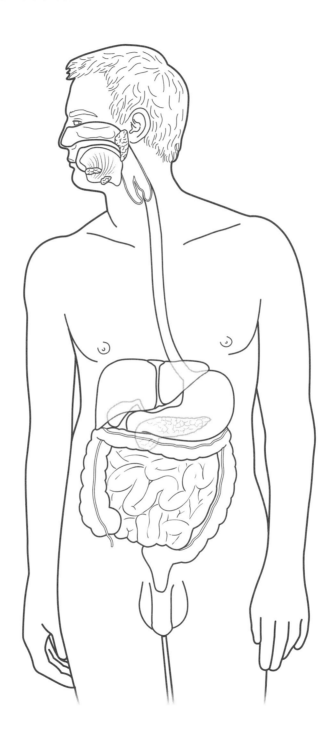

FIGURE **25.1** The organs of the digestive system

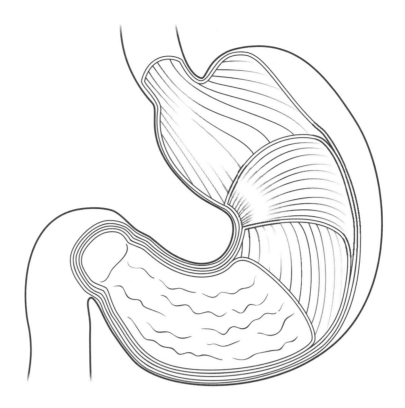

FIGURE **25.2** The stomach

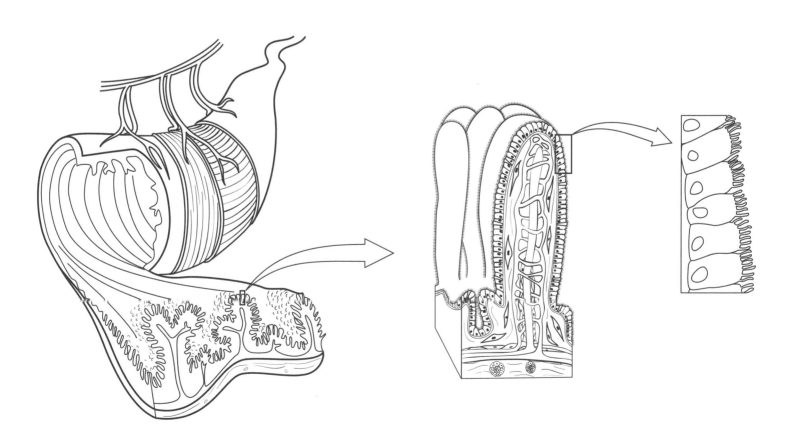

FIGURE **25.3** Gross and microscopic anatomy of the small intestine

EXPLORING ANATOMY & PHYSIOLOGY IN THE LABORATORY

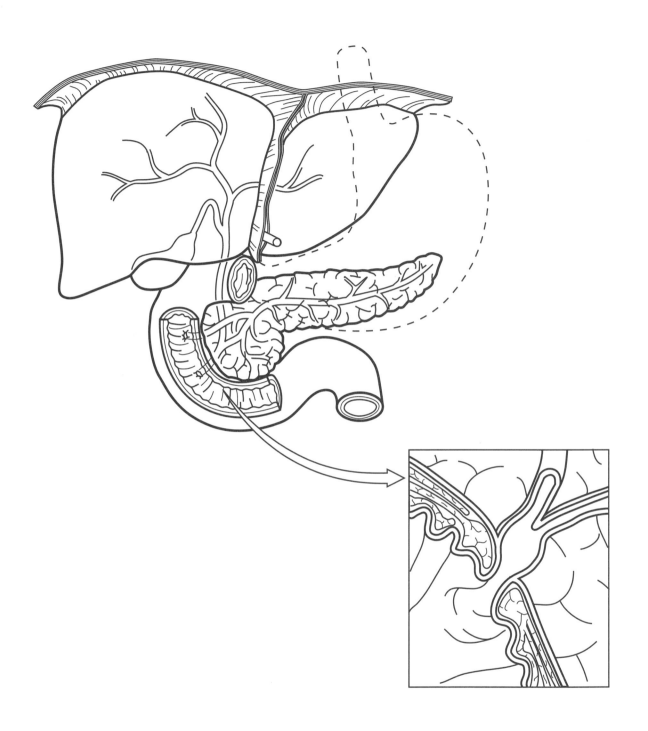

FIGURE **25.4** The liver, gallbladder, pancreas, and duodenum

 **Pre-Lab Exercise 3**

Digestive Enzymes

Chemical digestion is one of the main functions of the digestive system and is carried out with the help of numerous digestive enzymes. Record in Table 25.2 the organ that produces each enzyme and the function of each enzyme.

TABLE **25.2**	Digestive Enzymes	
Enzyme	**Source**	**Function**
Salivary amylase		
Pepsin		
Trypsin		
Chymotrypsin		
Carboxypeptidase		
Aminopeptidase		
Pancreatic lipase		
Pancreatic amylase		
Maltase		
Lactase		
Sucrase		
Nucleases		

Exercises

The food we eat contains nutrients that our cells use to build and repair body tissues and to make ATP. Food macro-molecules, however, are typically too large for the body to absorb and utilize, so the body must break them down into smaller molecules. This process of breaking foods down into smaller substances that can enter body cells is called **digestion** and is carried out by the **digestive system**. In general, the functions of the digestive system include taking in food, breaking food down mechanically and chemically into nutrients, absorbing these nutrients into the bloodstream, and eliminating indigestible substances.

We begin this unit with an introduction to the anatomy and histology of the organs of the digestive system. Next we examine the physiological processes of chemical digestion and emulsification. We conclude with a "big picture" view of digestion through a tracing exercise.

Exercise 1

Digestive System Anatomy

MATERIALS NEEDED

- Digestive system models
- Head and neck models
- Human torso models
- Human skulls with teeth
- Intestinal villus model
- Models of the tissue layers of the alimentary canal
- Digestive organ models (stomach, pancreas, liver, and duodenum)

The digestive system is composed of two types of organs: (1) the organs of the **alimentary canal**, also known as the **gastrointestinal** or **GI tract**, through which food travels, and (2) the **accessory organs**, which assist in mechanical or chemical digestion. The alimentary canal consists of the following organs (**Figure 25.5**):

1. **Mouth.** The alimentary canal begins with the mouth. In and around the mouth we find numerous accessory organs, including the **teeth**, the **tongue**, and the **salivary glands**. These accessory organs help the mouth to carry out mechanical digestion and chemical digestion.

2. **Pharynx.** The food next enters the pharynx, also known as the throat. The muscles surrounding the pharynx propel swallowed food into the next portion of the alimentary canal.

3. **Esophagus.** The esophagus is a narrow tube that is posterior to the heart and the trachea in the thoracic cavity. The esophagus contains both smooth and skeletal muscle fibers that contract via rhythmic contractions called **peristalsis,** which propels food into the stomach. A sphincter at the inferior end of the esophagus called the **cardioesophageal sphincter** (also known as the *lower esophageal sphincter*) prevents the contents of the stomach from regurgitating into the esophagus.

4. **Stomach.** The stomach, shown in **Figure 25.6**, has five regions:

   a. the dome-shaped **fundus,**

   b. the **cardia** near the cardioesophageal sphincter,

   c. the middle **body,**

   d. the inferior **pyloric antrum,** and

   e. the **pylorus.** A sphincter called the **pyloric sphincter** separates the stomach from the initial portion of the small intestine.

   Note in **Figure 25.6** that the stomach has interior folds called **rugae,** which allow it to expand considerably when it is filled with food. In addition, note that the wall of the stomach has three layers of smooth muscle (inner oblique, middle circular, and outer longitudinal layers) that work together to pummel ingested food into a liquid material called **chyme.**

5. **Small intestine.** The small intestine is the portion of the alimentary canal where most chemical digestion and absorption take place. It has three portions:
   a. the initial **duodenum,**
   b. the middle **jejunum,** and
   c. the terminal **ileum.**

   Of the three divisions, the duodenum is the shortest, measuring only about 10 inches. The jejunum and the ileum measure about 8 feet and 12 feet long, respectively.

6. **Large intestine.** The large intestine is named for its large diameter rather than its length, which only measures about 5.5 feet. Much of the large intestine features bands of longitudinal smooth muscle called **taeniae coli,** which have a drawstring effect on the large intestine that pulls it into pouches called **haustra.** Extending from the external surface of the large intestine are fat-filled pouches called **epiploic appendages.** The large intestine may be divided into five regions:
   a. the **cecum,** the pouch that receives contents from the ileum, from which it is separated by the **ileocecal valve,**
   b. the **vermiform appendix,** a blind-ended sac that extends from the cecum and contains lymphatic nodules,
   c. the **colon,** which itself has four divisions (the *ascending, transverse, descending,* and *sigmoid colon*),
   d. the **rectum,** the straight part of the large intestine that runs along the sacrum, and

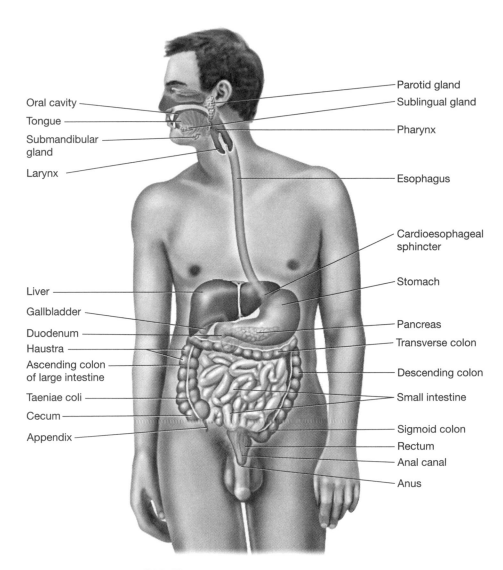

FIGURE 25.5 The organs of the digestive system

EXPLORING ANATOMY & PHYSIOLOGY IN THE LABORATORY

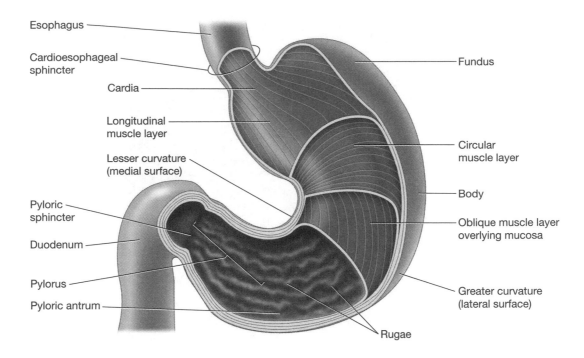

FIGURE **25.6** The stomach

    e. the **anal canal,** the terminal portion of the large intestine with two sphincters: the involuntary **internal anal sphincter** and the voluntary **external anal sphincter**.

The accessory digestive organs generally do not come into direct contact with the ingested food (the teeth and tongue are exceptions). Most of them instead secrete substances such as bile salts and enzymes that travel through a duct to the alimentary canal. The accessory digestive organs are the following:

1. **Teeth and tongue.** The teeth and the tongue are accessory organs located in the mouth that assist in mechanical digestion of ingested food. The tongue contains keratinized papillae called **filiform papillae,** which provide a rough surface that helps to break down food physically. Note that filiform papillae do not contain taste buds, which are located only on circumvallate, foliate, and fungiform papillae.

2. **Salivary glands.** The three types of salivary glands are located around the mouth and their secretions pass through a duct into the mouth.

    a. The largest salivary glands are the paired **parotid glands** that are superficial to the masseter muscle.

    b. The smaller **submandibular glands** are located just medial to the mandible.

    c. The **sublingual gland** is located under the tongue, as its name implies.

All three types of glands secrete saliva, which contains water, mucus, an enzyme called *salivary amylase,* and antimicrobial molecules such as *lysozyme.*

3. **Liver and gallbladder.** The liver and gallbladder are organs located on the right side of the abdominal cavity (**Figure 25.7**).

    a. The liver consists of four lobes—the large *right* and *left lobes* and the small, posterior *caudate* and *quadrate lobes.* The liver has a multitude of functions in the body, most of which are metabolic in nature. For example, recall from Unit 17 that all blood from the digestive organs and the spleen travels via the hepatic portal vein to the hepatic portal system of the liver. There, the absorbed nutrients and chemicals are processed before they enter the general circulation. One of the liver's main digestive functions is to produce a chemical called *bile,* required for the digestion and absorption of fats.

    b. Bile leaves the liver via a duct called the **common hepatic duct** after which much of it enters the gallbladder on the posterior side of the liver for storage. When stimulated by certain hormones, the gallbladder contracts and the bile contained within it is ejected through the **cystic duct**. Note in **Figure 25.7** that the cystic duct joins with the common hepatic duct to form the **common bile duct**, which empties into the duodenum at the **hepato-pancreatic ampulla**.

4. **Pancreas.** The pancreas is an exocrine and endocrine gland that sits posterior and inferior to the stomach. Its exocrine functions are digestive, whereas its endocrine functions are metabolic. The exocrine portion of the pancreas produces a fluid called **pancreatic juice**, which contains water, bicarbonate to neutralize the acid produced by the stomach, and a plethora of digestive enzymes. Pancreatic juice is released through the **pancreatic duct** and enters the duodenum at the hepatopancreatic ampulla.

Much of the alimentary canal and many of the accessory organs reside inside a cavity known as the **peritoneal cavity**. Like the pleural and pericardial cavities, the peritoneal cavity is found between a double-layered serous membrane, which secretes **serous fluid** that allows the organs to slide over one another without friction. The two peritoneal layers include the:

1. **Parietal peritoneum.** The outer parietal perito-neum is a thin membrane that is functionally fused to the abdominal wall and certain organs.

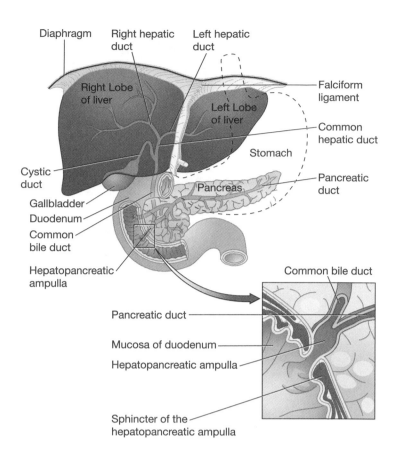

FIGURE 25.7 The liver, gallbladder, pancreas, and duodenum

2. **Visceral peritoneum.** The inner visceral peritoneum adheres to the surface of many digestive organs. The visceral peritoneum around the intestines folds over on itself to form a thick membrane known as the **mesentery**. The mesentery houses blood vessels, nerves, and lymphatic vessels, and anchors these structures and the intestines in place.

Procedure Model Inventory

Identify the following structures of the digestive system on models and diagrams, using your textbook and this unit for reference. As you examine the anatomical models and diagrams, record the name of the model and the structures you were able to identify on the model inventory in Table 25.3. After you have completed the activity, answer Check Your Understanding question 1 (p. 581).

Alimentary Canal

1. Mouth:
   a. Lips
   b. Cheeks
   c. Vestibule
   d. Oral cavity proper
   e. Hard palate
   f. Soft palate
   g. Uvula

2. Pharynx
   a. Oropharynx
   b. Laryngopharynx

3. Esophagus
   a. Cardioesophageal sphincter (or lower esophageal sphincter)

5. Stomach
   a. Cardia
   b. Fundus
   c. Body
   d. Pyloric antrum
   e. Pylorus
   f. Pyloric sphincter
   g. Rugae
6. Small intestine
   a. Duodenum
      (1) Hepatopancreatic ampulla
   b. Jejunum
   c. Ileum
   d. Ileocecal valve

7. Large intestine
   a. Haustra
   b. Taeniae coli
   c. Epiploic appendages
   d. Cecum
   e. Vermiform appendix
   f. Ascending colon
   g. Hepatic flexure
   h. Transverse colon
   i. Splenic flexure
   j. Descending colon
   k. Sigmoid colon
   l. Rectum
   m. Anal canal

Accessory Organs

1. Teeth
2. Tongue
   a. Filiform papillae
   b. Fungiform papillae
   c. Circumvallate papillae
   d. Foliate papillae
3. Salivary glands
   a. Parotid gland with parotid duct
   b. Submandibular gland
   c. Sublingual gland
4. Pancreas
   a. Pancreatic duct

5. Liver
   a. Right lobe, left lobe, caudate lobe, and quadrate lobe
   b. Falciform ligament
   c. Common hepatic duct
   d. Hepatic portal vein
   e. Hepatic arteries
   f. Hepatic veins
6. Gallbladder
   a. Cystic duct
   b. Common bile duct

Other Structures

1. Peritoneal cavity
2. Peritoneum
   a. Visceral peritoneum
   b. Parietal peritoneum
   c. Mesentery
3. Greater omentum

4. Blood vessels:
   a. Celiac trunk
   b. Left gastric artery
   c. Common hepatic artery
   d. Superior mesenteric artery
   e. Inferior mesenteric artery
   f. Superior mesenteric vein
   g. Inferior mesenteric vein
   h. Gastric veins
   i. Hepatic portal vein

Note: Your instructor may wish to omit certain structures included above or add structures not included in these lists. List any additional structures below:

TABLE **25.3**	Model Inventory for the Digestive System
Model	Digestive Structures Identified

EXPLORING ANATOMY & PHYSIOLOGY IN THE LABORATORY

Digestive System Histology

MATERIALS NEEDED

- Model of the layers of the alimentary canal
- Esophagus slide
- Stomach slide
- Duodenum slide
- Pancreas slide
- Liver slide
- Light microscope
- Colored pencils

The organs of the alimentary canal follow the same basic pattern that we have seen for other hollow organs: an inner layer of epithelial tissue that rests on connective tissue, a middle layer of smooth muscle, and an outer layer of supportive connective tissue. The tissue layers of the alimentary canal, shown in **Figure 25.8**, are named as follows:

1. **Mucosa.** The inner epithelial tissue lining of the alimentary canal is called the mucosa. It consists of simple columnar epithelium overlying a lamina propria and a thin layer of smooth muscle called the **muscularis mucosa**. The mucosa contains a collection of lymphoid nodules called **mucosal associated lymphoid tissue** (**MALT**), mucus-secreting goblet cells, and other glands that secrete products such as hydrochloric acid and enzymes. The mucosa is generally covered with a layer of mucus that helps to protect the underlying epithelium from the effects of the acid and digestive enzymes.

2. **Submucosa.** Deep to the mucosa we find the submucosa, which is a layer of connective tissue that houses blood vessels, nerves, lymphatics, and elastic fibers.

3. **Muscularis externa.** The muscularis externa actually contains two layers of smooth muscle—an inner circular layer and an outer longitudinal layer. These two layers contract alternately to produce the rhythmic contractions of **peristalsis**. Recall from the Exercise 1 that the muscularis externa of the stomach has three layers of smooth muscle, with an additional oblique layer.

4. **Serosa.** The outer connective tissue layer is called the serosa, which is partially composed of the visceral peritoneum throughout much of the alimentary canal. It is also known as the **adventitia**.

Although this pattern is followed throughout most of the alimentary canal, there are some notable differences, particularly in the esophagus (**Figure 25.9**). The mucosa of the esophagus is composed of stratified squamous epithelium with no goblet cells or thick mucus layer. The esophagus' muscularis externa changes as it progresses toward the stomach. The upper one-third of the esophagus is *skeletal muscle*, the middle one-third is about *half skeletal muscle and half smooth muscle*, and the lower one-third is *smooth muscle*. These differences in the muscularis externa allow you to determine the location of the section you are examining.

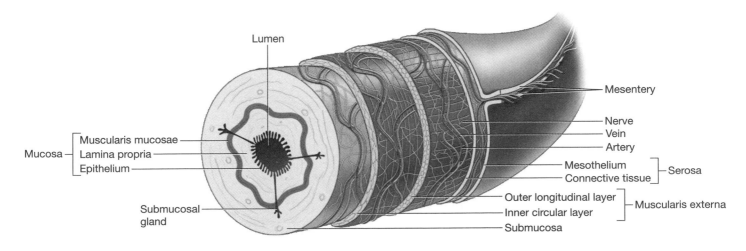

FIGURE 25.8 The tissue layers of the alimentary canal

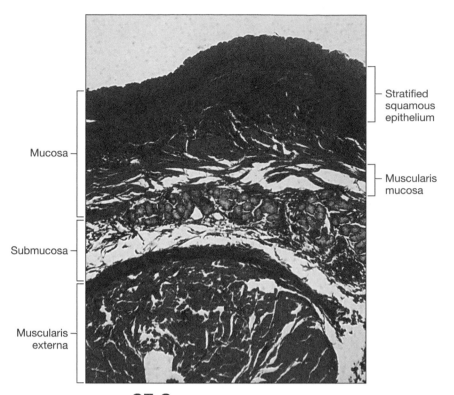

FIGURE **25.9** The esophagus, photomicrograph

Mucosa
— Stratified squamous epithelium

— Muscularis mucosa

Submucosa

Muscularis externa

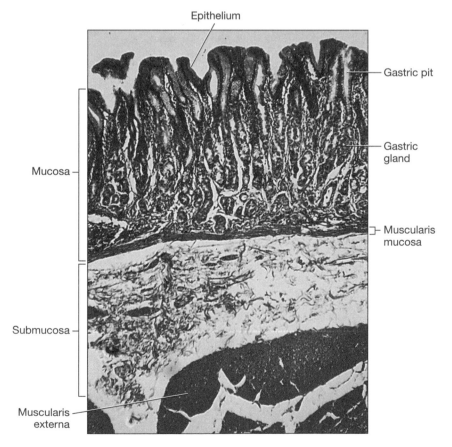

FIGURE **25.10** The stomach and gastric glands, photomicrograph

Epithelium

— Gastric pit

— Gastric gland

Mucosa

— Muscularis mucosa

Submucosa

Muscularis externa

The mucosae of both the stomach and the small intestine are folded into ridges. The stomach mucosa is heavily indented, which reflects the presence of numerous **gastric pits** that house **gastric glands** (**Figure 25.10**). These glands secrete products for digestion known collectively as **gastric juice**. Note in **Figure 25.10** that the stomach mucosa between the gastric pits contains a large number of goblet cells.

The small intestinal mucosa is also highly folded, but in a different manner and with a different purpose: They all increase the surface area available for absorption. The three sets of folds in the small intestine, shown in **Figure 25.11**, include the:

1. **Plicae circulares.** The plicae circulares, also known as *circular folds*, are folds of the mucosa and submucosa.

2. **Villi.** The fingerlike villi are the "fringe" on the plicae, and are composed of folds of mucosa. Note in **Figure 25.11** that between the villi are glands called **intestinal crypts**. Each villus is lined with intestinal cells called **enterocytes**, which surround a core containing blood vessels and a lymphatic vessel called a **lacteal**.

3. **Microvilli.** Microvilli are folds of the enterocytes' plasma membranes.

The two accessory digestive organs that we will examine in this exercise—the pancreas and the liver—are not hollow organs, so they do not follow the same histological pattern as the alimentary canal. The pancreas is a gland that consists of clusters of cells that produce both exocrine and endocrine secretions (**Figure 25.12**). The exocrine cells of the pancreas, called **acinar cells**, produce and secrete pancreatic juice (digestive enzymes, bicarbonate, and water) into ducts. The endocrine cells of the pancreas, called **pancreatic islets** or **islets of Langerhans,** secrete hormones such as insulin into the bloodstream. They are visible as small, circular groups of cells that stain less darkly than the surrounding acinar cells. We will reexamine these groups of cells in Unit 26, the Endocrine System.

EXPLORING ANATOMY & PHYSIOLOGY IN THE LABORATORY

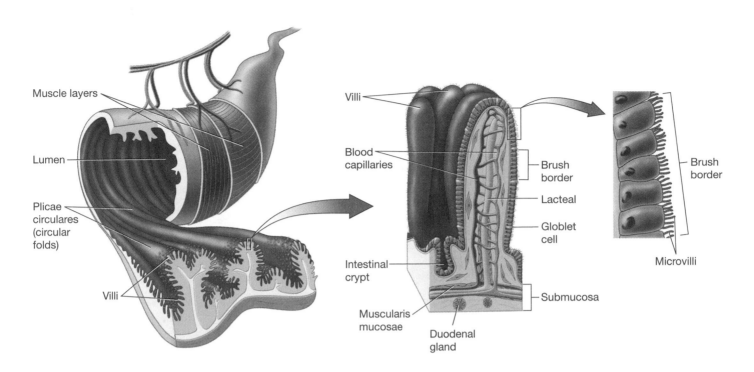

FIGURE **25.11** The folds of the small intestine

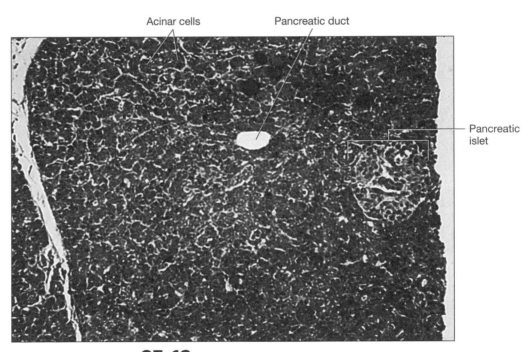

FIGURE **25.12** The pancreas, photomicrograph

The liver consists of hepatocytes organized into hexagonal plates of cells called **liver lobules (Figure 25.13)**. At the center of each lobule is a central vein, which will eventually drain into the hepatic veins. At each of the six corners of a liver lobule we find three small vessels collectively called **portal triads**. The three vessels are: (1) a **bile duct,** which carries bile made by hepatocytes and drains into the hepatic duct, (2) a **portal vein,** a tiny branch off the hepatic portal vein that delivers nutrient-rich blood to the liver for processing and detoxification, and (3) a **hepatic artery** (arteriole) that delivers oxygen-rich blood to the hepatocytes.

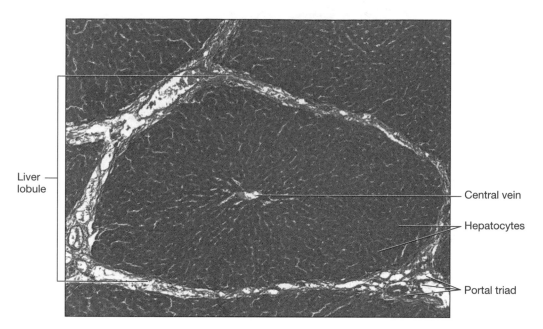

Liver lobule

Central vein

Hepatocytes

Portal triad

FIGURE 25.13
The liver, photomicrograph

Procedure Model Inventory

Identify on models and diagrams the following histologic structures of the digestive system, using your textbook and this unit for reference. As you examine the anatomical models and diagrams, record the name of the model and the structures that you were able to identify on the model inventory in Table 25.3 (note that this is the model inventory in Exercise 1).

1. Mucosa
   a. Epithelial tissue (simple columnar epithelium)
   b. Glands
   c. MALT
   d. Muscularis mucosae
2. Submucosa
   a. Blood vessels
   b. Nerves
   c. Lymphatic vessels

3. Muscularis externa
   a. Inner circular layer
   b. Outer longitudinal layer
4. Serosa
5. Intestinal villi
   a. Lacteal
   b. Enterocytes
   c. Microvilli

Procedure Microscopy of Digestive Organs

Obtain prepared slides of the esophagus, the fundus of the stomach, the duodenum, the pancreas, and the liver. Examine the slides initially with the low-power objective, and advance to higher-powered objectives to see more details. Use your colored pencils to draw what you see and label your drawing with the following terms.
When you have completed the activity, answer Check Your Understanding questions 2 through 4 (pp. 581–582).

Esophagus

1. Stratified squamous epithelium
2. Submucosa with blood vessels
3. Muscularis externa
   a. Smooth muscle
   b. Skeletal muscle
4. Serosa

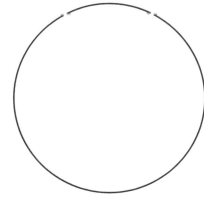

EXPLORING ANATOMY & PHYSIOLOGY IN THE LABORATORY

Fundus of the Stomach

1. Mucosa with goblet cells and gastric glands
2. Submucosa with blood vessels
3. Muscularis externa with smooth muscle layers
4. Serosa

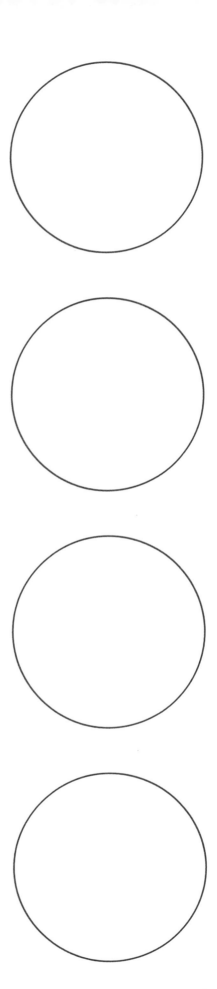

Duodenum

1. Villi
2. Mucosal-associated lymphoid tissue
3. Microvilli (visible as small "hairs" projecting from the cells)
4. Intestinal glands

Pancreas

1. Acinar cells
2. Pancreatic ducts
3. Pancreatic islets (islets of Langerhans)

Liver

1. Liver lobule
2. Portal triad
3. Central vein

Exercise 3

Digestion

MATERIALS NEEDED

- 15 glass test tubes
- Test tube rack
- Graduated cylinder
- Glass stirring rod
- Rubber stopper
- Detergent
- Distilled water
- Vegetable oil
- Sudan Red stain
- Small glass test tubes
- Graduated cylinders
- Starch solution
- Amylase enzyme
- Distilled water
- Water bath set to 37°C
- Boiling water bath
- Lugol's iodine solution
- Benedict's reagent
- Vegetable oil
- 0.1M Sodium hydroxide
- Bile
- Lipase
- Phenol red
- Egg white
- Pepsinogen in KCl buffer
- 0.5M Hydrochloric acid
- Biuret reagent

We now turn our attention to the physiology of food breakdown and absorption. Three major types of nutrients are broken down and absorbed in the alimentary canal: *carbohydrates*, *proteins*, and *lipids*. Carbohydrate digestion begins in the mouth with the enzyme **salivary amylase**, which catalyzes the reactions that digest polysaccharides into smaller oligosaccharides. The remaining polysaccharides and oligosaccharides are digested in the small intestine with the help of many enzymes, including pancreatic amylase and enzymes associated with the enterocytes of the small intestine called **brush border enzymes**.

Protein digestion begins in the stomach with the enzyme **pepsin**, secreted from chief cells in the gastric glands as the inactive pre-enzyme **pepsinogen**. Pepsinogen becomes the active enzyme pepsin when it encounters an acidic environment. Pepsin begins digesting proteins into polypeptides and some free amino acids. The remainder of protein digestion occurs in the small intestine with pancreatic enzymes such as **trypsin** and **chymotrypsin** and also brush border enzymes such as **carboxypeptidase** and **aminopeptidase**.

Lipid digestion does not begin until the lipids reach the small intestine. It is more complicated than protein or carbohydrate digestion because lipids are nonpolar molecules that do not dissolve in the water-based environment of the small intestine. This causes the lipids to clump together and form large fat "globules" in the small intestine. In this form, it is nearly impossible for the enzyme **pancreatic lipase** to catalyze the breakdown of lipids down into monoglycerides and free fatty acids because it has only a small surface area on which to work. Therefore, the first step in lipid digestion is to break up this large fat globule into smaller pieces, a process called **emulsification** (**Figure 25.14**).

Emulsification is accomplished by chemicals called **bile salts**, produced by the liver and stored in the gallbladder. Bile salts are amphipathic molecules that have both nonpolar and polar parts. Notice in **Figure 25.14** what happens to the fat globule when bile salts come into contact with it: The bile salts' nonpolar parts interact with the lipids, but their polar parts repel other lipids. The overall effect is that the fat globule is physically broken down into smaller fat pieces. This gives pancreatic lipase much more surface area on which to work, and the fats are digested into monoglycerides and free fatty acids.

Once chemical digestion is complete, all three nutrients are absorbed through the plasma membranes of the enterocytes lining the intestinal villi. Both monosaccharides and amino acids are absorbed into the enterocytes via facilitated diffusion or secondary active transport mechanisms. These nutrients then exit the other side of the enterocytes and enter the capillaries in the core of the villus. Remember from Exercise 1 that all blood from the small intestine and other abdominal organs is delivered to the hepatic portal system before being delivered to the general circulation. Therefore, both monosaccharides and amino acids travel first to the liver via the hepatic portal vein, after which they enter the blood in the hepatic veins and the inferior vena cava.

Lipid absorption, like lipid digestion, is more complicated because lipids are nonpolar and cannot be transported as free fatty acids through the water-filled plasma of the blood. In the small intestinal lumen, digested lipids remain associated with bile salts to form clusters of lipids and other nonpolar chemicals called **micelles**. Micelles escort the lipids to the enterocytes' plasma membranes, after which the lipids leave the micelles and enter the enterocytes by simple diffusion. Note that in the absence of bile salts, nearly all of the lipids will simply pass through the small intestine unabsorbed and be excreted in the feces.

Exploring Anatomy & Physiology in the Laboratory

In the enterocytes, lipids associate with other nonpolar substances and proteins to form protein-coated droplets called **chylomicrons**. Chylomicrons exit the enterocytes into the core of the villus by exocytosis, but they are too large to enter the blood capillaries. Instead, they enter the lacteal and join the lymph in the lymphatic system. For this reason, lipids are not delivered to the hepatic portal system like carbohydrates and amino acids. The chylomicrons travel within the lymphatic vessels until they enter the blood at the junction of the left subclavian vein and the left internal jugular vein.

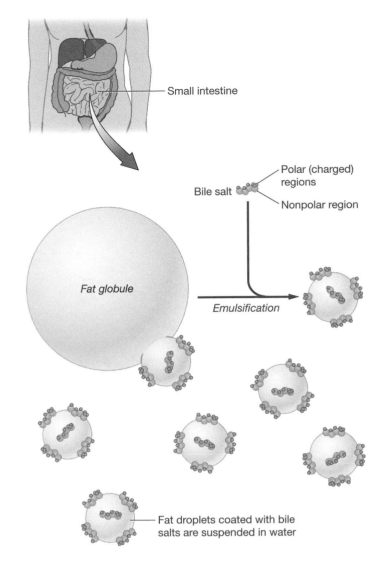

FIGURE **25.14** Emulsification of fats by bile salts

Procedure Test Carbohydrate Digestion

In the following procedure you will test for carbohydrate digestion using a starch solution and the enzyme amylase. You will check for digestion using two reagents: Lugol's iodine, which turns dark blue in the presence of starch, and Benedict's reagent, which forms a precipitate (solid) in the presence of simple sugars. You may interpret the results of your tests as follows:

▶ If the solution turns dark blue with Lugol's iodine, *no* or limited carbohydrate digestion occurred because starch is still present.

▶ If the solution develops a greenish-brownish red precipitate with Benedict's reagent, monosaccharides are present and carbohydrate digestion *did* occur.

Note: Safety glasses and gloves are required.

---

1 Obtain six glass test tubes and number them 1 through 6.

2 Place 3 ml of starch solution into tube 1 and tube 2.

3 Add 3 ml of amylase to tube 1.

4 Add 3 ml of distilled water to tube 2.

5 Set tubes 1 and 2 into a 37°C water bath and leave them in there for 30 minutes.

6 Remove the tubes from the water bath and mix the contents of each tube thoroughly with stirring rods (be certain to use two separate rods so you don't contaminate the contents of the tubes).

7 Divide the mixture in tube 1 equally into tubes 3 and 4 (each tube should contain about 3 ml of the mixture).

8 Divide the mixture in tube 2 equally into tubes 5 and 6 (each tube should contain about 3 ml of the mixture).

9 Add two drops of Lugol's iodine solution to tube 3 and tube 5. Record the results in Table 25.4.

10 Add 10 drops of Benedict's reagent to tube 4 and tube 6. Swirl the tubes gently to mix the contents.

11 Place tubes 4 and 6 into a boiling water bath for 3 minutes. *Use caution* to prevent the mixture from splattering and burning you or your classmates.

12 Remove the tubes from the boiling water, and record the results of the test in Table 25.4.

TABLE **25.4**	Results of Carbohydrate Digestion Experiment
Tube	**Reaction with Lugol's Iodine**
Tube 3 (starch + amylase)	
Tube 5 (starch + water)	
Tube	**Reaction with Benedict's Reagent**
Tube 4 (starch + amylase)	
Tube 6 (starch + water)	

13 Interpret your results:

   a. In which tube(s) did carbohydrate digestion occur? How do you know?

   b. In which tube(s) did no carbohydrate digestion occur? How do you know?

Procedure Test Protein Digestion

Now you will examine protein digestion with the enzyme pepsin. Recall that pepsin is produced by the chief cells as the inactive enzyme *pepsinogen*, and that pepsinogen requires certain conditions to become active. You will assess for protein digestion with the chemical Biuret reagent, which turns purple-blue in the presence of undigested proteins and lavender-pink in the presence of amino acids. When you have completed the activity, answer Check Your Understanding question 5 (p. 582).

Note: Safety glasses and gloves are required.

---

1 Obtain four test tubes and number them 1 through 4.

2 Add about 3 ml of egg white (or albumin) to each test tube.

3 Add 3 ml of pepsinogen in 0.1M NaOH (pH about 8) to tube 1.

4 Add 3 ml of pepsinogen in 0.5M hydrochloric acid (pH about 2–3) to tube 2.

5 Add 3 ml of 0.1M NaOH to tube 3.

6 Add 3 ml of 0.5M hydrochloric acid to tube 4.

7 Place each tube in a 37°C water bath for one hour.

8 Remove the tubes and add 5 drops of Biuret reagent to each tube. Record the color change that you observe in each tube in Table 25.5.

TABLE 25.5	Results of Protein Digestion Experiment
Tube	**Color of Mixture after Adding Biuret Reagent**
Tube 1 (egg white + pepsinogen + NaOH)	
Tube 2 (egg white + pepsinogen + HCl)	
Tube 3 (egg white + NaOH)	
Tube 4 (egg white + HCl)	

9 Interpret your results:
   a. In which tube(s) did protein digestion occur? How do you know?

   b. In which tube(s) did no protein digestion occur? How do you know?

c. What effect does pH have on protein digestion with pepsin? Why?

Procedure Demonstrate Lipid Emulsification

Before we examine the chemical digestion of lipids, let's examine what happens during the process of emulsification. In this exercise you will use four compounds to observe emulsification in action:

▶ *Lipids*: The source of lipids for this exercise is vegetable oil.

▶ *Emulsifying agent*: Detergents are considered to be emulsifiers because they have both polar parts and nonpolar parts, similar to bile salts. The emulsifying agent in this exercise, therefore, will be a liquid detergent.

▶ *Distilled water.*

▶ *Red Sudan stain*: This is a stain that binds only to lipids. The sole purpose of this stain is to make the lipids more visible during the procedure.

When you have completed the activity, answer Check Your Understanding question 6 (p. 582).

Note: Safety glasses and gloves are required.

---

1 Obtain a glass test tube and add approximately 2 ml of distilled water to it.

2 Add about 2 ml of vegetable oil to the water.

3 Place a rubber stopper into the tube and shake it vigorously for 15 seconds. Allow it to stand for 2 minutes. What happens to the oil and water?

4 Add three or four drops of red Sudan stain, and shake the tube again for 15 seconds. What color is the oil? What color is the water?

5 Add about 1 ml of liquid detergent to the mixture and shake the tube vigorously for 15 seconds. Allow the tube to stand for 2 minutes. What has happened to the solution? Is it still two distinct colors? Explain your results.

Procedure Test Lipid Digestion

In the following procedure you will be comparing the ability of four solutions to digest vegetable oil: lipase and bile, lipase alone, bile alone, and water. You will check for the presence of digestion using an indicator called **phenol red**. Phenol red appears pink at an alkaline (basic) pH, changes to an orange-red color at a neutral pH, and changes to a yellow color when the pH becomes acidic. If fat digestion has occurred, fatty acids will be released that will decrease the pH of the contents of your tube and turn them orange-red or yellow. You may interpret your results in the following way:

Pink color = the pH is basic and no (or limited) fat digestion occurred

Red-orange color = the pH is neutral and some fat digestion occurred

Yellow color = the pH is acidic and significant fat digestion occurred

---

Note: Safety glasses and gloves are required.

---

1 Obtain four glass test tubes and number them 1 through 4.

2 Add 3 ml of vegetable oil to each test tube.

3 Add 8–10 drops of the pH indicator phenol red to each test tube. The oil now should appear pink. If it does not appear pink, add drops of 0.1M NaOH to each tube until the indicator turns pink.

4 Add 1 ml of lipase and 1 ml of bile to tube 1.

5 Add 1 ml of lipase to tube 2.

6 Add 1 ml of bile to tube 3.

7 Add 1 ml of distilled water to tube 4.

8 Place each tube in a 37°C water bath for 30 minutes.

9 Remove the tubes from the water bath and record the color of each tube in Table 25.6.

TABLE **25.6**	Results of Lipid Digestion Experiment	
Tube Number	**Color**	**pH (Acidic, Neutral, or Alkaline)**
1 (oil + lipase + bile)		
2 (oil + lipase)		
3 (oil + bile)		
4 (oil + water)		

10 Interpret your results:

   a. In which tube(s) did lipid digestion occur? How do you know?

   b. In which tube(s) did limited or no lipid digestion occur? How do you know?

   c. What effect does bile have on lipid digestion? Why?

 Exercise 4

Time to Trace!

MATERIALS NEEDED

- Laminated outline of the human body
- Water-soluble marking pens

Now it's time to put all of the digestive anatomy and physiology together to get a "big picture" view of the digestive system. In this exercise you will trace the pathway that three different nutrients take from their ingestion at the mouth to their arrival at the heart. You will trace a cookie (primarily carbohydrates), an egg (primarily protein), and greasy fried food (primarily lipids).

Along the way, detail the following for each:

1. The *anatomical pathway* that each takes, from ingestion through its passage through the alimentary canal, to its absorption into the blood, and finally to its passage through the blood until it reaches the heart.

2. The *physical* and *chemical processes* that break down each substance, including enzymatic breakdown, churning, chewing, and emulsification.

   Some hints:

- Don't forget that carbohydrates and proteins travel through the hepatic portal system before they enter the general circulation.

- Remember that digestion and absorption are quite different for lipids. For example, fats are not absorbed into the same structures as proteins and carbohydrates.

- Use the text in Exercise 3 and your list of enzymes that you completed in Pre-Lab Exercise 3 for reference.

- Refer to the tracing exercises from Unit 17 (p. 403) and Unit 20 (p. 465) to review the pathway of blood and lymph flow through the body.

Tracing Steps:

1. Cookie: **Start:** mouth $\longrightarrow$

      heart: **End**

2. Egg: **Start:** mouth $\longrightarrow$

      heart: **End**

3. Greasy fried food: **Start:** mouth $\longrightarrow$

      heart: **End**

NAME _____

SECTION _____ DATE _____

Check Your Recall

1 Label the following structures on **Figure 25.15**.

- Parotid gland
- Transverse colon
- Esophagus
- Pancreas
- Appendix
- Sublingual gland
- Liver
- Cecum
- Sigmoid colon
- Gallbladder
- Submandibular gland
- Anal canal

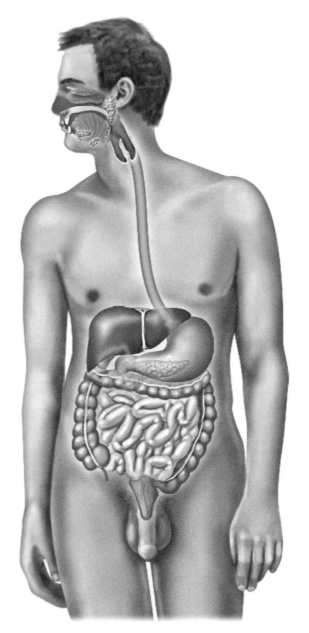

FIGURE **25.15** The organs of the digestive system

2 Label the following structures on Figure 25.16.

- Cardioesophageal sphincter
- Fundus
- Cardia
- Duodenum
- Rugae
- Pyloric antrum
- Body
- Pylorus
- Esophagus
- Pyloric sphincter

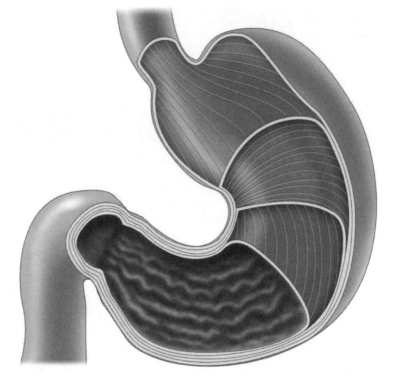

FIGURE **25.16** The stomach

3 Label the following structures on Figure 25.17.

- Liver
- Gallbladder
- Pancreatic duct
- Cystic duct
- Duodenum
- Hepatopancreatic ampulla
- Pancreas
- Common bile duct
- Common hepatic duct

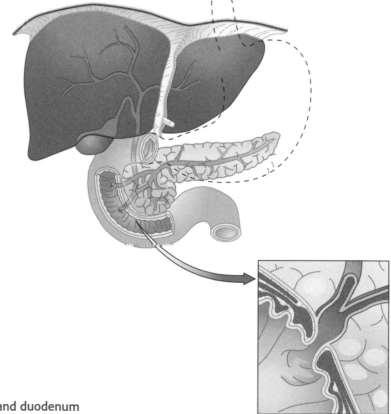

FIGURE **25.17** The liver, gallbladder, pancreas, and duodenum

4 Which of the following organs is *not* part of the alimentary canal?

a. Esophagus
b. Gallbladder
c. Cecum
d. Ileum

5 Mark the following statements as true (T) or false (F). If the statement is false, correct it so it becomes a true statement.

_____ The peritoneal cavity is located between the visceral peritoneum and the mesentery.

_____ The longest segment of the small intestine is the duodenum.

_____ The stomach has three layers of smooth muscle that contract to churn food into chyme.

_____ The gallbladder produces and stores bile.

_____ The small intestine features three sets of progressively smaller folds that increase surface area for absorption.

_____ The liver consists of plates of hexagonal liver lobules.

6 Label the following structures on **Figure 25.18**.

• Circular layer of smooth muscle
• Lumen
• Serosa
• Submucosa
• Muscularis mucosae
• Longitudinal layer of smooth muscle
• Mucosa

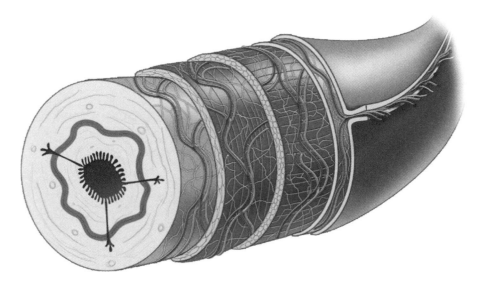

FIGURE 25.18 The tissue layers of the alimentary canal

7 Fill in the blanks: The exocrine cells of the pancreas are called _____ and secrete _____.

The endocrine cells of the pancreas are called _____ and secrete _____.

8 Matching: Match the following with the correct definition.

a. Salivary amylase _____   A. Protein-coated lipid droplets that are absorbed

b. Bile salts _____   B. Digestive enzyme(s) associated with the enterocytes of the small intestine

c. Micelles _____   C. Emulsifies / emulsify fats

d. Lacteal _____   D. Protein-digesting enzyme(s) produced by the pancreas

e. Brush border enzymes _____   E. Required to activate pepsinogen

f. Pancreatic lipase _____   F. Begin(s) carbohydrate digestion in the mouth

g. Hydrochloric acid _____   G. The structure(s) into which lipids are absorbed

h. Trypsin _____   H. Protein-digesting enzyme(s) produced by the stomach

i. Chylomicrons _____   I. Clusters of bile salts and digested lipids

j. Pepsin _____   J. Enzyme(s) that digest(s) lipids into free fatty acids and monoglycerides

9 Why is the digestion and absorption of lipids more complex than the digestion and absorption of carbohydrates and proteins?

10 How does the absorption of lipids differ from the absorption of carbohydrates and proteins?

Check Your Understanding | **Critical Thinking and Application Questions**

1 *Peritonitis* is an infection of the fluid in the peritoneal cavity. Why do you think this condition might be difficult to treat? Why would this condition have wide-ranging effects on the abdominal organs?

2 One of the common consequences of gallstones is blockage of the common bile duct, which prevents bile from being emptied into the duodenum. Predict the possible consequences of this condition.

3 Explain how the forms of the small intestine, stomach, and esophagus follow their functions.

4 The condition known as heartburn is most often caused by acid regurgitating from the stomach into the esophagus. Why do you think the acid tends to burn the esophagus and produce pain but does not similarly burn the stomach?

5 Many dietary supplements contain digestive enzymes that the manufacturers claim are necessary to digest food properly. What will happen to these enzymes in the stomach? (*Hint*: Enzymes are proteins.) Will the enzymes continue to function once they have reached the small intestine? Why or why not?

6 Explain why detergents are able to remove both water-based and oil-based stains.

Endocrine System

OBJECTIVES

Once you have completed this unit, you should be able to:

- Identify endocrine organs and structures.

- Identify microscopic structures of endocrine organs.

- Trace the functions, stimulus for secretion, and target tissues of various hormones.

- Solve clinical endocrine "mystery cases."

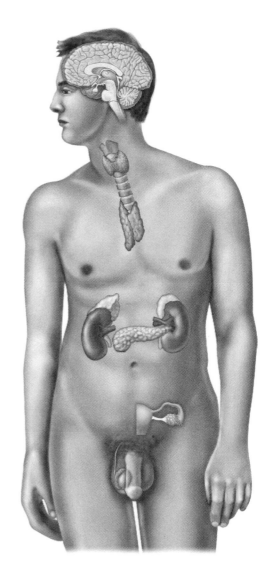

Pre-Lab Exercise 1
Key Terms

Table 26.1 lists the key terms with which you should be familiar before coming to lab. Please note that key hormones are covered in Pre-Lab Exercise 3.

TABLE 26.1	Key Terms

Term   **Definition**

General Terms

Endocrine organ (gland) _____

_____

Hormone _____

_____

Target tissue _____

_____

Negative feedback _____

_____

Endocrine Organs

Hypothalamus _____

_____

Anterior pituitary _____

_____

Posterior pituitary _____

_____

Thyroid gland _____

_____

Parathyroid glands _____

_____

Pineal gland _____

_____

Thymus _____

_____

Pancreas _____

_____

Adrenal cortex _____

_____

Adrenal medulla _____

_____

Ovaries _____

_____

Testes _____

_____

Endocrine System Anatomy

Label and color the structures of the endocrine system depicted in **Figure 26.1** with the terms from Exercise 1. Use your text and Exercise 1 in this unit for reference.

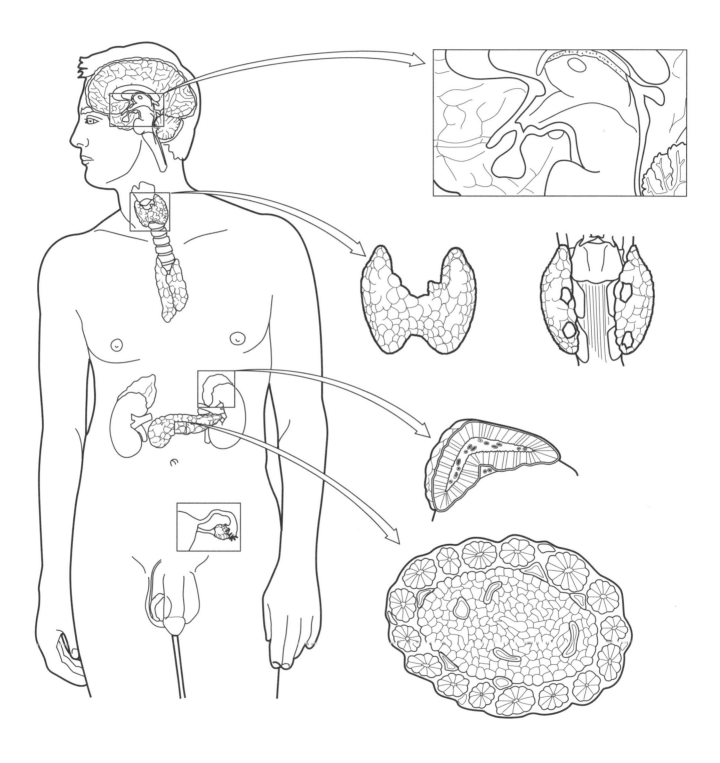

FIGURE 26.1 The endocrine system

Pre-Lab Exercise 3

Hormones: Target Tissues and Effects

Fill in Table 26.2 with the organ that secretes each hormone, the stimulus for that hormone's secretion, and the hormone's target tissue.

TABLE **26.2** Properties of Hormones			
Hormone	**Organ that Secretes the Hormone**	**Stimulus for Secretion**	**Target Tissue**
Antidiuretic hormone			
Oxytocin			
Thyroid-stimulating hormone			
Adrenocorticotropic hormone			
Growth hormone			
Prolactin			
Melatonin			
Thyroxine and triiodothyronine (T4 and T3)			
Calcitonin			
Thymosin and thymopoietin			
Parathyroid hormone			
Cortisol			
Aldosterone			
Epinephrine and norepinephrine			
Insulin			
Glucagon			

Exercises

The **endocrine system** is a diverse group of ductless glands that plays a major role in regulating the body's homeostasis. It works closely with the other system that maintains homeostasis—the nervous system. Although these two systems both work to maintain the body's homeostasis, you will notice that the methods by which they do so differ. The nervous system works via action potentials (nerve impulses) and release **neurotransmitters** that directly affect target cells. The effects are nearly immediate, but they are temporary. In contrast, the endocrine system works via secretion of **hormones**—chemicals secreted into the bloodstream that typically act on distant targets. The effects of hormones are not immediate, but they are longer-lasting than those of nerve impulses.

In general, hormones function to regulate the processes of other cells, including inducing the production of enzymes or other hormones, changing the metabolic rate of the cell, and altering permeability of the plasma membranes. You might think of hormones as the "middle managers" of the body, because they communicate the messages from their "bosses" (the endocrine glands) and tell other cells what to do. Some endocrine glands (e.g., the thyroid and anterior pituitary glands) secrete hormones as their primary function. Others, however, secrete hormones as a secondary function, examples of which are the heart (atrial natriuretic peptide), adipose tissue (leptin), kidneys (erythropoietin), and stomach (gastrin).

This unit introduces you to the anatomy, histology, and physiology of the endocrine organs and hormones. To close out this unit, you will play "endocrine detective" and try to solve three "endocrine mysteries" involving our favorite cast of characters—Ms. Magenta and company.

Exercise 1

Endocrine System Anatomy

MATERIALS NEEDED

- Endocrine system models
- Human torso models
- Head and neck models

The ten organs in the body that have hormone secretion as a primary function are the hypothalamus, the pituitary gland, the pineal gland, the thyroid gland, the parathyroid glands, the thymus, the adrenal gland, the pancreas, and the ovaries or testes (**Figure 26.2**). We discuss the first eight organs in this unit. The testes and ovaries are discussed in Units 27 and 28.

1. **Hypothalamus.** The hypothalamus, the inferior part of the diencephalon, can be likened to the endocrine system's chief executive officer (CEO). It releases hormones that inhibit and stimulate secretion—called **inhibiting** and **releasing hormones**, respectively—from the anterior pituitary gland. The anterior pituitary in turn releases hormones that stimulate other endocrine and exocrine glands in the body. In addition to inhibiting and releasing hormones, the hypothalamus makes the hormone **oxytocin**, which triggers uterine contraction and milk ejection from the mammary gland, and **antidiuretic hormone** (**ADH**), which causes water retention from the collecting ducts of the kidneys.

2. **Pituitary gland.** Note in **Figure 26.2** that the pituitary gland is actually two separate structures:
   a. The **anterior pituitary gland,** or **adenohypophysis,** is composed of glandular epithelium and secretes a variety of hormones that impact other tissues in the body. Examples include **thyroid-stimulating hormone** (**TSH**), which stimulates growth of and secretion from the thyroid, **growth hormone** (**GH**), which increases the rate of cell division and protein synthesis in all tissues, **prolactin,** which stimulates milk production from mammary glands, **adrenocorticotropic hormone** (**ACTH**), which stimulates secretion from the adrenal cortex, and two reproductive hormones, **luteinizing hormone** and **gonadotropin-releasing hormone,** which will be discussed in the next two units.
   b. The other component of the pituitary gland is the **posterior pituitary** or **neurohypophysis,** which is actually composed of nervous tissue rather than glandular tissue. In fact, the posterior pituitary doesn't produce any hormones at all, and functions merely as a place for storing the oxytocin and ADH produced by the hypothalamus.

3. **Pineal gland**. Recall from Unit 13 (p. 285) that the tiny pineal gland is located in the posterior and superior diencephalon. It secretes the hormone **melatonin** in response to decreased light levels, and acts on the reticular formation of the brainstem to trigger sleep.

4. **Thyroid gland**. The thyroid gland, located in the anterior and inferior neck superficial to the larynx, is composed of hollow spheres called **thyroid follicles**. The cells that line the thyroid follicles are simple cuboidal cells called **follicle cells,** and they surround a gelatinous, iodine-rich substance called **colloid**. The follicle cells respond to TSH from the anterior pituitary by secreting a chemical into the colloid that reacts with iodine to produce two different hormones: **thyroxine** or **T4,** which has four iodine molecules, and **triiodothyronine** or **T3,** which has three iodine molecules. T3 is the most active of the two hormones and acts on essentially all cells in the body to increase the metabolic rate, increase protein synthesis, and regulate the heart rate and blood pressure, among other things. About 10 times as much T4 is produced as is T3, and the body converts T4 to T3 when T3 levels in the blood drop.

   In between the follicles we find another cell type called the **C cells** or **parafollical cells**. These cells produce the hormone **calcitonin,** one of the body's main hormones that regulate calcium homeostasis. Calcitonin is secreted when calcium levels in the blood rise, and it decreases blood calcium by triggering osteoblast activity and bone deposition.

5. **Parathyroid glands**. As you can see in **Figure 26.2,** the small parathyroid glands are located on the posterior thyroid gland. They secrete the hormone **parathyroid hormone** (**PTH**), which works with calcitonin to maintain calcium homeostasis. PTH is secreted in response to decreased levels of calcium in the blood, and it triggers osteoclast activity and resorption of bone tissue, increased calcium absorption from the gut, and increased calcium reabsorption from the kidneys. Hormones such as PTH and calcitonin that have opposite actions are called **antagonists**.

6. **Thymus**. The thymus gland sits in the superior mediastinum. It is largest and most active in infancy and early childhood, during which time it secretes the hormones **thymosin** and **thympoietin**. Both of these hormones stimulate the development of T lymphocytes within the thymus. In adults most of the thymic tissue is gradually replaced by fat and other connective tissue.

7. **Adrenal gland**. As the name implies, the adrenal glands sit atop the superior pole of each kidney. Like the pituitary gland, the adrenal gland is actually two separate glands.
   a. The superficial region consists of glandular tissue called the **adrenal cortex,** and it secretes **steroid hormones** in response to stimulation by ACTH. The outermost zone of the adrenal cortex secretes steroids called **mineralocorticoids** such as **aldosterone,** which regulate fluid and electrolyte homeostasis. The middle zone of the adrenal cortex secretes steroids called **glucocorticoids** such as **cortisol,** which regulate the stress response, blood glucose, fluid homeostasis, and inflammation. The innermost zone secretes glucocorticoids and steroids called **gonadocorticoids** that impact the gonads and other tissues.
   b. The deep region of the adrenal gland, called the **adrenal medulla,** consists of modified postsynaptic sympathetic neurons that secrete **epinephrine** and **norepinephrine** in response to sympathetic stimulation.

8. **Pancreas**. In Unit 25 you learned about the exocrine functions of the pancreas, and now we turn to its endocrine functions. These functions are carried out by cells in small, round "islands" called **pancreatic islets,** or **islets of Langerhans**. The cells within the pancreatic islets secrete the hormones **insulin** and **glucagon,** which play a major role in regulating blood sugar levels. Insulin triggers the uptake of glucose by cells, which decreases blood glucose, and glucagon triggers the release of stored glucose from the liver, which increases blood glucose. Note that glucagon and insulin are antagonists.

Let's now examine these structures on models and charts. Note that human torso models are typically a good place to start when studying the endocrine system, as most of the organs are easy to find. The one exception is the thymus; many torsos and models choose not to show this structure because it is fairly inactive in adults.

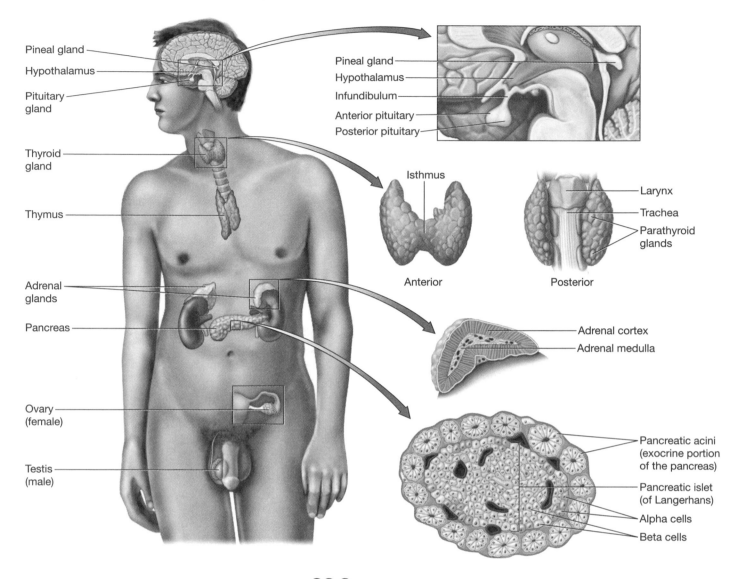

Pineal gland
Hypothalamus
Pituitary gland

Thyroid gland

Thymus

Adrenal glands

Pancreas

Ovary (female)

Testis (male)

Pineal gland
Hypothalamus
Infundibulum
Anterior pituitary
Posterior pituitary

Isthmus

Anterior

Larynx
Trachea
Parathyroid glands

Posterior

Adrenal cortex
Adrenal medulla

Pancreatic acini (exocrine portion of the pancreas)
Pancreatic islet (of Langerhans)
Alpha cells
Beta cells

FIGURE **26.2** The endocrine system

Procedure Model Inventory

Identify the following structures of the endocrine system on models and diagrams, using your textbook and this unit for reference. As you examine the anatomical models and diagrams, record on the model inventory in Table 26.3 the name of the model and the structures you were able to identify. After you have completed this activity, answer Check Your Understanding questions 1 and 2 (p. 603).

Endocrine Glands

1. Hypothalamus
   a. Infundibulum
2. Pituitary gland
   a. Anterior pituitary
   b. Posterior pituitary
3. Pineal gland
4. Thyroid gland
   a. Isthmus

5. Parathyroid glands
6. Thymus gland
7. Adrenal glands
   a. Adrenal cortex
   b. Adrenal medulla
8. Pancreas
9. Ovaries
10. Testes

TABLE **26.3** Model Inventory for the Endocrine System	
Model/Diagram	**Structures Identified**

 Exercise 2

Endocrine Organ Histology

MATERIALS NEEDED

- Thyroid gland slide
- Adrenal gland slide
- Pancreas slide
- Light microscope
- Colored pencils

In this exercise you will examine the histology of three endocrine organs: the thyroid gland, the adrenal gland, and the pancreas. (We will examine the ovary and testis in the next unit.) Following are keys to identification of each tissue:

1. **Thyroid gland.** Thyroid follicles are perhaps one of the easiest structures to identify (**Figure 26.3**). Look for a ring of simple cuboidal epithelium surrounding the reddish-brown, acellular colloid. In between the follicles, note that you can see large C cells.

2. **Adrenal gland.** The adrenal cortex may be subdivided into three zones: the outer *zona glomerulosa*, a thin layer where the cells are arranged in clusters; the middle *zona fasciculata*, where the cells are stacked on top of one another and resemble columns; and the inner *zona reticularis*, where the cells stain more darkly and are tightly packed (**Figure 26.4**). The innermost adrenal medulla is distinguished from the zones of the cortex by its numerous, loosely arranged blood vessels. The regions of the adrenal gland are best viewed on low or medium power. Switch to high power to get more detail of the individual cells.

3. **Pancreas.** In Unit 25 you examined the exocrine portion of the pancreas, the enzyme-secreting *acinar cells*. Most of the pancreas is composed of acinar cells; however, note the small "islands" of tissue in **Figure 26.5**. These islands are the insulin- and glucagon-secreting pancreatic islets. Typically, the islets are lighter in color than the surrounding acinar cells. On high power, the differences between the two tissues are easily seen: The acinar cells are slightly cuboidal and are arranged around a duct, whereas the islet cells have no distinctive arrangement (there also may be larger, stain-free spaces between the islet cells).

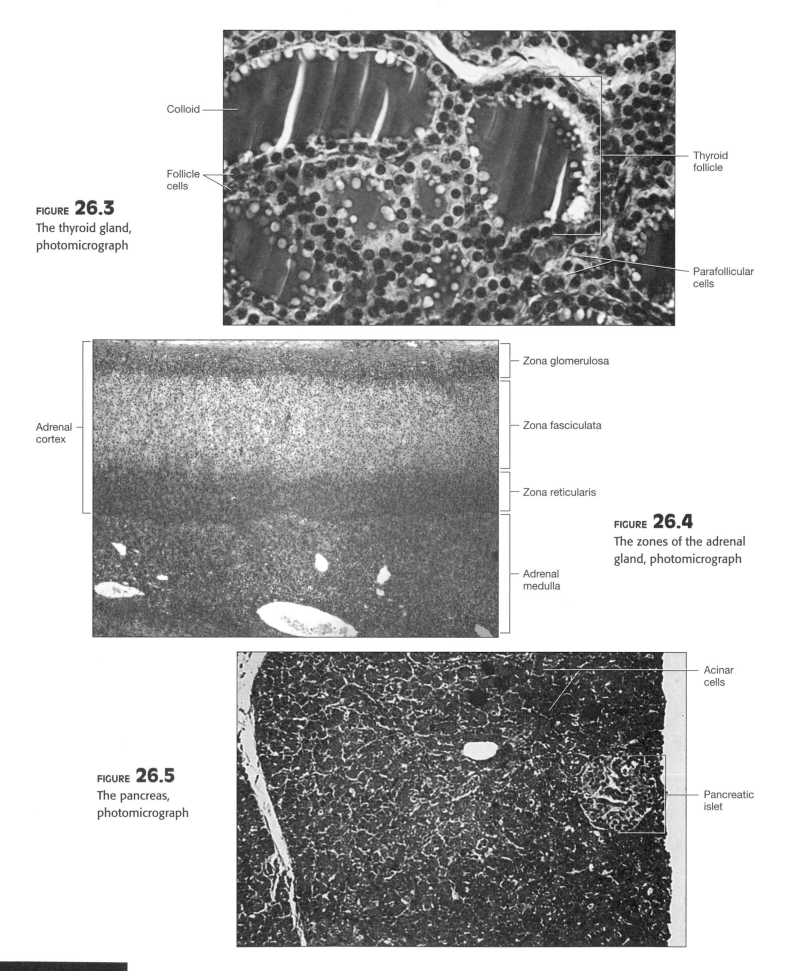

Colloid

Follicle cells

FIGURE 26.3
The thyroid gland, photomicrograph

Thyroid follicle

Parafollicular cells

Adrenal cortex

Zona glomerulosa

Zona fasciculata

Zona reticularis

FIGURE 26.4
The zones of the adrenal gland, photomicrograph

Adrenal medulla

Acinar cells

FIGURE 26.5
The pancreas, photomicrograph

Pancreatic islet

EXPLORING ANATOMY & PHYSIOLOGY IN THE LABORATORY

Procedure Microscopy

Obtain prepared slides of the thyroid gland, the adrenal gland, and the pancreas. Place each slide on the stage of the microscope and scan it on low power. Advance to higher power to see the cells and associated structures in greater detail. Use your colored pencils to draw what you see in the field of view and label your drawing with the terms indicated.

Thyroid

1. Follicle cells
2. Colloid
3. C (parafollicular) cells

Adrenal Gland

1. Zones of the adrenal cortex:
   a. Zona glomerulosa
   b. Zona fasciculata
   c. Zona reticularis
2. Adrenal medulla

Pancreas

1. Acinar cells (exocrine portion)
2. Pancreatic islets (endocrine portion)

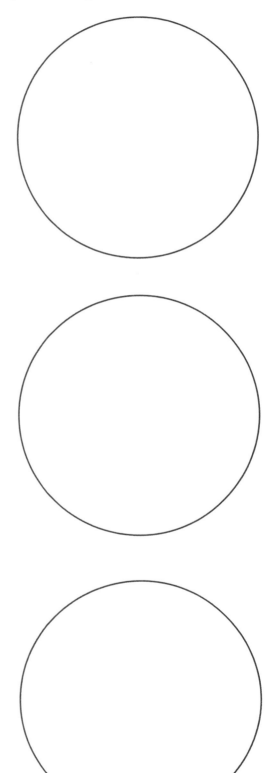

Exercise 3

Time to Trace: Negative Feedback Loops

Earlier in this unit we pointed out that each hormone has its own stimulus for secretion. The stimulus for secretion is generally a disturbance of homeostasis such as a change in body temperature, a change in the concentration of blood glucose or electrolytes, or a stressor. The hormone's response is to act on distant target cells to cause changes that restore homeostasis. When homeostasis is restored, the concentration of the hormone and activity of the glands declines. This type of response is called a **negative feedback loop.**

In this exercise you will be tracing a hormone's negative feedback loop from the initial homeostatic disturbance through the hormone's effects on its target cells to restore homeostasis. Following is an example:

Start: blood calcium rises $\longrightarrow$ calcitonin is released from C cells of the thyroid gland $\longrightarrow$ calcitonin levels in the blood rise $\longrightarrow$ calcitonin triggers osteoblast activity and calcium deposition into bone $\longrightarrow$ blood calcium decreases $\longrightarrow$ calcitonin levels in the blood decline $\longrightarrow$ **End**

Some of the negative feedback loops are slightly more complex as they involve multiple organs, such as the following:

Start: body temperature and/or metabolic rate drops $\longrightarrow$ thyrotropin-releasing hormone is released from the hypothalamus $\longrightarrow$ thyroid-stimulating hormone is released from the anterior pituitary $\longrightarrow$ T3 and T4 are produced and released from the thyroid gland $\longrightarrow$ T4 and T3 concentrations rise in the blood $\longrightarrow$ T3 increases the metabolic rate of cells and increases heat production $\longrightarrow$ body temperature and metabolic rate return to normal $\longrightarrow$ levels of thyrotropin-releasing hormone in the blood decrease $\longrightarrow$ levels of thyroid-stimulating hormone in the blood decrease $\longrightarrow$ levels of T3 and T4 in the blood decrease $\longrightarrow$ **End**

Now it's your turn! Note that sometimes the effects of the hormones cross over, so I've given you a hint at the start of each negative feedback loop, as well as the general number of steps in each loop. After you have completed the activity, answer Check Your Understanding questions 3 through 6 (pp. 603–604).

1. **Start:** the concentration of the blood increases (i.e., there is *inadequate water* in the blood) $\longrightarrow$ the hypothalamus

   releases _____ and stores it in the _____ $\longrightarrow$

   _____ $\longrightarrow$ _____ $\longrightarrow$

   _____ $\longrightarrow$ _____ **End**

2. **Start:** blood glucose increases $\longrightarrow$ the pancreas releases _____ $\longrightarrow$

   _____ $\longrightarrow$ _____ $\longrightarrow$

   _____ $\longrightarrow$ _____ **End**

3. **Start:** blood calcium decreases $\longrightarrow$ the parathyroid glands release _____ $\longrightarrow$

   _____ $\longrightarrow$ _____ $\longrightarrow$

   _____ $\longrightarrow$ _____ **End**

4. **Start:** blood potassium and hydrogen ion concentrations increase → the adrenal cortex produces

_____ → _____ →

_____ → _____ →

_____ **End**

5. **Start:** the body is exposed to a stressor → the hypothalamus releases _corticotropin-releasing hormone_ →

_____ → _____

→ _____ → _____

→ the body adapts to the stressor → _____

End

 ## Exercise 4

Endocrine "Mystery Cases"

In this exercise you will be playing the role of "endocrine detective" to solve endocrine disease mysteries with our favorite cast of characters. In each case you will have a victim who has suddenly fallen ill with a mysterious malady. You will be presented with a set of "witnesses," each of whom will give you a clue as to the nature of the illness. Other clues will come from samples that you send off to the lab for analysis. You will solve the mystery by providing the victim with a diagnosis. You may wish to use your textbook for assistance with these cases.

Case 1: The Cold Colonel

You are called upon to visit the ailing Col. Lemon. Before you see him, you speak with three witnesses who were with him when he fell ill.

Witness statements:

▶ _Ms. Magenta_: "Colonel Lemon has been hot-blooded for as long as I've known him. But I noticed that he couldn't seem to keep warm. He kept complaining about being cold. . . ."

▶ _Mr. Olive_: "Just between you and me, I've noticed that the old chap has put on quite a bit of weight lately."

▶ _Professor Purple_: "The Colonel and I used to go on major expeditions together. Now he just doesn't seem to have the energy to do much of anything."

What are your initial thoughts about the witnesses' statements? Does one hormone come to mind that may be the cause?

You see the Colonel and collect some blood to send off to the lab. The analysis comes back as follows:

T3 (triiodothyronine): 0.03 ng/dl (normal: 0.2–0.5 ng/dl)

T4 (thyroxine): 1.1 µg/dl (normal: 4–7 µg/dl)

TSH (thyroid-stimulating hormone): 86 mU/l (normal: 0.3–4.0 mU/l)

Analyze the results. Why do you think the T3 and T4 are low and the TSH is elevated?

Based upon the witness statements and the laboratory analysis, what is your final diagnosis?

Case 2: The Bloated Ms. Blanc

Your next call is to the home of Ms. Blanc. As before, you have three witnesses to interview.

Witness statements:

▶ *Ms. Magenta*: "Between you and me, she's really let herself go. She has fat deposits in all of these strange places, like around her face and trunk, and a weird hump on her back. If you ask me, some exercise may do her some good!"

▶ *Col. Lemon*: "Not that I want to talk about such things, but she's seemed bloated, a bit swollen, lately. She told me that even her blood pressure has gone up!"

▶ *Mrs. Feather*: "The poor dear. She's been sick so much lately with all kinds of infections. Personally, I think her immune system needs boosting. But my herbal teas don't seem to help."

What are your initial thoughts, based upon the witnesses' statements? Does one hormone come to mind that could produce these effects?

Your next step is to speak with Ms. Blanc. During the interview you notice a bottle of pills on her nightstand. You log the pills as evidence and send them off to the lab for analysis. The lab report shows the pills to be the medication *prednisone*, which you know to be a glucocorticoid similar to the hormone cortisol. Is this finding significant? Why or why not?

Based upon the witness statements and the laboratory analysis, what is your final diagnosis?

Case 3: The Parched Professor

Your last call is to the aid of Professor Purple. Three witnesses are present from whom to take statements.

Witness statements:

▶ *Mr. Olive*: "I swear that I saw him drink a full glass of water every half an hour today. He kept saying how thirsty he was!"

▶ *Ms. Blanc*: "He must be going to the . . . well, you know, the little boys' room, two or three times every hour!"

▶ *Mrs. Feather*: "He's been saying lately that his mouth is dry and that he feels weak. Personally, I think he's just not following a healthy diet! He should be drinking some of my herbal teas!"

Based upon the witnesses' statements, what are your initial thoughts? Does one hormone come to mind that could produce these effects?

You interview Professor Purple and collect blood and urine specimens to be sent off to the lab for analysis. The lab reports that the urine osmolality is 150 mOsm/kg, which means that the urine is overly dilute (too *much* water in the urine). The blood osmolality is 300 mOsm/kg, meaning that the blood is overly concentrated (too *little* water in the blood). The lab also reports that his blood glucose is completely *normal*. What is the significance of these clues?

Based upon the witness statements and the laboratory analysis, what is your final diagnosis? (*Hint:* Think of the hormone that is supposed to trigger water retention from the kidneys. Is there a disease where this hormone is deficient?)

NAME _____

SECTION _____ DATE _____

Check Your Recall

1 Label Figure 26.6 with the terms below.

- Hypothalamus
- Anterior pituitary
- Posterior pituitary

- Thyroid gland
- Parathyroid glands
- Adrenal cortex

- Adrenal medulla
- Pancreas
- Pancreatic islet

- Ovary
- Testis

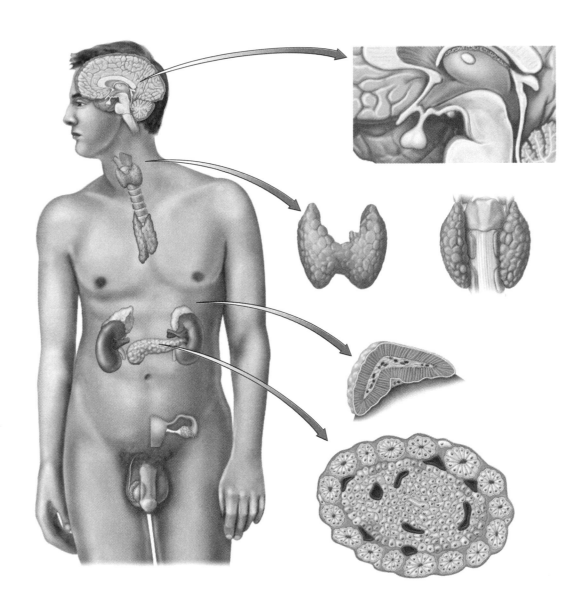

FIGURE **26.6** The endocrine system

2 Fill in the blanks: The nervous system works through secretion of _____, whereas the endocrine system works via secretion of _____.

3 Which of the following is *not* a function of the hypothalamus?
  a. Produces antidiuretic hormone and oxytocin
  b. Stimulates production and release of hormones from the anterior pituitary
  c. Stimulates production and release of hormones from the posterior pituitary
  d. Inhibits the production and release of hormones from the anterior pituitary

4 Which of the following endocrine organs are part of the diencephalon?
  a. Hypothalamus
  b. Pineal gland
  c. Thymus
  d. Both a and b are correct.
  e. Both b and c are correct.

5 Which of the following sets of hormones are antagonists?
  a. T3 and T4
  b. Calcitonin and parathyroid hormone
  c. Epinephrine and cortisol
  d. Growth hormone and thyroxine

6 Matching: Match the following endocrine organs with the hormone(s) that each secretes.

  a. Pineal gland   _____   A. Parathyroid hormone

  b. Thyroid gland   _____   B. Insulin and glucagon

  c. Pancreas   _____   C. Steroid hormones

  d. Thymus   _____   D. Epinephrine and norepinephrine

  e. Parathyroid glands   _____   E. Thyroxine, triiodothyronine, and calcitonin

  f. Adrenal cortex   _____   F. Thyroid-stimulating hormone, growth hormone

  g. Anterior pituitary   _____   G. Melatonin

  h. Adrenal medulla   _____   H. Thymosin and thymopoietin

7 Which of the following is *not* true regarding endocrine organ histology?
  a. The thyroid gland consists of rings of simple cuboidal follicle cells surrounding colloid.
  b. The pancreas has an exocrine portion consisting of pancreatic islets and an endocrine portion consisting of acinar cells.
  c. The adrenal cortex has three zones of cells that secrete three different types of hormones.
  d. The adrenal medulla is modified nervous tissue of the sympathetic nervous system.

8 What is a negative feedback loop? Cite an example of a negative feedback loop in the endocrine system.

9 List the stimulus for secretion for each of the following hormones (note that some hormones are stimulated by other hormones):

a. Glucagon _____

b. Calcitonin _____

c. Parathyroid hormone _____

d. Thyroxine and triiodothyronine (T4 and T3) _____

e. Cortisol _____

f. Antidiuretic hormone _____

10 List the target tissue and effects of each of the following hormones:

a. Insulin:

b. Adrenocorticotropic hormone (ACTH):

c. Growth hormone:

d. Prolactin:

e. Oxytocin:

f. Thymosin:

NAME _____

SECTION _____ DATE _____

 Check Your Understanding | **Critical Thinking and Application Questions**

1 Endocrine glands are often called "ductless glands" because their products are secreted directly into the bloodstream. How is this an example of structure following function?

2 How are the endocrine system and nervous system similar? How do they differ?

3 Tumors of the parathyroid gland often result in secretion of excess parathyroid hormone. Considering the function of this hormone, predict the effects of such a tumor.

4 The hormone calcitonin is prescribed to treat the disease osteoporosis. Explain why this hormone would help to reduce bone loss in patients affected with this disease.

5 Drugs known as *potassium-sparing diuretics* work by blocking the effects of aldosterone on the kidney. What effect would these drugs have on the amount of urine produced, on the blood pressure, and on potassium concentrations in the blood? Could these drugs have a negative impact on the acid–base balance of the body? Explain.

6 The disease *diabetes mellitus, type I* is characterized by destruction of the cells that produce insulin in the pancreatic islets. Predict the symptoms and effects of this disease.

Reproductive System

OBJECTIVES

Once you have completed this unit, you should be able to:

- Identify structures of the male and female reproductive system.

- Identify the stages of meiosis and describe how meiosis differs from mitosis.

- Describe the histology of the ovary, seminiferous tubules, and spermatozoa.

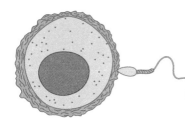

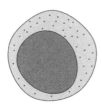

Pre-Lab Exercise 1
Key Terms

Table 27.1 lists the key terms with which you should be familiar before coming to lab.

TABLE **27.1**	Key Terms

Term | **Definition**

Structures of the Male Reproductive System

Testes _____
_____

Seminiferous tubules _____
_____

Epididymis _____
_____

Vas deferens _____
_____

Spermatic cord _____
_____

Seminal vesicle _____
_____

Prostate gland _____
_____

Bulbourethral gland _____
_____

Penis _____
_____

Corpus spongiosum _____
_____

Corpora cavernosa _____
_____

Structures of the Female Reproductive System

Ovaries _____

_____

Ovarian follicles _____

_____

Uterine tube _____

_____

Uterus _____

_____

Endometrium _____

_____

Myometrium _____

_____

Perimetrium _____

_____

Cervix _____

_____

Vagina _____

_____

Vulva _____

_____

Mammary glands _____

_____

Gametogenesis Terms

Meiosis I _____

_____

Meiosis II _____

_____

Haploid _____

_____

Spermatogenesis _____

_____

Oogenesis _____

_____

Polar body _____

_____

 Pre-Lab Exercise 2

Male Reproductive Anatomy

Label and color the structures of the male reproductive system depicted in **Figure 27.1** with the terms from Exercise 1. Use your text and Exercise 1 in this unit for reference.

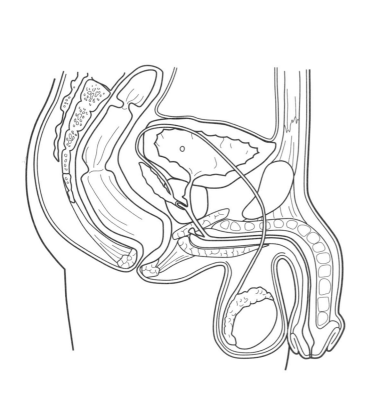

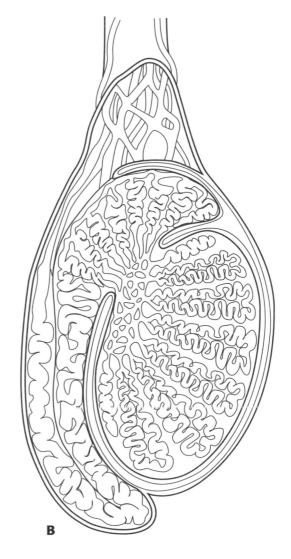

A

B

FIGURE 27.1 Male reproductive organs: (A) midsagittal section of the male pelvis; (B) midsagittal section through the testis

Pre-Lab Exercise 3

Female Reproductive Anatomy

Label and color the structures of the female reproductive system depicted in **Figure** 27.2 with the terms from Exercise 2. Use your text and Exercise 2 in this unit for reference.

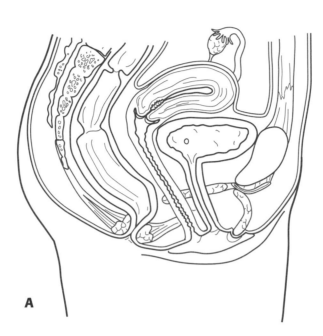

A

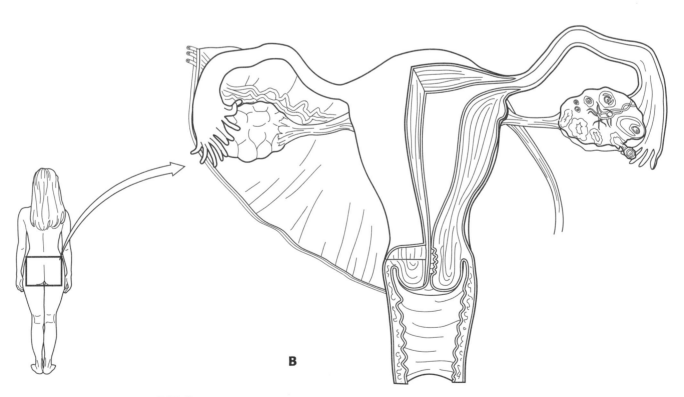

B

FIGURE **27.2** Female reproductive tract: (A) midsagittal section of the female pelvis; (B) posterior view of the female reproductive organs

Pre-Lab Exercise 4
Stages of Mitosis

In this unit we will discuss the process of meiosis. Meiosis proceeds in a fashion similar to mitosis, so it's a good idea to review the discussion of mitosis before you begin the lab. Fill in Table 27.2 with the events that are occurring in the cell in each of the stages of mitosis.

TABLE 27.2	Stages of Mitosis
Stage of Mitosis	**Events Occurring in the Cell**
Prophase	
Metaphase	
Anaphase	
Telophase	

Exercises

The other organ systems in the human body that we have discussed all function in some manner to help maintain homeostasis of the body. The reproductive system, however, plays little role in maintaining homeostasis and instead functions to perpetuate the species. The main organs of the reproductive system are the **gonads**—the testes and the ovaries—which produce **gametes,** or sex cells, for reproduction.

We begin this unit with the anatomy of the male (Exercise 1) and female (Exercise 2) reproductive systems. Then we turn to the main functions of these organs: **gametogenesis,** the formation of new gametes.

Exercise 1

Male Reproductive Anatomy

MATERIALS NEEDED

▸ Male reproductive models
▸ Human torso models

The **testes,** the gamete-producing organs of the male, are situated outside the body in a sac of skin and connective tissue called the **scrotum** (Figure 27.3). They are located externally because sperm production requires a temperature lower than body temperature of about 34°C (about 94°F). The testes are surrounded by an outer connective tissue sheath called the **tunica vaginalis** and an inner sheath called the **tunica albuginea,** which divides the interior of the testes into **lobules.** Each lobule contains a tightly coiled **seminiferous tubule** where **spermatogenesis** takes place.

The seminiferous tubules converge near the superior part of the testis to form a structure called the **rete testis.** The rete testis exits the testis to join the first segment of the duct system of the male reproductive tract, the **epididymis** (Figures 27.3 and 27.4). Immature sperm produced by the seminiferous tubules migrate to the epididymis to finish their maturation,

EXPLORING ANATOMY & PHYSIOLOGY IN THE LABORATORY

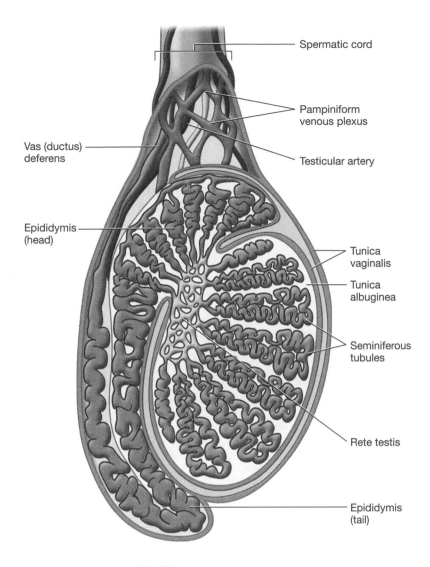

FIGURE 27.3 Midsagittal section through the testis

- Spermatic cord
- Pampiniform venous plexus
- Vas (ductus) deferens
- Testicular artery
- Epididymis (head)
- Tunica vaginalis
- Tunica albuginea
- Seminiferous tubules
- Rete testis
- Epididymis (tail)

after which they exit via a long tube called the **vas (or ductus) deferens**. The vas deferens travels superiorly through the **spermatic cord**, a structure that also carries the testicular artery, testicular veins (the pampiniform plexus), and nerves.

Note in **Figure 27.4** that once the vas deferens enters the pelvic cavity, it crosses superiorly and posteriorly over the urinary bladder to join a gland called the **seminal vesicle**. At this point, the vas deferens merges with the duct from the seminal vesicle to form the **ejaculatory duct**. This duct passes through the **prostate gland**, where it joins with the **prostatic urethra**. The prostatic urethra becomes the **membranous urethra** as it exits the prostate, and then becomes the **spongy urethra** as it enters the corpus spongiosum of the penis.

The male reproductive tract consists of three exocrine glands: the prostate gland, seminal vesicles, and **bulbourethral glands**. The seminal vesicles and the prostate gland together produce about 90% of the volume of **semen**, a fluid that contains chemicals to nourish and activate sperm. The smaller bulbourethral glands produce an alkaline secretion that is released prior to the release of sperm during ejaculation.

The **penis** is composed of three erectile bodies: the single **corpus spongiosum** and the paired, dorsal **corpora cavernosa**. The corpus spongiosum, which surrounds the spongy urethra, enlarges distally to form the **glans penis**. All three bodies consist of vascular spaces that fill with blood during an erection.

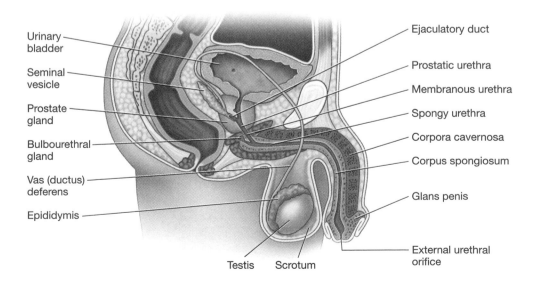

- Urinary bladder
- Seminal vesicle
- Prostate gland
- Bulbourethral gland
- Vas (ductus) deferens
- Epididymis
- Ejaculatory duct
- Prostatic urethra
- Membranous urethra
- Spongy urethra
- Corpora cavernosa
- Corpus spongiosum
- Glans penis
- External urethral orifice
- Testis
- Scrotum

FIGURE 27.4 Midsagittal section through the male pelvis

Procedure Model Inventory

Identify the following structures of the male reproductive system on models and diagrams using your textbook and this unit for reference. As you examine the anatomical models and diagrams, record on the model inventory in Table 27.3 the name of the model and the structures you were able to identify. When you have completed the activity, answer Check Your Understanding questions 1 through 3 (p. 625).

1. Scrotum
2. Testes
   a. Tunica albuginea
   b. Tunica vaginalis
   c. Seminiferous tubules
   d. Rete testis
3. Epididymis
4. Spermatic cord
   a. Testicular arteries
   b. Pampiniform plexus

5. Cremaster muscle
6. Vas deferens
7. Ejaculatory duct
8. Urethra
   a. Prostatic urethra
   b. Membranous urethra
   c. Spongy urethra
   d. External urethral orifice

9. Glands:
   a. Seminal vesicle
   b. Prostate gland
   c. Bulbourethral gland
10. Penis
    a. Corpora cavernosa
    b. Corpus spongiosum
       (1) Glans penis

TABLE 27.3 Model Inventory for the Male Reproductive System	
Model/Diagram	Structures Identified

Exercise 2

Female Reproductive Anatomy

MATERIALS NEEDED

- Female reproductive models
- Human torso models

The female reproductive organs lie in the pelvic cavity, with the exception of the almond-shaped **ovaries**, which are located in the peritoneal cavity (**Figures 27.5** and **27.6**). The ovaries, the female gonads, produce gametes (**oocytes,** or eggs) that travel through the reproductive tract to be fertilized. The ovaries are held in place by several ligaments, including the **ovarian ligament,** a part of the **broad ligament** called the **mesovarium,** and the **suspensory ligament**. The suspensory ligaments extend from the lateral body wall and carry with them the ovaries' blood supply.

EXPLORING ANATOMY & PHYSIOLOGY IN THE LABORATORY

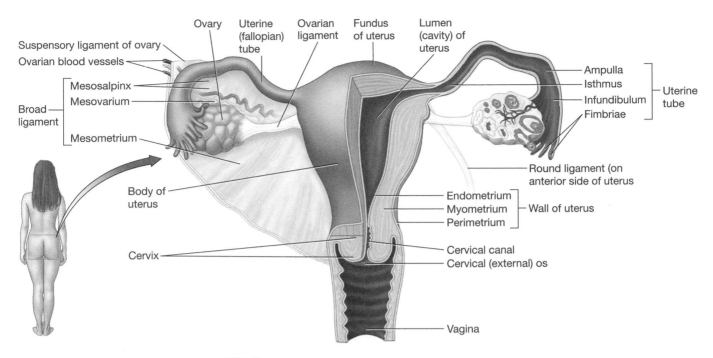

FIGURE **27.5** Posterior view of the female reproductive organs

Developing oocytes within the ovaries are encased in structures called **follicles** (**Figure 27.7**). Follicles are present in various stages, ranging from the immature **primordial follicles** to the mature **vesicular** (or **Graafian**) **follicle** from which an oocyte is released during ovulation. Notice in **Figure 27.5** that, unlike the male reproductive tract, the tubule system of the female reproductive tract is not continuous. Therefore, the oocyte actually is released into the pelvic cavity and the fingerlike extensions of the **uterine** (**fallopian**) **tube,** called **fimbriae,** must "catch" the oocyte and bring it into the uterine tube.

The uterine tubes join the superolateral portion of the **uterus.** The uterus is situated between the urinary bladder and the rectum. It is held in place by the broad ligament and an anterior ligament called the **round ligament.** The uterus consists of three portions: the dome-shaped

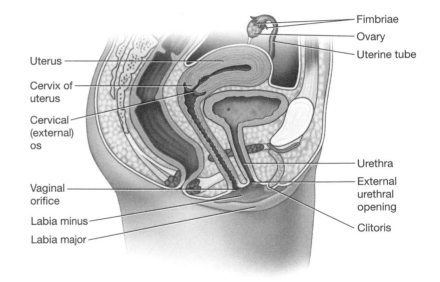

FIGURE **27.6** Midsagittal section of the female pelvis

fundus, the central **body,** and the narrow **cervix.** The wall of the uterus has three layers: the inner epithelial and connective tissue lining called the **endometrium,** in which a fertilized ovum implants; the middle, muscular **myometrium;** and the outermost connective tissue lining, the **perimetrium** (which is actually just the visceral peritoneum).

The inferiormost portion of the cervix is the **cervical** (**external**) **os.** The **vagina** extends inferiorly from the cervical os and terminates at the vaginal orifice. Flanking the vaginal orifice are the **greater vestibular** (Bartholin's) **glands,** which secrete mucus to lubricate the vaginal canal during coitus.

The external anatomy of the female is collectively called the **vulva,** and consists of the **labia majora** and **labia minora,** the **clitoris,** and the urethral and vaginal orifices. The labia are paired skinfolds, analogous to the scrotum in the male. They enclose a region called the **vestibule,** which houses the clitoris and the urethral and vaginal orifices.

The **mammary glands** are not true reproductive organs but do have an associated reproductive function in milk production (**Figure 27.8**). Mammary glands are present in both males and females (males can produce milk too), but their

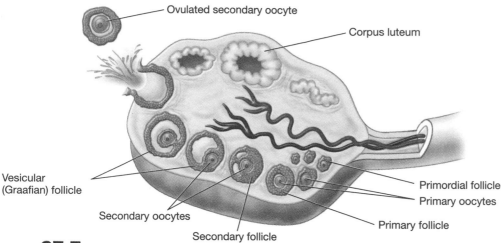

FIGURE **27.7** The ovary

anatomy is most appropriately discussed with female anatomy. Internally, mammary glands consist of 15–25 **lobes,** each of which has smaller **lobules** that contain milk-producing **alveoli.** Milk leaves the alveoli through **alveolar ducts,** which join to form storage areas called **lactiferous sinuses.** Milk leaves through the **nipple,** which is surrounded by a darkly pigmented area called the **areola.**

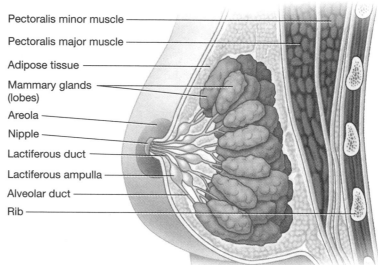

FIGURE **27.8** The mammary gland

 Procedure **Model Inventory**

Identify the following structures of the female reproductive system on models and diagrams, using your textbook and this unit for reference. As you examine the anatomical models and diagrams, record on the model inventory in Table 27.4 the name of the model and the structures you were able to identify. After you have completed the activity, answer Check Your Understanding questions 4 and 5 (p. 626).

1. Ovary
2. Ovarian follicles
   a. Primordial follicle
   b. Primary follicle
   c. Secondary follicle
   d. Graafian (vesicular) follicle
3. Uterine (fallopian) tube
   a. Fimbriae
4. Uterus
   a. Fundus
   b. Body
   c. Cervix
   d. Layers
      (1) Endometrium
      (2) Myometrium
      (3) Perimetrium
5. Vagina
6. Vulva
   a. Vestibule
   b. Labia majora
   c. Labia minora
   d. Clitoris
   e. Urethral orifice
7. Ligaments
   a. Broad ligament
      (1) Mesometrium
      (2) Mesovarium
      (3) Mesosalpinx
   b. Round ligament
   c. Suspensory ligament of the ovary
   d. Ovarian ligament
8. Mammary glands
   a. Areola
   b. Nipple
   c. Lobe
   d. Alveolar duct
   e. Lactiferous sinus
   f. Adipose tissue

TABLE **27.4**	Model Inventory for the Female Reproductive System
Model/Diagram	Structures Identified

 Exercise 3

Meiosis

MATERIALS NEEDED

- Meiosis models
- Mitosis models
- 4 colors of pipe cleaners

As you may recall from Unit 4, somatic cells divide by a process called mitosis. During mitosis, a cell replicates its DNA and divides its DNA and organelles into two identical daughter cells (see Pre-Lab Exercise 4 for a review of the stages). Each new **diploid** cell is identical to the original cell. **Gametogenesis** (oogenesis and spermatogenesis), however, must proceed in a different way for two reasons:

1. If each gamete were to have the same genetic material, we all would be genetically identical to our siblings.

2. If each gamete were to have two sets of chromosomes, our offspring would have four sets of chromosomes and we would be tetraploid.

For this reason, gametes undergo a process known as **meiosis** rather than mitosis.

Meiosis is also known as **reduction division** because gametes proceed through two rounds of cell division and each gamete ends up with only one set of chromosomes (a **haploid** cell). As you can see in **Figure 27.9**, meiosis begins in a manner similar to mitosis: The chromosomes replicate, and for a brief period the gametes have twice the normal genetic material (92 chromosomes). The cell then goes through **meiosis I**, which is similar to mitosis except for one important difference that occurs during prophase I: Replicated chromosomes line up next to one another so closely that they overlap in several places—a phenomenon called **synapsis**.

Each synapse has four chromosomes, and therefore the entire structure is called a **tetrad**. As the tetrads align themselves on the equator of the cell and prepare to attach to spindle fibers during metaphase I, the areas of the chromosomes that overlap form points called **chiasmata**, or **crossover**. As anaphase I begins, the chromosomes exchange pieces of genetic material at the points of crossover.

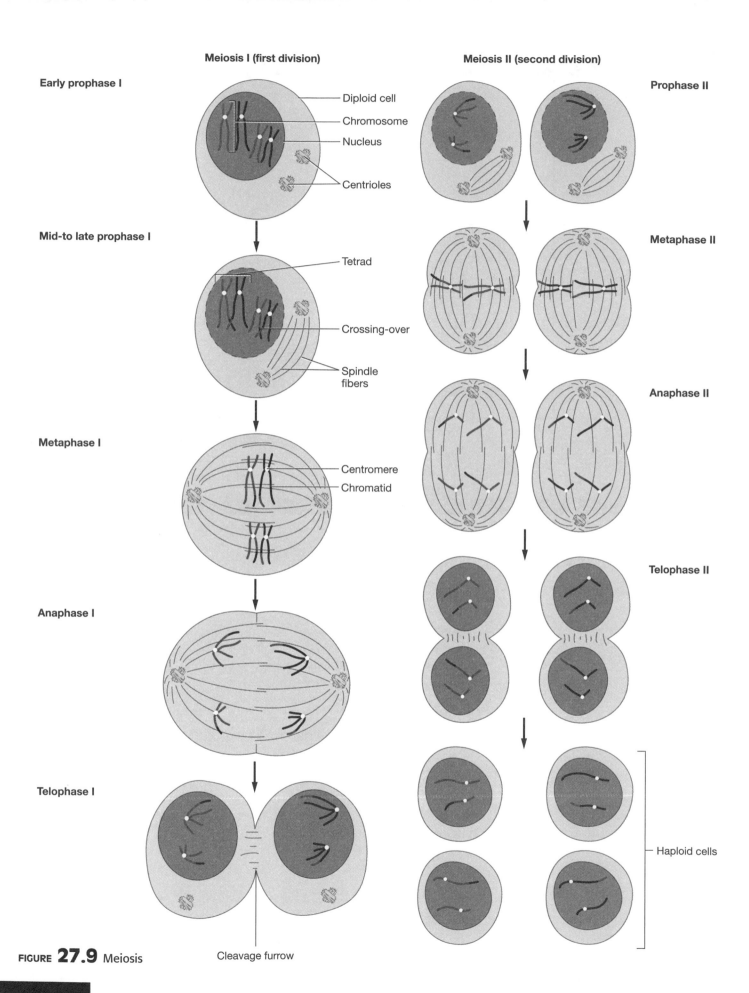

Meiosis I (first division)

Meiosis II (second division)

Early prophase I

Diploid cell
Chromosome
Nucleus

Centrioles

Prophase II

Mid-to late prophase I

Tetrad

Crossing-over

Spindle fibers

Metaphase II

Metaphase I

Centromere
Chromatid

Anaphase II

Anaphase I

Telophase II

Telophase I

Haploid cells

Cleavage furrow

FIGURE **27.9** Meiosis

616

Once telophase I and cytokinesis are complete, the gametes have 46 chromosomes. They then go through the second round of division, **meiosis II**. Meiosis II is similar to meiosis I with one major exception: The DNA does *not* replicate prior to the start of prophase II. At the end of telophase II, each gamete has only 23 chromosomes and is a haploid cell.

Procedure Model Meiosis

In the following procedure you will model meiosis with either a set of cell meiosis models or with pipe cleaners. When you have completed the activity, answer Check Your Understanding question 6 (p. 626).

1 Obtain a set of meiosis and mitosis models.

2 Arrange the mitosis models in the correct order.

3 Now arrange the meiosis models in the correct order.

4 Compare the models of the two processes. List all differences you can see between meiosis and mitosis below.

As an alternative procedure, make model chromosomes with pipe cleaners. Use yellow and green pipe cleaners for the chromosomes undergoing meiosis, and red and blue pipe cleaners for the chromosomes undergoing mitosis. Show how the chromosomes duplicate and divide in the two processes, and show how the chromosomes undergoing meiosis form points of crossover.

Exercise 4

Spermatogenesis and Oogenesis

MATERIALS NEEDED

- Testis slide
- Sperm slide
- Ovary slide
- Light microscope
- Colored pencils

Gametogenesis is really two separate processes: **spermatogenesis**, which takes place in the seminiferous tubules of the testes, and **oogenesis**, which takes place in the ovary and uterine tube. Spermatogenesis, shown in **Figure 27.10**, begins with stem cells located at the outer edge of the seminiferous tubules called **spermatogonia**. Before puberty, these diploid cells undergo repeated rounds of mitosis to increase their numbers. As puberty begins, each spermatogonium divides into two different cells—one cell that stays a spermatogonium and another cell that becomes a **primary spermatocyte**.

The primary spermatocyte then undergoes meiosis I and divides into two **secondary spermatocytes**, which migrate closer to the lumen of the tubule. The two secondary spermatocytes undergo meiosis II and give rise to four haploid **spermatids**. The spermatids then move to the epididymis to mature into functional gametes. Mature sperm contain three parts: the **head**, in which the DNA resides, the **midpiece**, which contains an axomere and mitochondria, and the **flagellum**.

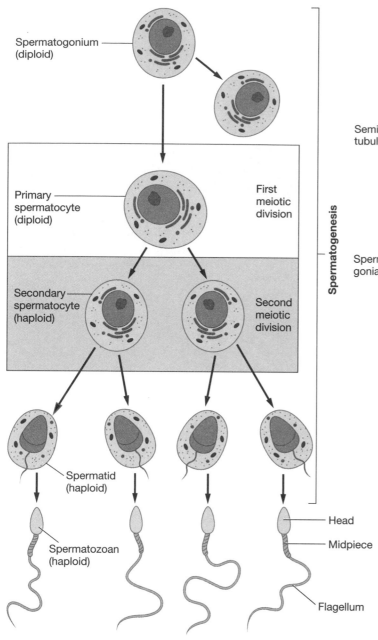

FIGURE **27.10** Spermatogenesis

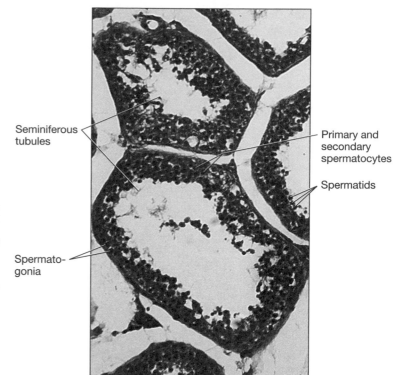

FIGURE **27.11** Seminiferous tubules, photomicrograph

Note in **Figure 27.11** that most of the stages of spermatogenesis are identifiable on a microscope slide of the seminiferous tubules. On the inner edge of the tubule near the lumen are small, round cells with little cytoplasm. These are the spermatids. Note that the spermatids lack the characteristic organelle seen in mature sperm—the flagellum. As we move to deeper layers of the tubule wall, we can see primary and secondary spermatocytes, and on the outer edge we can see the cuboidal spermatogonia. In between the tubules are small clusters of cells called **interstitial** (or **Leydig**) **cells**. These cells make testosterone, which is required for spermatogenesis to take place.

Like spermatogenesis, oogenesis proceeds through meiosis to yield a haploid gamete, the **ovum** (**Figure 27.12**). But the two processes differ in some notable ways:

▶ *The number of oocytes is determined before birth.* During the fetal period, stem cells called **oogonia** undergo mitosis, increasing their numbers to about 500,000 to 700,000. This is the total number of oocytes that a woman will ever produce. This is in sharp contrast to spermatogenesis, which begins at puberty and continues throughout a male's lifetime.

▶ *Meiosis I begins during the fetal period but is arrested.* Still during the fetal period, the oogonia become encased in a **primordial follicle**, enlarge, and become **primary oocytes**. The primary oocytes begin prophase I but are arrested at this stage. Meiosis I does not resume until puberty.

▶ *The first meiotic division results in one secondary oocyte and one polar body.* At puberty, one primary oocyte enlarges and becomes encased in a **primary follicle**. This primary oocyte then completes meiosis I to produce a **secondary oocyte** and a small bundle of nuclear material called a **polar body**. The formation of a polar body allows the oocyte to

EXPLORING ANATOMY & PHYSIOLOGY IN THE LABORATORY

conserve cytoplasm, which will have to sustain the cell if fertilization occurs. The secondary oocyte is initially encased in a **secondary follicle**, which enlarges to form a **vesicular** or **Graafian follicle**.

▶ *Meiosis II completes only if fertilization takes place.* The secondary oocyte begins meiosis II and is released when the vesicular follicle ruptures during ovulation. Note that the ruptured follicle then becomes an endocrine organ called a **corpus luteum**. But the secondary oocyte only completes meiosis II to form an ovum and a second polar body if fertilization occurs. If fertilization does not occur, the secondary oocyte degenerates without ever completing meiosis II.

The ovary offers a nice opportunity to view oogenesis and meiosis in action (**Figure 27.13**). Though you typically cannot see the chromosomes of the oocytes, you can determine the stage of the oocyte by looking at the surrounding follicle: Primary oocytes are encased in primordial and primary follicles, and secondary oocytes in secondary and vesicular follicles. Note that we cannot see oogonia in the ovary because these cells begin meiosis I and become primary oocytes during the fetal period.

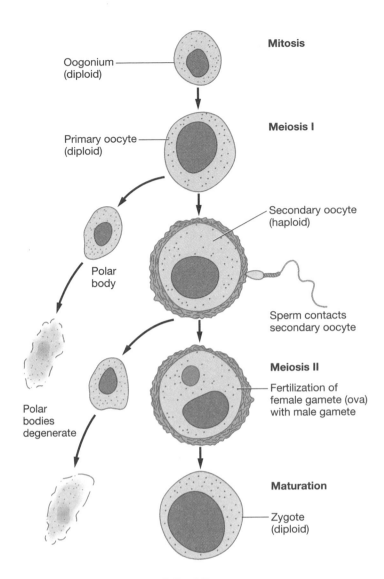

FIGURE **27.12** Oogenesis

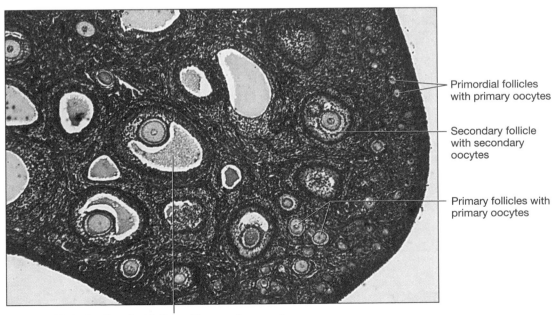

FIGURE **27.13**
Ovary, photomicrograph

Vesicular (Graafian) follicle with secondary oocyte

Procedure Microscopy of Male Reproductive Structures

In this exercise you will examine prepared slides of the testes and a sperm smear. You will want to examine both slides on high power, and you may wish to examine the sperm smear with an oil immersion lens. Because sperm cells are so small, the slide will likely have a thread or some other marker to help you find them. As you examine the slides of the seminiferous tubules and the sperm smear, use colored pencils to draw what you see, and label your drawings with the terms indicated.

Testes

1. Seminiferous tubule
2. Spermatogonia
3. Primary spermatocytes
4. Spermatids
5. Interstitial cells

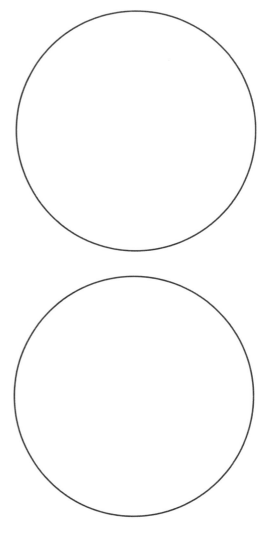

Sperm

1. Head
2. Midpiece
3. Flagellum

Procedure Microscopy of the Ovary

Obtain a prepared slide of an ovary. Use your colored pencils to draw what you see and label your drawing with the following terms. Please keep in mind that you may have to examine multiple slides to see all the follicular stages.

1. Primordial follicle
2. Primary follicle
   a. Primary oocyte
3. Secondary follicle
4. Graafian (vesicular) follicle
   a. Secondary oocyte
   b. Antrum

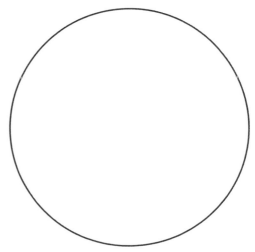

EXPLORING ANATOMY & PHYSIOLOGY IN THE LABORATORY

NAME _____

SECTION _____ DATE _____

 Check Your Recall

1 Label the following structures on **Figure 27.14**.

- Seminiferous tubules
- Tunica albuginea
- Tunica vaginalis
- Rete testis
- Epididymis
- Spermatic cord
- Vas deferens

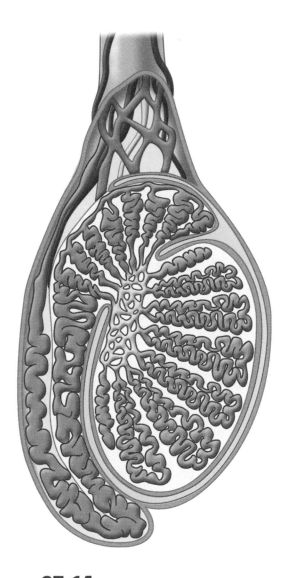

FIGURE **27.14** Midsagittal section through the testis

2 Label the following structures on **Figure 27.15.**

- Epididymis
- Testis
- Scrotum
- Vas deferens
- Prostate gland
- Seminal vesicle
- Ejaculatory duct
- Bulbourethral gland
- Corpora cavernosa
- Corpus spongiosum
- Glans penis

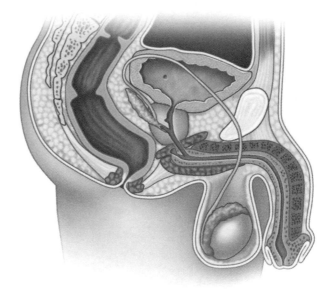

FIGURE **27.15** Midsagittal section through the male pelvis

3 Label the following structures on **Figure 27.16.**

- Uterus
- Cervix
- Cervical os
- Vagina
- Uterine tube
- Ovary
- Clitoris
- Labia major
- Labia minus

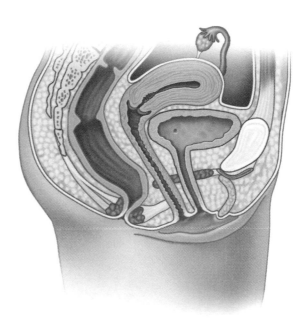

FIGURE **27.16** Midsagittal section of the female pelvis

EXPLORING ANATOMY & PHYSIOLOGY IN THE LABORATORY

4 Label the following structures on **Figure 27.17**.

- Fundus of uterus
- Body of uterus
- Cervix
- Endometrium
- Myometrium

- Perimetrium
- Uterine tube
- Fimbriae
- Broad ligament

- Ovary
- Ovarian follicle
- Round ligament
- Suspensory ligament

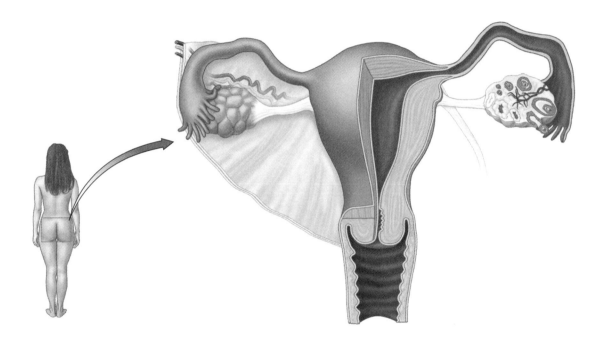

FIGURE **27.17** Posterior view of the female reproductive organs

5 Which of the following statements about meiosis is false?
a. Gametes proceed through two rounds of cell division.
b. Crossover occurs during meiosis, which increases genetic diversity.
c. The DNA replicates before both meiosis I and meiosis II.
d. Meiosis is carried out only in the gonads to produce gametes.

6 Fill in the blanks: The daughter cells produced by mitosis are _____ with _____

set(s) of chromosomes, whereas the gametes produced by meiosis are _____ with _____

set(s) of chromosomes.

7 Number the following events of spermatogenesis in the proper order, with number 1 next to the first event and number 4 next to the final event.

_____ Primary spermatocytes undergo meiosis I to produce two secondary spermatocytes.

_____ The two secondary spermatocytes undergo meiosis II to produce four haploid spermatids.

_____ Spermatogonia divide into more spermatogonia and primary spermatocytes.

_____ Spermatogonia undergo repeated rounds of mitosis.

8 Which of the following statements about spermatogenesis and oogenesis is *false*?
a. Meiosis II does not complete in oogenesis unless fertilization takes place.
b. Meiosis I begins during the fetal period but is arrested in oogenesis.
c. Spermatogenesis begins at puberty and continues throughout the male's lifetime, whereas the total number of oocytes a woman will produce is determined before birth.
d. Spermatogenesis results in one spermatid and two polar bodies, whereas oogenesis results in four ova.

9 Spermatids migrate to the _____ to mature.
a. vas deferens
b. epididymis
c. seminal vesicle
d. prostate gland

10 Label the following structures on **Figure 27.18**.
- Corpus luteum
- Primary follicle
- Secondary follicle
- Vesicular follicle
- Primary oocyte
- Secondary oocyte

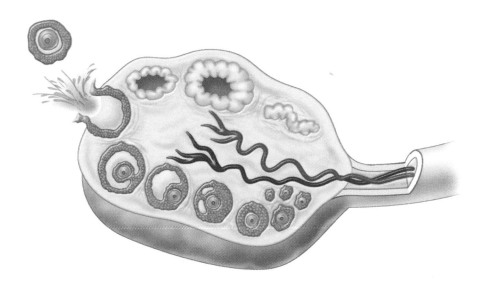

FIGURE 27.18 Ovary

EXPLORING ANATOMY & PHYSIOLOGY IN THE LABORATORY

NAME _____

SECTION _____ DATE _____

1 A condition called *testicular torsion* results when the spermatic cord becomes twisted. Why would this condition be a surgical emergency?

2 What structure is sectioned in a vasectomy? What are the effects of this procedure? Will it impact the number of sperm or the volume of semen produced?

3 The condition *benign prostatic hypertrophy*, in which the prostate is enlarged, often results in urinary retention—the inability to completely empty the bladder. Considering the anatomy of the male genitourinary tract, explain this symptom.

4 A tubal (or ectopic) pregnancy results from implantation of a fertilized ovum in the uterine tube instead of the uterus. Why is this dangerous?

5 One of the most common complaints of pregnant women is the need to urinate often. Explain this, considering the arrangement of the pelvic organs.

6 What happens if one of the chromosomes doesn't separate fully during meiosis II? Give an example of a common disorder caused by this phenomenon. What are the symptoms of this disorder?

Human Development and Heredity

OBJECTIVES

Once you have completed this unit, you should be able to:

- Identify and describe structures and membranes associated with the fetus.

- Identify and describe the stages of development and structures of the fetal circulation.

- Trace a developing zygote (or conceptus) from fertilization to implantation.

- Define the terms phenotype and genotype, and perform analyses to determine a subject's phenotype and genotype.

- Describe dominant, recessive, and sex-linked traits.

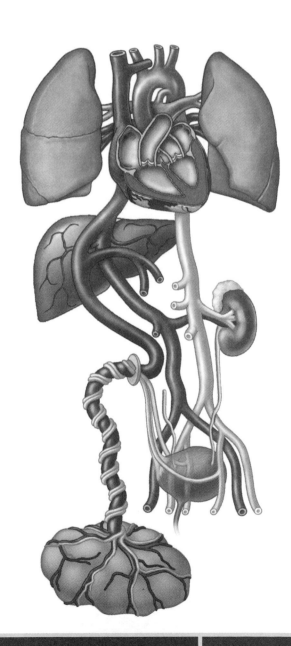

Pre-Lab Exercises | Complete the following exercises prior to coming to lab, using your textbook and lab manual for reference.

Pre-Lab Exercise 1
Key Terms

Table 28.1 lists the key terms with which you should be familiar before coming to lab.

TABLE 28.1	Key Terms

Term	Definition
Development	
Fertilization	
Zygote	
Morula	
Blastocyst	
Implanting blastocyst	
Embryo	
Fetus	
Fetal structures	
Chorion	

Chorionic villi _____

_____

Amnion _____

_____

Placenta _____

_____

Inheritance

Heredity _____

_____

Allele _____

_____

Phenotype _____

_____

Genotype _____

_____

Autosomal dominant _____

_____

Autosomal recessive _____

_____

Sex-liked trait _____

_____

Pre-Lab Exercise 2
Fetal Membranes

Label and color the fetal membranes and structures depicted in **Figure 28.1** with the terms from Exercise 2. Use your text and Exercise 2 in this unit for reference.

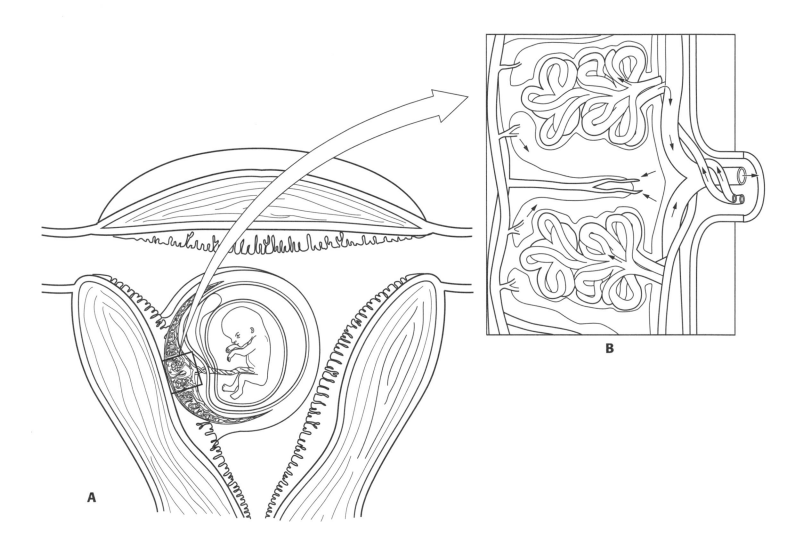

FIGURE 28.1 Embryo and membranes (A) embryo in uterus; (B) chorionic villi

Fetal Cardiovascular Anatomy

Label and color the structures of the fetal cardiovascular system depicted in **Figure 28.2** with the terms from Exercise 2. Use your text and Exercise 2 in this unit for reference.

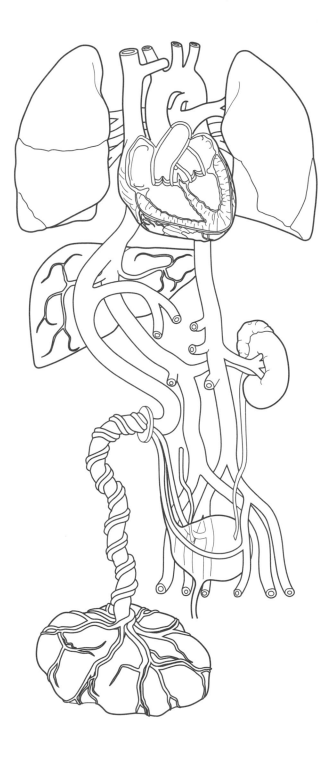

FIGURE 28.2 Fetal cardiovascular anatomy

Exercises

When two gametes unite in a process known as **fertilization,** a diploid **zygote** forms and the incredible process of development begins. In this unit we examine human development and the unique anatomy and structures of the developing human. We conclude by exploring the basics of **heredity**—the passing of traits from parents to offspring.

 ## Exercise 1

Fertilization and Implantation

The first event in human development is **fertilization,** which takes place in the uterine tube when sperm cells encounter a secondary oocyte. Fertilization begins with a process called the **acrosomal reaction,** in which the acrosomes in the heads of the sperm release digestive enzymes by exocytosis. The acrosomal enzymes digest the corona radiata and the zona pellucida, which are the secondary oocyte's outer coverings.

When the zona pellucida has been cleared, the head and midpiece of a single sperm enter the oocyte. As the sperm enters, the secondary oocyte completes meiosis II, yielding an ovum and a second polar body. The sperm and ovum nuclei swell to become **pronuclei,** and their nuclear membranes rupture. The two sets of exposed chromosomes then fuse to form a single-celled structure called a **zygote** (**Figure 28.3**).

The zygote, also called a **conceptus,** undergoes multiple changes during the period of time known as **gestation.** Gestation is considered to be a 40-week period extending from the mother's last menstrual period to birth. The changes that the conceptus undergoes can be divided into three processes:

1. an increase in cell number,
2. cellular differentiation, and
3. the development of organ systems.

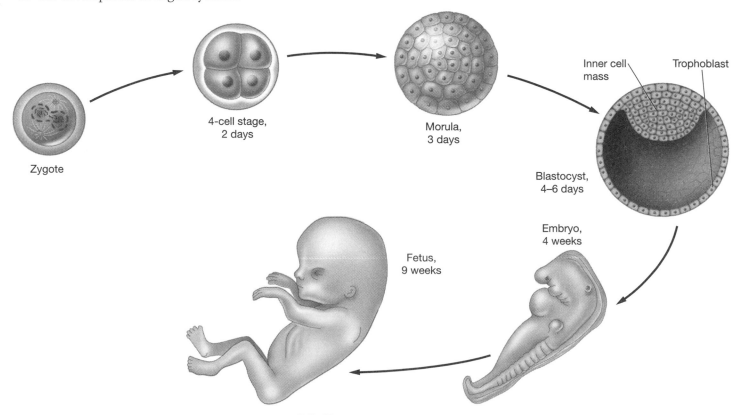

FIGURE 28.3 The stages of development

EXPLORING ANATOMY & PHYSIOLOGY IN THE LABORATORY

The increase in cell number begins about 30 hours after fertilization as the zygote travels down the uterine tube. At this time a process called **cleavage** occurs, in which the cell undergoes a series of mitotic divisions to produce two cells, then four, and so on. At about day 3, the conceptus reaches the uterus as a 16-cell ball called a **morula** (which means "little mulberry"). The morula floats around the uterus for another 2 to 3 days and continues to divide until it becomes a hollow sphere called a **blastocyst**. Note in **Figure 28.3** that the blastocyst has two populations of cells: the rounded **inner cell mass** and an outer layer of cells called the **trophoblast**.

At about day 6, the blastocyst adheres to the endometrium and begins the process of **implantation**. The implanting blastocyst secretes digestive enzymes that eat away the endometrial lining, and the endometrium reacts to the injury by growing over and enclosing the blastocyst. The process of implantation is generally complete by the second week, about the time a woman's menstrual period would begin.

Procedure Time to Trace

In this exercise you will be tracing the gametes from their production in the male and female gonads to the point at which they meet to form a zygote. You will also trace the zygote's development to the point at which it implants into the endometrium. You may want to review the Unit 27 exercises on male and female reproductive anatomy and meiosis before you begin.

Step 1: Male Gamete

Trace the male gamete from its earliest stage—the spermatogonium in the seminiferous tubule—through the stages of spermatogenesis to a mature sperm, and then through the male reproductive tract until it exits the body and enters the uterine tube of the female reproductive tract. Trace the pathway using **Figure 28.4** and also fill in the space below.

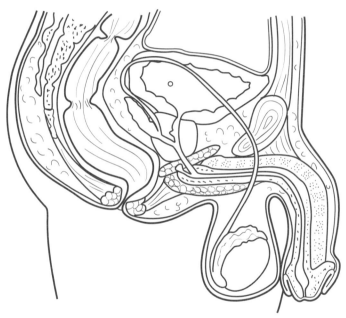

FIGURE **28.4** Male reproductive tract

Start: Spermatogonium

End: Uterine tube

Step 2: Female Gamete

Trace the female gamete from its earliest stage, the oogonium in a primordial follicle, to the time at which it is ovulated and enters the uterine tube. Trace the pathway using **Figure 28.5** and also fill in the space below.

Start: Oogonium

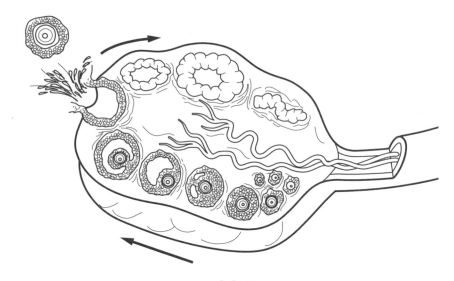

FIGURE **28.5** Ovary and follicles

End: Uterine tube

Step 3: Fertilization and Implantation

Trace the sperm and the ovum from the time they meet in the uterine tube to become a zygote until implantation occurs in the endometrium. Trace the pathway using **Figure 28.6** and also fill in the space below.

Start: Sperm and ovum meet in uterine tube

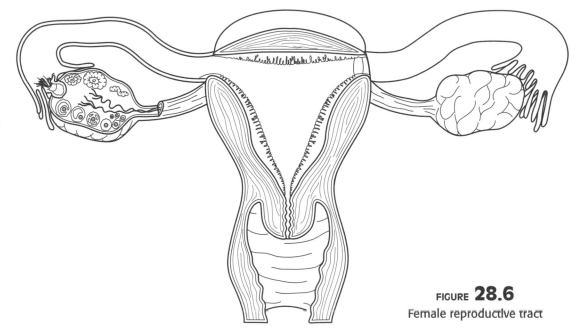

FIGURE **28.6**
Female reproductive tract

End: Implanted blastocyst

Exercise 2

Embryogenesis, Fetal Development, and Fetal Cardiovascular Anatomy

MATERIALS NEEDED

- Sequence of embryonic development model
- Female reproductive system with fetus model
- Fetal circulation model

As the blastocyst implants, it begins a process known as **embryogenesis** during which cellular differentiation takes place. During embryogenesis, the inner cell mass differentiates into three primary tissue layers known as **germ layers**, from which all other tissues arise. The outermost germ layer is called **ectoderm**; the middle germ layer is called **mesoderm**; and the innermost germ layer is called **endoderm**. The three primary germ layers are formed by the end of the second week, at which point the conceptus is considered an **embryo**.

The next 6 weeks of the embryonic period are marked by development of the **placenta**, formation of the extraembryonic membranes, and differentiation of the germ layers into rudimentary organ systems (**Figure 28.7**). The placenta begins to form at day 11, when the trophoblast forms the **chorion**, the fetal part of the placenta. Note in **Figure 28.7** that the chorion develops elaborate projections called the **chorionic villi**, which eat into uterine blood vessels to create a space filled with maternal blood called the **placental sinus**. Nutrients and oxygen from maternal blood diffuse from the placental sinus to the chorionic villi and

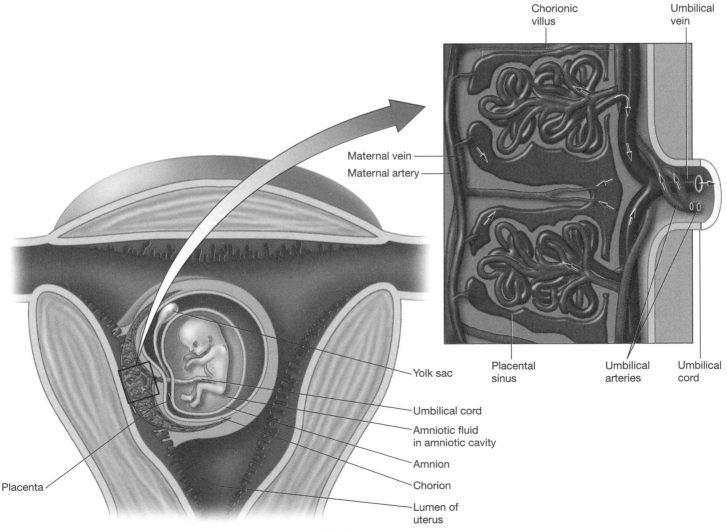

FIGURE 28.7 Embryo in uterus

are delivered to the embryo by the large **umbilical vein.** Note that the umbilical vein is red, which reflects the fact that it carries oxygenated blood. Wastes are drained from the embryo via the paired **umbilical arteries.** All three vessels travel through the **umbilical cord.**

During this period the extraembryonic membranes develop. The innermost membrane is called the **amnion.** The amnion completely surrounds the embryo (and later the fetus) and suspends it in **amniotic fluid** within the **amniotic cavity.** The amniotic fluid protects the embryo from trauma by allowing it to remain buoyant and also protects it from fluctuations in temperature. In addition, the cavity provides the embryo with ample space in which to move, which is critical to early muscle development.

As the placenta and extraembryonic membranes form, the primitive organ systems develop. This process completes by end of the eighth week, at which point the conceptus is considered a **fetus.** For the duration of gestation, the organ systems become progressively more specialized and the fetus continues to grow and develop.

The fetus' cardiovascular system differs significantly from that of the neonate. Within the fetal cardiovascular system are three **shunts** that bypass the relatively inactive lungs and liver and reroute the blood to other, more metabolically active organs. The three shunts include the following (**Figure 28.8**):

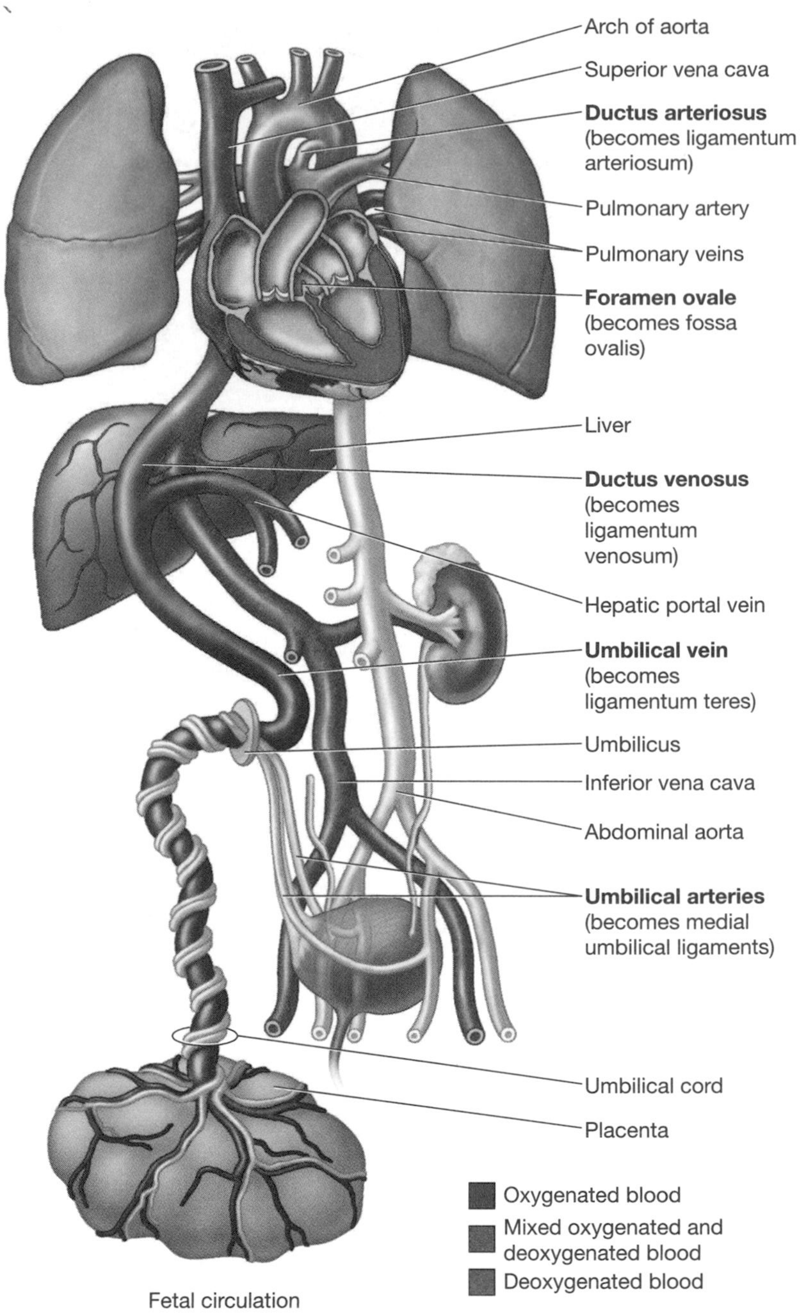
Arch of aorta
Superior vena cava
Ductus arteriosus (becomes ligamentum arteriosum)
Pulmonary artery
Pulmonary veins
Foramen ovale (becomes fossa ovalis)
Liver
Ductus venosus (becomes ligamentum venosum)
Hepatic portal vein
Umbilical vein (becomes ligamentum teres)
Umbilicus
Inferior vena cava
Abdominal aorta
Umbilical arteries (becomes medial umbilical ligaments)
Umbilical cord
Placenta
Oxygenated blood
Mixed oxygenated and deoxygenated blood
Deoxygenated blood
Fetal circulation

FIGURE 28.8 Fetal circulatory system

1. **Ductus venosus.** The ductus venosus is a shunt that bypasses the liver. The umbilical vein delivers a small amount of blood to the liver but sends the majority of the blood through the ductus venosus to the fetal inferior vena cava. It closes around the time of birth and becomes the ligamentum venosum.
2. **Foramen ovale.** The foramen ovale is a hole in the interatrial septum that shunts blood from the right atrium to the left atrium. This allows the blood to bypass the collapsed fetal lungs. The foramen ovale closes about the time of birth, leaving a permanent indentation in the interatrial septum called the **fossa ovalis.**
3. **Ductus arteriosus.** The ductus arteriosus is a vascular bridge between the pulmonary artery and the aorta that bypasses the pulmonary circuit. It also closes around the time of birth and leaves behind a remnant called the **ligamentum arteriosum.**

Procedure Model Inventory

Identify the following structures of development on models and diagrams, using your textbook and this unit for reference. As you examine the anatomical models and diagrams, record on the model inventory in Table 28.2 the name of the model and the structures you were able to identify. When you have completed the activity, answer Check Your Understanding questions 1 through 4 (pp. 649–650).

1. Stages of development
   a. Zygote
   b. Morula
   c. Blastocyst
   d. Implanted blastocyst
   e. Embryo
   f. Fetus
2. Fetal membranes
   a. Amnion
      (1) Amniotic cavity
      (2) Amniotic fluid
   b. Chorion
      (1) Chorionic villi
   c. Yolk sac

3. Vascular structures
   a. Placenta
      (1) Placental sinus
      (2) Maternal arteries and veins
   b. Umbilical cord
      (1) Umbilical arteries
      (2) Umbilical vein
   c. Foramen ovale
   d. Ductus venosus
   e. Ductus arteriosus

TABLE 28.2	Model Inventory for the Stages of Development and Fetal Structures
Model	Structures Identified

Heredity

MATERIALS NEEDED

- Well plate
- Simulated blood of Ms. Rose, Ms. Magenta, Col. Lemon, Professor Purple, and Mr. Olive
- Anti-A, anti-B, and anti-Rh antisera
- PTC tasting papers
- Ishihara chart for color blindness

Heredity is defined as the passing of traits from parents to offspring. We refer to two properties of the cell when discussing traits:

1. **Phenotype** is the physical expression of a specific trait, such as hair color or blood type. Traits such as "blonde hair," "brown eyes," and "type A blood" refer to an individual's phenotype.

2. **Genotype** is the genetic makeup of an organism, or what a cell contains in its DNA. When someone performs DNA analysis, he or she is examining the genotype. We have two copies of each of our genes—one from the maternal side and one from the paternal side. The varying forms of a gene are known as **alleles**. The genotype is described using letters to stand for each of the alleles that are inherited. It is generally not possible to determine a person's precise genotype without doing DNA typing; however, if we know the phenotype, we can determine *possible* genotypes. Let's look at an example using blood groups:

a. From each parent, we can inherit one of the following ABO genes:
   (1) A allele: codes for A antigens
   (2) B allele: codes for B antigens
   (3) O allele: codes for neither A nor B antigen (please keep in mind that there is no "O" antigen)

b. From each parent, we can inherit one of the following Rh factor genes:
   (1) + allele: codes for the Rh antigen
   (2) − allele: does not code for the Rh antigen

We can put all of this information into a table to determine which phenotypes accompany which possible genotypes. Fill in the remainder of Table 28.3, following the pattern from the first two examples.

A person who inherits the same gene from both parents is said to be **homozygous** (e.g., the genotype AA is homozygous). A person who inherits different genes from each parent is said to be **heterozygous** (e.g., the genotypes AB and BO are heterozygous).

TABLE 28.3 Phenotype and Genotype	
Genotype(s)	**Phenotype**
AA++, AA+−, AO++, AO+−	A+
AA−−, AO−−	A−
BB++, BB+−, BO++, BO+−	
	B−
OO++, OO+−	
	O−
	AB+
AB−−	

Procedure Determining Paternity with Blood Typing

Note: Safety glasses and gloves are required.

Blood typing is a simple and quick way to determine a child's potential heredity. Following is an example:

Child's Phenotype: B+

Mother's Phenotype: O−

Potential Father 1's Phenotype: AB+

Potential Father 2's Phenotype: O−

We start by first determining the possible genotypes of each person:

⬗ Child's potential genotypes: BB++, BB+−, BO++, BO+−
⬗ Mother's potential genotypes: OO−−
⬗ Father 1's potential genotypes: AB++, AB+−
⬗ Father 2's potential genotypes: OO−−

Now what we have to do is attempt to determine which genes the child received from his mother. For the ABO gene, the mother can only give her child an O allele. For the Rh gene, the mother can only give her child a − allele. So, we can say that the child received an O allele and a − allele from mom.

The final step is to determine which gene(s) the child got from his father. The child has a B allele and a + allele that mom couldn't have given him, so these B and + alleles had to come from dad. Now let's look at both fathers' potential genotypes. We see that father 1 does have a B allele but father 2 does not. We see also that father 1 has a + allele, but father 2 does not. Therefore:

We cannot exclude father 1 as a potential father of the child.

We *can* exclude father 2 as a potential father of the child.

Notice here that we can't say for certain that father 1 is indeed the child's father. All we can say is that he cannot be excluded and that further testing is necessary. We can say for certain, however, that father 2 is completely excluded and does not have to be tested further.

For this procedure we return to our favorite cast of characters to sort out some sticky paternity issues using the genetics of blood typing. While investigating the murders from Unit 19, the investigators discovered that Ms. Magenta was hiding a secret: She had a 20-year-old daughter named Ms. Rose. The investigators are uncertain as to Ms. Rose's paternity, but they are certain that there are three possible fathers—Col. Lemon, Mr. Olive, and Professor Purple. You must determine if any of these three men may be ruled out as Ms. Rose's father on the basis of their blood types.

Following is the procedure for testing the blood samples. When you have completed the activity, answer Check Your Understanding question 5 (p. 650).

1 Obtain a well plate, anti-A, anti-B, and anti-Rh antisera, and samples of blood from Ms. Magenta, Ms. Rose, Col. Lemon, Mr. Olive, and Professor Purple.

2 Place two drops of Ms. Magenta's blood into three separate wells.

3 Place one drop of anti-A antiserum into the first well and watch for a reaction. Record your results in Table 28.4.

4 Place one drop of anti-B antiserum into the second well and watch for a reaction. Record your results in Table 28.4.

5 Place one drop of anti-Rh antiserum into the third well and watch for a reaction. Record your results in Table 28.4.

6 Repeat this procedure with the blood of Ms. Rose, Col. Lemon, Mr. Olive, and Professor Purple. Record your results in Table 28.4.

TABLE 28.4 | Blood Type Results

Blood Sample	Reaction with Anti-A?	Reaction with Anti-B?	Reaction with Anti-Rh?
Ms. Magenta			
Ms. Rose			
Col. Lemon			
Mr. Olive			
Prof. Purple			

7 Use your data from Table 28.4 to complete Table 28.5.

TABLE 28.5 | Phenotypes and Potential Genotypes

Blood Sample	Blood Type (Phenotype)	Potential Genotypes
Ms. Magenta		
Ms. Rose		
Col. Lemon		
Mr. Olive		
Prof. Purple		

8 Interpret your results:

a. Can any of the men be excluded as Ms. Rose's potential father? Explain.

b. Can any of the men *not* be excluded as Ms. Rose's potential father? Explain.

Procedure Testing the Ability to Taste PTC

The genetic effect known as **dominance** refers to the effect that different alleles have on an individual's phenotype. An allele that is expressed as the phenotype is known as the **dominant allele**, and an allele that is not expressed and has little to no measurable effect on phenotype is known as the **recessive allele**. Dominant alleles are nearly always represented with capital letters and recessive alleles with lower-case letters.

The example of free and attached earlobes can be used show the effect of dominant and recessive genes on phenotype. The gene that codes for free earlobes is dominant and written as *E*, whereas the gene that codes for attached earlobes is recessive and written as *e*. Three different combinations of genes and two phenotypes are possible:

EE, or homozygous dominant individuals, will have free earlobes;

Ee, or heterozygous individuals, will have free earlobes;

ee, or homozygous recessive individuals, will have attached earlobes.

Another dominant allele is one that determines the ability to taste **PTC** (phenylthiocarbamide), a bitter chemical found in foods such as Brussels sprouts and coffee. About 75% of the population has the dominant allele, *T*, and can therefore taste PTC. The remaining 25% of the population is homozygous for the recessive allele, *t*, and cannot taste PTC. In the following procedure, you will test your ability and that of your classmates to taste PTC. You will then pool your data to determine the percentage of tasters and non-tasters in your class.

1 Obtain a piece of PTC paper and place it on your tongue.

2 If you can taste the PTC right away, you may remove the paper and dispose of it. If you do not taste it right away, leave the paper on your tongue for up to 1 minute. If you cannot taste the PTC after 1 minute, remove the paper and dispose of it.

Were you able to taste PTC?

3 Pool the data from your class and record it below:

_____ Number of tasters   _____ Number of non-tasters

_____ Percent of class that are tasters   _____ Percent of class that are non-tasters

4 Interpret your results:
a. Which genotype(s) do the tasters have?

b. Which genotype(s) do the non-tasters have?

c. How do the results from your class compare to the general population?

Procedure Testing Color Vision

Recall that humans have 23 pairs of chromosomes. Chromosomes 1–22 are called **autosomes,** and the last pair are the **sex chromosomes.** The two sex chromosomes are known as **X** and **Y.** Females have the genotype XX and males have the genotype XY. Until now we have been discussing **autosomal traits,** or those that are passed down on autosomes. Certain traits, however, are passed down on the sex chromosomes, and these are known as **sex-linked traits.**

The inheritance of autosomal traits is not generally influenced by gender; however, this is not the case with the inheritance of sex-linked traits. We can use the example of the blood disease hemophilia to see how this works. Hemophilia is the result of a recessive allele (*h*) found on the X chromosome. The normal allele is the dominant allele *H*.

▶ A female with the genotype X^HX^H will *not* have the disease.

▶ A female with the genotype X^HX^h will *not* have the disease; however, she will be a "carrier" of the disease who is capable of passing the allele to her offspring.

▶ A female with the genotype X^hX^h will have the disease.

▶ A male with the genotype X^HY will *not* have the disease.

▶ A male with the genotype X^hY will have the disease.

Notice here that because males have only one X chromosome, they require only one copy of the recessive allele to express the trait and have the disease. Conversely, females still require two copies of the recessive allele to express the trait. For this reason, hemophilia is much more common in males than in females.

Another example of a sex-linked trait is color blindness, or the inability to distinguish between certain colors. The allele for the disease is a recessive allele *c* and the allele for normal color vision is the dominant allele *C*. The following procedure will allow you to test your vision for abnormalities in color vision, which may result from the presence of this allele. The test uses a chart called an *Ishihara chart* that has a number embedded among a group of colored discs. A person with normal color vision will be able to read the number, whereas a person with color blindness will only be able to see randomly arranged colored dots.

1 Obtain an Ishihara chart. Hold the chart at a comfortable reading distance. Record what you see below:

Was the test negative or positive for color blindness? (Note that "negative" means that your color vision was normal.)

2 Pool the results from your class and record the data below:

_____ Number of females with normal color vision

_____ Number of females with impaired color vision

_____ Number of males with normal color vision

_____ Number of males with impaired color vision

3 Interpret your results:

    a. What is/are the genotype(s) for females with normal color vision?

    b. What is the genotype for females with impaired color vision?

    c. What is the genotype for males with normal color vision?

    d. What is the genotype for males with impaired color vision?

    e. How did the number of males with impaired color vision compare to the number of females with impaired color vision? Explain this finding.

NAME _____

SECTION _____ DATE _____

Check Your Recall

1 Matching: Match the following terms with the correct definition.

a. Chorion _____   A. A single-celled structure formed after fertilization

b. Blastocyst _____   B. The end of a sperm that contains digestive enzymes

c. Fetus _____   C. A hollow sphere of cells that implants into the endometrium

d. Gestation _____   D. The conceptus from 2 weeks to 8 weeks of development

e. Zygote _____   E. A 16-cell ball that reaches the uterus

f. Morula _____   F. The fetal membrane that contributes to the placenta

g. Amnion _____   G. The early tissues that will go on to form all tissues in the body

h. Embryo _____   H. The period that extends from the mother's last menstrual period to birth

i. Acrosome _____   I. The innermost fetal membrane

j. Germ layer _____   J. The conceptus from 9 weeks to birth

2 Label **Figure** 28.9 with the terms below.
- Umbilical vein
- Umbilical arteries
- Placenta
- Foramen ovale
- Ductus venosus
- Ductus arteriosus

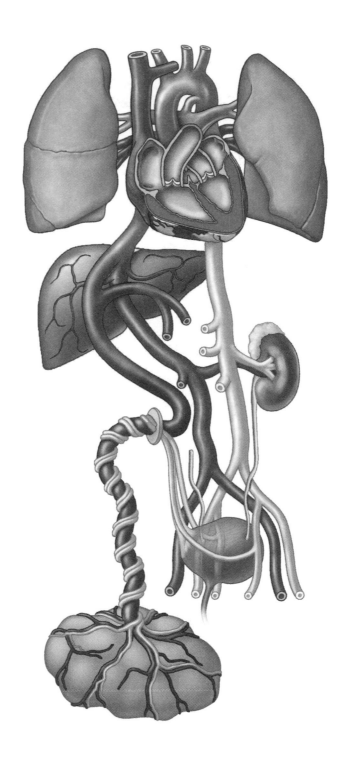

FIGURE **28.9** Fetal circulatory system

EXPLORING ANATOMY & PHYSIOLOGY IN THE LABORATORY

3 Label **Figure 28.10** with the terms below.

- Amnion
- Amniotic cavity
- Chorion
- Placenta

- Yolk sac
- Chorionic villi
- Placental sinus

- Maternal arteries and veins
- Umbilical vein
- Umbilical arteries

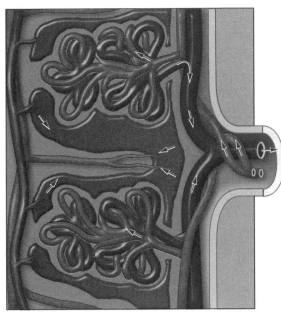

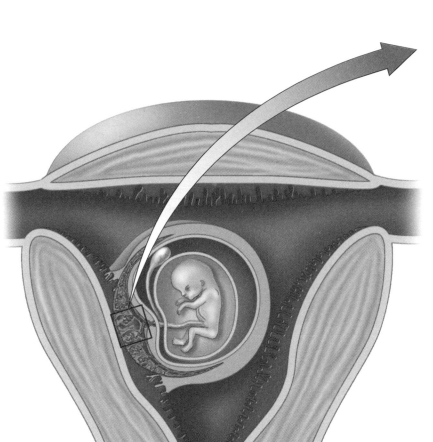

FIGURE **28.10** Embryo in uterus

4 Which of the following is *not* one of the three basic changes that a conceptus undergoes during development?
a. An increase in cell number
b. Cellular differentiation
c. Development of organ systems
d. Neoplasia

5 The placenta is formed from

    a. the chorion.
    b. maternal tissues.
    c. the amnion.
    d. Both a and b are correct.
    e. Both b and c are correct.

6 Fill in the blanks: The umbilical _____ carries oxygenated blood to the fetus, and the

_____ carries deoxygenated blood away from the fetus. The _____

bypasses the fetal liver, and the _____ and _____ bypass the fetal lungs.

7 Variant forms of a gene are known as

    a. genotypes.
    b. alleles.
    c. phenotypes.
    d. antigens.

8 Fill in the blanks: The _____ is the physical manifestation of a trait, and the _____
is the genetic makeup of an organism.

9 How do dominant and recessive alleles differ?

10 Which of the following statements is false?

    a. Sex-linked disorders tend to affect females more than males.
    b. Chromosomes 1–22 are called autosomes.
    c. A person with two identical genes is homozygous.
    d. Blood typing may be used to conclusively exclude paternity, but it cannot conclusively prove paternity.

Check Your Understanding | Critical Thinking and Application Questions

1 Predict the effects of failure of the foramen ovale and/or the ductus arteriosus to close shortly after birth.

2 The use of certain medications, such as nonsteroidal anti-inflammatories (e.g., ibuprofen), in the seventh to ninth month of pregnancy increases the risk for premature closure of the ductus arteriosus. What problems might this potentially cause for the fetus and the neonate?

3 A condition called *placenta previa* occurs when the placenta is too low in the uterus. Why do you think this is a cause for concern?

4 Active labor generally is considered to begin about the time the "water breaks." What is the "water," and which membranes rupture to release it? What could be the potential consequences if the membranes were to rupture too early?

5 Further investigation has revealed that Ms. Magenta is in fact *not* Ms. Rose's mother. Because of a freak baby-switching accident, Mrs. Feather is really Ms. Rose's biological mother. Mrs. Feather's blood type is determined to be AB+. Col. Lemon, Professor Purple, and Mr. Olive remain Ms. Rose's potential fathers.

a. What is Mrs. Feather's phenotype? What are her possible genotypes?

b. Can any of the potential fathers be excluded? Explain.

c. What if Ms. Rose's blood type were O+? How would this change the scenario, both *maternally* and *paternally*?

Physiologic Connections

OBJECTIVES

Once you have completed this unit, you should be able to:

- Find physiologic connections between different organs and systems in the body.

- Review the physiology of the organ systems that we have covered.

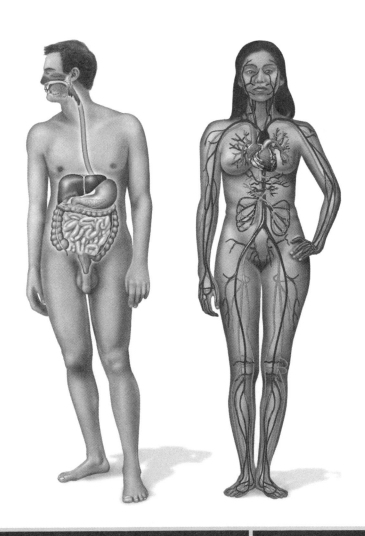

651

In anatomy and physiology the anatomical connections between organs and system are usually easy to see. All organs are connected by the cardiovascular system; bones are united by ligaments and cartilage; and accessory organs of the digestive system are joined to the alimentary canal by common ducts. What is sometimes difficult to see is the *physiologic* connections that exist between all organs and organ systems.

At this point in our study of anatomy and physiology, we have examined each organ system and its individual organs and have studied how the parts fit together to make the whole. Along the way, we also have developed an appreciation for how the form of the system, organ, tissue, and cell all follow the primary functions of each part. Now, in this last unit, we will develop an appreciation for how the physiologic functions of each organ complement one another.

Exercises

Exercise 1

Physiologic Connections

MATERIALS NEEDED

- Anatomical models: human torso
- Laminated outline of human body
- Water-soluble marking pens

Your final task in this lab manual is to find physiologic connections between seemingly unrelated groups of organs. The physiologic connection between the salivary glands, stomach, and duodenum is easy to find, but the connection between the heart, bone, and thyroid may be a bit less obvious. Following are some hints, and an example, to make this task easier.

- Begin each exercise by listing the functions of each organ listed. Some functions are obvious, and others may require a bit of digging on your part. Take, for example, the skin. The skin has the obvious functions of protection and thermoregulation. But don't forget about its other tasks: The skin also synthesizes vitamin D when it is exposed to certain wavelengths of ultraviolet radiation; it excretes small amounts of metabolic wastes; and it and serves immunologic functions. These less-discussed functions are easy to overlook but are vitally important.

- Beware of the tendency to fall back to anatomical connections. When given the combination "lungs, kidneys, and blood vessels," you may be tempted to say that the lungs breathe in oxygen, which is delivered by the blood vessels to the kidneys. That's all true, but this is actually an anatomical connection. With some creative thinking on your part, you could come up with something much better. Remember that lung tissue makes angiotensin-converting enzyme, and, having said that, the connection between the blood vessels and the kidneys becomes more obvious (at least I hope it does).

- When in doubt, look to the endocrine system. Remember that many different organs, not just the organs discussed in the endocrine unit, make hormones. The heart, kidneys, and even adipose tissue all function in some respects as endocrine organs. For the organs that have a primary endocrine function, don't overlook some of the hormones they make. For example, the thyroid gland makes T3 and thyroxine but also calcitonin, which students tend to forget.

Here's an example of physiologic connection:

Lungs, Kidney, and Bone

Connection: The lungs have the function of *gas exchange*. When this exchange is inadequate and oxygen intake is insufficient, the kidneys detect the decreased partial pressure of oxygen in the blood. The kidneys, in turn, make the hormone *erythropoietin*, which acts on the bone marrow to stimulate increased *hematopoiesis* and an increased rate of red blood cell maturation.

Wasn't that easy?

Now it's your turn.

Part 1: Heart, Parathyroid Glands, Bone

Hint: Think electrophysiology of the heart.

Connection:

Part 2: Small Intestine, Lymphatic Vessels, Liver

Hint: How does the body process dietary fats?

Connection:

Part 3: Spleen, Liver, Large Intestine

Hint: Think about the functions of the spleen with respect to erythrocytes.

Connection:

Part 4: Adrenal Glands, Blood Vessels, Kidney

Hint: What do the different regions of the adrenal gland make?

Connection:

Part 5: Pancreas, Sympathetic Nervous System, Liver

You're on your own for this one!!

Connection:

EXPLORING ANATOMY AND PHYSIOLOGY IN THE LABORATORY

Congratulations!

*You've done what you probably thought was impossible—
completed this manual! I truly hope it has been fun
(at least a little!) and that you have enjoyed it as much
as I have. Best wishes with your future plans,
down whichever path they may take you.*

Photo Credits

Chapter 1

Figures 1.7, 1.8: Van De Graaff, Morton, and Crawley, *A Photographic Atlas for the Anatomy and Physiology Laboratory, 6e* © Morton Publishing

Chapter 5

Figures 5.7, 5.16, 5.21: Leboffe, *A Photographic Atlas of Histology* © Morton Publishing

Chapter 10

Figures 10.1, 10.2: Leboffe, *A Photographic Atlas of Histology* © Morton Publishing

Chapter 12

Figure 12.2: Leboffe, *A Photographic Atlas of Histology* © Morton Publishing

Chapter 13

Figures 13.12, 13.13, 13.14: Van De Graaff, Morton, and Crawley, *A Photographic Atlas for the Anatomy and Physiology Laboratory, 6e* © Morton Publishing

Chapter 15

Figures 15.8, 15.9, 15.10: Van De Graaff, Morton, and Crawley, *A Photographic Atlas for the Anatomy and Physiology Laboratory, 6e* © Morton Publishing

Chapter 23

Figure 23.8: Van De Graaff, Morton, and Crawley, *A Photographic Atlas for the Anatomy and Physiology Laboratory, 6e* © Morton Publishing

Chapter 25

Figure 25.4: Leboffe, *A Photographic Atlas of Histology* © Morton Publishing

Chapter 26

Figure 26.2: Leboffe, *A Photographic Atlas of Histology* © Morton Publishing

Index

EXPLORING ANATOMY & PHYSIOLOGY IN THE LABORATORY

reactions, testing, 36–38
solubilities, 38–41
chemosenses / chemoreceptors, 350
chiasmata, 615
chlorine (Cl), 30–31
cholesterol, 62
chondrocytes, 90, 136
chordae tendinae, 370
chorionic villi, 630, 635
chromatin, 64
chromosomes, 72, 615, 619
sex, 642
chylomicrons, 569
cilia, 62, 84, 87
circle of Willis, 394, 397, 415
circulation, fetal, 646. *See also* blood flow;
cardiovascular
cisterna chyli, 457
clavicle, 161
climbing stairs exercise, 249
clitoris, 613
CNS. *See* central nervous system
coccyx, 157
cochlea, 346, 347
coitus, 613
collagen, 90, 131, 340, 408
collagen vascular diseases, 416
colloid, 589
osmotic pressure, 457
color blindness, 642
color vision, testing, 642–643
concentration gradient, 67
conceptus, 632
condyles, 164
connective tissue, 82–83, 90–94, 190, 408
bone, 131
in digestive system, 563
in female anatomy, 613
in heart, 368
hints & tips, 95
in lymphatic system, 461
in male anatomy, 610
microscopy of, 95
nerve, 311
respiratory, 486, 487
skin, 115, 116
in urinary system, 518
cooperative learning approach, 155, 166
COPD (chronic obstructive pulmonary
disease), 504
coracoid and coronoid processes, 161
cornea, 340
coronary arteries, 370
corpus cavernosa / corpus spongiosum, 611
corpus luteum, 619
Corti, organ of, 347
cortisol, 589
covalent bonds, 38, 39
cranial / cranium
bones, 151, 152, 154, 156
cavity, 8
meninges, 288

nerves, 306–307, 311, 312–313, 314
testing, 315–319
cribriform plate, 154, 350
cricoid cartilage, 481
cricothyroid ligament, 481
cricothyroidotomy, 489–490
crista galli, 154
crossbridge cycle, 213, 214, 215
cutaneous sensation, 352
cystic duct, 560
cytokinesis, 72–73, 617
cytoplasm, 62, 72, 208, 436, 619
cytoskeleton, 62, 263
cytosol, 62, 63, 70, 211

daughter cells, 72, 615
dendrites, 101, 264
depolarization, 213, 261, 268, 377, 378
dermis structures / dermis, 112–113, 115
layers, 116
muscles in, 219
desmosomes, 376
detoxification, 63
development. *See* human development
diabetes mellitus, 604
diaphragm, 239, 240, 502
diaphysis, 136
diastolic blood pressure, 424
diencephalon, 283, 285, 588, 589
diffusion, 67–69, 70, 568
DiGeorge syndrome, 471
digestion / digestive system, 568
accessory organs, 557, 559, 560, 561,
564, 652
anatomy, 553–555, 557
carbohydrate test, 569–570
chemical, 557, 568
enzymes, 63, 556, 559, 560, 563, 633
of fats, 36
histology, 551–552
lipid, 568, 573
mechanical, 557
organs, 553, 558, 559, 566, 577
physiology, 552
protein, 568, 571
structures, 550–551
tracing exercises, 574–575
directional terms, 5
diuretics, potassium-sparing, 604
DNA (deoxyribonucleic acid), 64, 615, 638
extraction procedure, 43
Doppler ultrasound, 428
ductus arteriosus, 636
duodenum, 555, 558, 560, 561, 578, 652
dura mater, 286
dynamometer, 218

ear, 335, 338, 346, 356
model inventory, 347
earlobe example of phenotype, 641
ectoderm, 635
edema, 471

ejaculation and ejaculatory duct, 611
elastin, 90
elbow joint, 162, 246
electronegativity, 32, 38
electrons, 30–32, 38
electrophysiological events in heart, tracing,
378
embryo / embryogenesis, 630, 635, 647
emphysema, 511
emulsification, 568, 572, 574
endocrine system, 285, 590, 599, 652
anatomy, 586, 588
and disease mystery cases, 595–598
organs, 584–585, 591, 619
and pancreas, 589
endoderm, 635
endometrium, 613, 633, 634
endomysium, 98, 208
endoplasmic reticulum, 63, 208, 263
endosteum, 132
endothelium / endothelial cells, 408, 539.
See also epithelial tissue
enterocytes, 564, 568, 569
enzymes, 32, 36–38
acrosomal, 632
angiotensin-converting, 652
brush border, 568
digestive, 63, 556, 559, 560, 563, 564,
574, 633
hormone-producing, 588
osteoclasts secreting, 131
pepsin, 571
eosinophils, 436
epidermal structures and layers / epidermis,
112, 115, 116
epididymis, 610, 611, 617
epiglottis, 480
epimysium, 208
epinephrine, 324, 589
epiphyses, 136
epiploic appendages, 558
epithalamus, 285
epithelial tissue / epithelium, 82, 84–88,
115, 131, 486, 538, 563
digestive, 563
glandular, 588
hints & tips, 88
microscopy, 88–89
olfactory, 264, 350, 351
pseudostratified ciliated columnar, 486
simple columnar, 563
simple cuboidal, 486, 591
simple squamous, 487, 520, 525
skin, 115
stratified squamous, 480
transitional, 521, 525, 526
equilibrium, 346, 349–350
error of localization, 352
erythroblastosis fetalis, 452
erythrocytes, 91, 436, 439, 443, 444, 445,
542, 543
erythropoietin, 652

EXPLORING ANATOMY & PHYSIOLOGY IN THE LABORATORY

thorax muscles, 230–231, 240
throat, 479, 557
thymosin, 589
thymus, 457, 588, 589
thyroid
    cartilage, 481
    connections to other organs, 652
    follicles, 589
    gland, 588, 589, 591, 592
tibia, 164, 166
    arteries, 394
tissue. *See also* connective tissue
    adipose, 90, 131, 132, 518, 588
    of alimentary canal, 563, 579
    blood as, 91
    bone / osseous, 91, 131–132, 133, 589
    elastic, 501
    epithelial, 82, 84–88, 115, 131, 486, 563
    eye, 340
    heart layers, 368
    kidney, 519, 525, 526
    lung, 502, 652
    lymph, 457
    muscle, 83, 98–99, 206, 207, 208, 209, 210–211, 219–220, 221
        cardiac, 376
        microscopy of, 100–101, 222
    nervous, 83, 101–102, 115, 131, 263
    subcutaneous, 115
    thymic, 589
tongue, 351, 559, 561
tonicity, 70, 71
tonsils, 456, 457, 458, 459
total lung capacity (TLC), 504
touch
    receptor, mapping, 120
    sense of, 338, 352
trabeculae, 132
trachea, 482, 486, 487
transfusion reaction, 443
transverse process and foramina, 157
trochlear notch, 161, 162
trophoblast, 633
tropomyosin, 208, 213, 215
troponin, 208, 213
trunk muscles, 240, 241, 244, 252–253
TSH (thyroid-stimulating hormone), 588, 589
tuberosity, deltoid, 161
tubular reabsorption and secretion, 537, 538
tunica interna and externa, 408, 409, 416
tunica vaginalis, 610
tunics, 340

two-point discrimination test, 352–353
tympanic
    membrane, 346
    region (skull), 151

ulna, 151, 162, 163
ultrasound, 428
umbilical vein, 636
universal blood donor, 445
urea, 542, 544
ureter, 521, 522
urethra, 521, 522
    female, 613
    male, 611
urethral sphincter, 521, 522
urinalysis, 540–542
urinalysis reagent test strips, 539, 540
urinary bladder, 521, 522, 525, 526
    microscopy, 527
urinary system, 514–515, 516, 518–521.
        See also kidneys
    model inventory, 522
    organs, 525–526
urine, 518, 525, 540
    osmolality, 597
uterus
    embryo in, 630, 635, 647
    and uterine tube, 613, 614, 617, 632, 633, 634
uvea, 340

vagina, 613
valves, heart, 362, 370, 419–420
vasectomy, 625
vascular
    disease, 416, 423
    structures, 482
vas deferens, 611
vasoconstriction / vasodilation, 424, 464
vein(s), 368, 392, 398–400, 408, 409, 412, 542
    of abdomen and hepatic portal system, 393
    brachial, 400
    of head and neck, 388–389, 401
    hepatic, 559, 565, 568
    interlobar, 519
    jugular, 398, 457, 569
    of lower limbs, 389, 401
    renal, 400, 519, 522
    splenic, 400
    subclavian, 400, 457, 569
    systemic, 370
    testicular, 611

of trunk, 388, 401
    umbilical, 636
    of upper limbs, 389, 401
    vertebral, 157
vena cava, 363, 368, 398, 400, 568
    fetal, 636
ventilation, pulmonary, 498–499
    model, 502
    pH and, 506
ventral (anterior) cavity, 8
ventral rami of spinal nerves, 320, 321, 329
ventricles, 368, 370, 394
vertebrae, 157, 178
vertebral
    artery, 387
    cavity and column, 8, 158–159, 176, 240, 288, 289
    foramen and foramina, 157, 288
vesicular follicle, 619
vestibular apparatus (brain), 349
vestibule (female), 613
vestibulocochlear nerve, 347
villi
    chorionic, 630
    intestinal, 564, 566, 569
visceral
    layer, 8
    motor division of ANS, 311
    pericardium, 368
    peritoneum, 560, 563
vision, 339. *See also* eye
    peripheral, 340
    rods and cones, comparing, 344
vitamin D, 140
vitreous humor, 340
vocal cords / folds, 481
voice box, 480, 481
Volkmann's canals, 132
vomer, 155
vulva, 613, 614

walking in place exercise, 248–249
water (H_2O), 31–32
Weber test, 348
well plate diagram, 440
white blood cells, 436
wrist, 161, 162, 241, 244

xiphoid process, 157

zona pellucida, 632
zygomatic processes / bones, 154, 155
zygote, 632, 633